BRAIN THEORY
SPATIO-TEMPORAL ASPECTS OF BRAIN FUNCTION

BRAIN THEORY

SPATIO-TEMPORAL ASPECTS OF BRAIN FUNCTION

Edited by

A. AERTSEN
Institut für Neuroinformatik
Rühr-Universität Bochum
Bochum, Germany

1993
ELSEVIER
AMSTERDAM - LONDON - NEW YORK - TOKYO

ELSEVIER SCIENCE PUBLISHERS B.V.
Sara Burgerhartstraat 25
P.O. Box 211, 1000 AE Amsterdam, The Netherlands

ISBN: 0 444 89839 5

This book is printed on acid-free paper.

Transferred to digital printing 2005

PREFACE

Modern theories of brain function are increasingly concerned with dynamics. The task of organizing perception and behaviour in a meaningful interaction with the external world prompts the brain to recruit its various resources in a properly coordinated manner. Vis-a-vis the complexity and the multitude of the dynamics involved, a careful orchestration of the various processing components, distributed over space and time, is essential. Hence, it should come as no surprise that a number of recent developments in brain science have emphasized the aspect of spatio-temporal coordination. This holds for experimental and theoretical investigations alike. Experimental studies of the brain at different levels of resolution have established a multitude of distributed physiological mechanisms, ranging from interactions at the level of individual ionic channels, via the integrative processes governing the behaviour of single neurons, all the way up to the dynamics within and across populations of neurons or even entire regions of the brain. Similarly, theoretical investigations of the behaviour of model neurons and computational networks of neuron-like elements have revealed a number of intriguing dynamical phenomena which acquire their meaning only if the entire spatio-temporal complexity is considered. Thus, much of recent theoretical work in brain science is, in fact, concerned with the development of an adequate conceptual framework for the space-time dynamics exhibited by natural as well as by artificial brains.

The present collection of papers intends to capture these various developments in the brain sciences. It brings together new insights and concepts from various branches of experimental and theoretical neuroscience, partly in the form of review chapters, partly in short, focussed contributions, or critical essays. Starting point of this enterprise were the presentations and discussions at the Fourth International Meeting on Brain Theory held at the Istituto per la Ricerca Scientifica e Tecnologica (IRST) in Trento (Italy) on April 11-13, 1992. This meeting, organized by Valentino Braitenberg, Werner von Seelen, Luigi Stringa and Ad Aertsen, set out to explore "the problems of the processing of the temporal dimension of sensory input and of the generation of space-time patterns in the motor output, as well as the intervening storage and transformation of temporal patterns in nerve nets". The meeting was the fourth in a series, starting in 1984 at the International Center for Theoretical Physics in

Trieste [1], and continuing in 1986 in Bad Homburg [2] and in 1990 at Schloss Ringberg [3].

The meeting lasted three full days, providing a natural segmentation into three subheadings. On the first day, spatio-temporal aspects of brain function were discussed in the context of processing of sensory input and perception. Similarly, the third day focussed on spatio-temporal aspects of brain function at the output end: planning and control of movement. These two sessions flanked the middle one, which was dedicated to the intervening level of neuronal activity in the working brain, and the various dynamics observed at different levels of resolution in space and time. We adopted essentially the same tripartition to organize the present book. A fourth part combines contributions that transcend this scheme. A declared goal of these various meetings was "to raise an interest in theoretical models that actively seek confrontation with experimental data from the functioning brain, and by a didactic effort aimed at experimentalists to present their data in a format that makes them more amenable to theory". The same goal was pursued in the composition of the present book. The proof of the pudding is left as an exercise to the reader. The Fourth International Meeting on Brain Theory was jointly sponsored by the European Commission, the Gertrud Reemtsma Stiftung, the Istituto per la Ricerca Scientifica e Tecnologica (Trento), the Accademia degli Agiati, as well as by contributions from the various institutions who financed the participation of their delegates. Splendid hospitality, together with most efficient organization was provided by IRST, Trento. The last afternoon was hosted by the Accademia degli Agiati in the neighbouring town of Rovereto. The generous support by all these institutions is most gratefully acknowledged.

Ad Aertsen

References

[1] Palm G, Aertsen A (eds): Brain Theory. Berlin: Springer, 1986.

[2] von Seelen W, Leinhos U, Shaw G (eds): Organization of Neural Networks: Structures and Models. Weinheim: VCH Verlag, 1987.

[3] Aertsen A, Braitenberg V (eds): Information Processing in the Cortex: Experiments and Theory. Berlin: Springer, 1992.

CONTENTS

III. Space-Time Dynamics of Neuronal Activity in the Working Brain

IV. The Brain: The Organization of Perception and Action

Part One

Space-Time Aspects of Brain Function: Sensory Perception

Brain Theory / A. Aertsen (Editor)

Embodiments of geometry

J.J. Koenderink

Utrecht Biophysics Research Institute (UBI), Buys Ballot Laboratory, Princetonplein 5, 3584 CC Utrecht, the Netherlands.

Abstract This tutorial is meant to offer a broad overview of the importance of geometrical language for the description of the structures encountered in front end visual systems. Since the paper is of a tutorial nature, technicalities are suppressed and quite a few liberties with mathematical rigour are taken.

The brain of any optically guided agent has to be able to deal with geometrical problems as a matter of routine. Yet the notion that formal geometries might be powerful and apt tools in the description of functional brain structure is an alien one to neurophysiologists and psychophysicists alike. Yet I will argue for this very possibility.

The discussion will focus on the structure of the "front end" visual system in neurophysiology and the "visual field" in the psychology of perception. Several striking examples of empirically disclosed structures that map almost immediately upon certain formal geometries are discussed. Of course many research problems remain, those relevant to the issue of the brain as a "geometry engine" are summarily discussed.

1 GEOMETRY AND THE BRAIN

For the neuroanatomist the brain exists primarily *in space*, it is an incredibly complicated three dimensional shape. Good neuroanatomists have to be excellent intuitive geometers. Yet the notion that parts of the brain can advantageously be described as *embodiments of space* is probably in need of some explanation. The problem is that most nonmathematicians tend to hold an extremely limited view of geometry. My first aim in this introduction is to set the scene by trying to broaden this view.

One may distinguish (at least) three different views of "geometry":

logician's view: a geometry is a categorical system defined via an arbitrary set of axioms, stating all relations that may hold among a set of arbitrary items[1]. Some of such systems are (for historical reasons) known as "geometries";

algebraist's view: a geometry is the set of algebraic relations admitted by some fiducial structure, *e.g.*, all relations invariant under the action of some arbitrary fiducial group of transformations[2]. Various examples of such systems are conventionally known as "geometries";

(synthetic) geometer's view: geometries are about certain primitive objects ("point", "line") that are abstractions of certain physical entities (pin pricks, rulers) and carry intuive connotations[3]. In a geometry one studies the properties of the configurations that can be constructed using a predetermined set of ideal operations (*e.g.*, connecting two points with a line).

We may also distinguish the *physicist's*, or *engineer's* view of geometry[4,5], which is totally different from all of the above, because the physicist's geometry is *about* certain entities that exist as material systems (pin pricks, rulers).

The "geometry" one learns at school (in the tradition of Euclid's Elements) is (ideally) of the synthetic variety, although usually with a physicist's flavour. If this is your background you are bound to think of the notion of describing functional brain structures *as* geometry as preposterous.

The abstract logician's view of geometries as sets endowed with additional structure at first sight appears to be "about nothing". However, when you realize that a description of the functional order of parts of the brain can be little else but a description of the formal relations that exist or can exist among possible brain states, it becomes clear that abstractions such as the logics, geometries, *etc.*, are exactly the formal structures one needs. Of course one needs at least one thing in addition to the pure logic: Some way to interprete or relate the abstract entities to actual observations (neuroanatomical, physiological or psychophysical).

Geometrical descriptions of functions of brain and mind are indeed common. In psychophysics one describes the internal structure of sensory modalities via multidimensional Riemann spaces (*e.g.*, color space[6,7] or binocular space[8]), in physiology one often describes the empirically determined correlation between signals recorded from different structures in geometrical terms ("common input", "overlapping receptive fields"), in neuroanatomy one quite naturally describes the common "projections" geometrically. Indeed, one often uses the geometrical description as an "explanation" of certain phenomena. For instance, "local sign" is most often explained away (though quite mistakingly I think) by pointing out the retinotopical organization of certain cortical areas[9]. Attempts to interprete brain structures as "embodiments" of geometrical entities have been comparatively rare though. Perhaps the most common instance is that of an identification of mechano– or photo–receptors or cutanous or visual receptive fields as "points" in a rather literal sense. A more sophisticated (and rather extreme) example is the identification of certain dendritic tree structures with the Lie group generators of movements in the visual field[10].

In most cases the identification is essentially *ad hoc* and very teneous. In order to arrive at something more objective one has to provide *operational definitions* of geometrical entities for which physiologically plausible implementations can be pointed out. Such need not necessarily be very complicated at all. Consider a—argueably the simplest—example. According to Euclid a point is "that which has no parts", *i.e.*, which has either no discernable inner structure or whose inner structure we choose to ignore. (*E.g.*, as an astronomer may treat our galaxy as a "point" in a somewhat more global context.) How can one operationalize such a definition? An example is to consider the photoreceptors as points, but such an identification doesn't lead to exciting consequences, it is hardly more than a metaphorical way of speaking. A more viable approach[11] is to define a point as an *operator* that returns a minimal description of what "is at it". An example would be an operator that operates on retinal illuminance patterns and returns the illuminance at some location. By recording the output of such an operator for various inputs one characterizes the point. (For instance, the operator will only be responsive to a restricted set of punctate stimuli, or by noting the response to linear illuminance gradients in two orthogonal directions we obtain the "Cartesian coordinates" of the point and so forth.) For such an operator one easily imagines a physical embodiment, *e.g.*, a flat photovoltaic cell, a receptive field. There is no need for the physical operator to have "zero size" (if that were possible at all!), but one should turn a blind eye to its inner structure. (Thus the photovoltaic cell returns only a number, the incident flux per integration period.)

Of course the definition of geometrical objects is only a start. One will attempt to define *relations* between the objects in an operational manner. If succesful, one may use geometry as a *language* that describes the functional brain structure[12]. This has obvious advantages: because geometries are among the best understood mathematical objects one is then in a position to predict as yet unnoticed relations, and so forth. This is exactly the manner in which many natural phenomena have been formalized (*e.g.*, symplectic geometry is often used as a model of classical mechanics). It is just that brain science has largely ignored such obvious possibilities.

An additional issue I haven't mentioned is that some sensory modalities (haptics, optics, to a lesser extent acoustics) and motorical competences (grasping, manipulation, locomotion, ...) are most naturally described in geometrical language in the first place, this holds for both the stimulus as well as the response domains. The full process of optically, ...guided behavior can often be treated as a sequence of mappings between geometries.

2 GENERAL ISSUES

It will not be possible to discuss all (or even many) of the geometrical aspects of optically, ...guided behavior and the related brain processes and structures. I will focus on the optical domain ("vision") and merely pick out a few goodies. Because I can't assume that all of you will be thoroughly conversant with issues of abstract geometry I need first to spend some time on explaining various important distinctions and domains.

2.1 Local sign

The problem of "local sign" has been latently present in philosophical discussions since the classical period, but was first formulated precisely by Hermann Lotze[13] in the second half of the nineteenth century. His reasoning is fairly subtle. Lotze is concerned with conscious perception. Of course "perception" is by definition an aspect of consciousness, but I want to stress the fact that local sign does not apply to mere sensorimotor coordination. The latter can always be explained as an—be it incredibly complicated—map from the receptors to the skeletomuscular system for which one may think of network implementations, that is essentially a triviality. The point is that the perceiver *knows* where the photoreceptors (or rather retinal receptive fields) are in the visual field. That such is not trivial, but is somehow established during ontogenesis, is clear from the fact that cases exist in which this process fails (cases of amblyopia sometimes termed "tarachopia", or "scrambled visual field"). That the brain has a knowledge of *where* parts of the body are is not a trivial feat, as is well illustrated by the familiar phenomenon of "referred pain".

That somatotopy doesn't provide an "explanation" may become clear when you perform the *Gedanken* experiment of spatially permuting two cortical pyramidal cells leaving all the wiring intact: will this permute the locations in the visual field? If you believe so you'd better skip the remainder. Hermann Lotze had no need for homunculi and neither have I.

2.2 Symmetries

Another important point has to do with the issue of symmetries, or the "universality" of the human visual system. When someone explains a theory of the visual system usually a person from the audience will rise and object that the human visual system is not at all like the theory because .... Such a remark is almost certainly factually correct, but most often somewhat besides the point. I'm not claiming here that theories should be immune from empirical verification! It is just that a healthy pragmatic attitude increases potential pay offs. A theory that doesn't explain the universe may still be useful, as long as one is clear about its (always present) limits. For instance, strictly speaking all theories of physics are false except for some that are too hard to verify (superstrings?). For instance, geometrical optics or classical mechanics have been proven "false" a century ago, yet everybody uses them in daily laboratory situations: it would be sheer madness to discard anything as useful merely because one perceives obvious limits to the domain of applicability. Yet a similar practice is customary among brain scientists.

Most theories of vision start with symmetry assumptions, such as homogeneity, isotropy and selfsimilarity of the visual field. All these assumptions are obviously false[14,15]: the peripheral visual field differs from the central part, the vertical differs from the horizontal (not to speak of "oblique effects"), size invariance doubtless fails at very small or extremely large sizes. Does this doom the theories from the very start? I don't think so. For such a practice would lead one to abandon theory altogether, and one would be left with phenomenological descriptions of specific systems. Then "science" would become mere butterfly collecting and effectively cease to exist. This can't be tolerated.

Suppose one had to design a visual system for a robot[16] fit for an unspecified environment. Such a system cannot be committed to any specific world. One has to assume that one place will (on the average) be like the other, that there are no special directions, or any preferred size. The system would have to be designed on the assumptions of full symmetry: homogeneity, stationarity (in time), isotropy and selfsimilarity. Once one knows something about the prospective environment, one would of course use these constraints. (For instance, if all objects the robot ever had to deal with would be known to be of a single size, one would abandon selfsimilarity.) It is the same in the animal kingdom, where evolution has fitted species for specific (very generalized) biotopes. As these biotopes differ, so will the visual systems. From a theoretical point of view it is advantageous to view all visual systems as *specializations* of the fiducial totally uncommitted (tailored for an extremely varied biotope) system. Then one can build a general theory[17] and study the hierarchy of possible specializations of it. Although *no particular* visual system will be exactly like the fiducial universal prototype, yet they are more easily described as (in itself lawful) deviations from the ideal than as completely independent entities. Nontrivial comparisons between the visual systems of different varieties depend on the existence of such a genetic tree of theories.

In this paper I have no room to discuss many possible specializations and will largely concentrate on the fiducial, totally uncommitted system. Such a system is only constrained by the expectation that the structure of the world is basically equally important at all places, in all directions, at all sizes, at any epoch or timescale. Evolution will weed out systems that overly specialize in such a world. This principle turns out to be much more powerful than one might expect at first blush.

2.3 Local, multilocal and global structure

The concepts "local", "multilocal" (or neigborhood) and "global" are often used in a loose sense. In mathematics these terms have obtained a precise meaning[18], and are sometimes extended by further terms of the same nature (*e.g.*, "punctal"). I will use a fairly strict convention in this paper[19]. A property will be denoted as

punctal if it applies to a single point. *Example*: barometric pressure at Trento;

local if it applies to an infinitesimal neighborhood of a point. By that I mean that such geometrical objects as vectors, or tensors, which may be defined by operators ("receptive fields") at a single location, may enter the description. *Example*: direction into which the water will flow at some point;

multilocal if it applies to more than one distinct (though usually close) point. *Example*: height difference between Trento and Amsterdam;

global if it applies to the space as a whole. *Example*: the surface of the globe has no boundary.

Punctal and global properties are typically of little interest for our pursuits. It is very important to understand the vital differences between local and multilocal fully though.

The all important difference between the two is that *locally* all geometrical expertise can be encoded into the structure of the receptive fields. There is no need to be able to find the distance between receptive fields (they're all at the same location anyway) for instance. In contradistinction, *multilocally* the geometrical expertise has to be encoded into the relations between the receptive fields. For instance, when are three simple cells collinear (an extended local sign problem)? When are two simple cells at different locations parallel (problem of "parallel transport", or "connection")? Whereas local processing does not even need local sign, for multilocal processing one has to postulate some kind of calibration procedure.

3 TOPOLOGY

"Topology" is a branch of geometry that abstracts from all metrical (distance, angle, ...) and projective (collinearity, convexity, ...) properties. It is merely concerned with coincidence, overlap, boundary of regions, *etc.* Although this may seem very insignificant, exactly because of these restrictions topology is actually the all important backbone (or "glue") common to all geometries. Without topology little is left to do. For our purposes we practically equate "topology" with "local sign".

In order to start with *something*, I assume that we have a collection of "points". These will be operationally defined in the way indicated before. Let's consider relations between the points. Such relations will have to be operationalized too of course. A very simple example will be *coincidence*, or equality of points. I consider two points to be coincident if and only if they never fail to produce identical responses to all stimuli I try. The physiologist can distinguish the locations where he puts his probe, but *functionally* such points are equal, and one cannot distinguish between them. (Notice that the location where the physiologist puts the probe never enters the operational definition of the point. The brain *itself* cannot know this location, or one would have another local sign problem!).

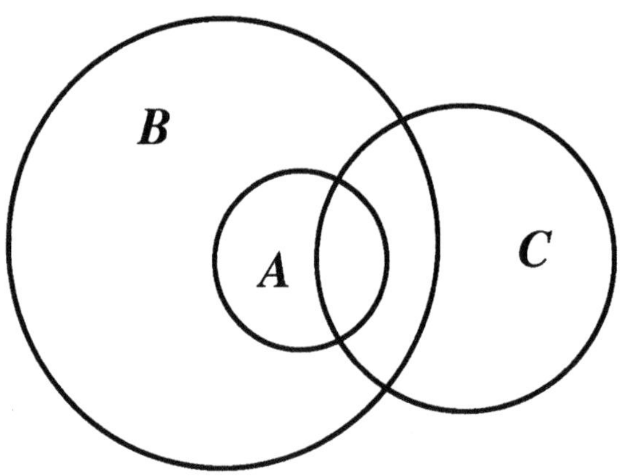

Figure 1: Functional definition of "inclusion": a receptive field (A say) is included in another (B say) if and only if any receptive field (C say) that correlates with it (A with C) also correlates with the other (B with C).

That correlated neural activity leads to the notion of "being at the same place" in the mind is vividly illustrated by a curious phenomenon first described (and explained in our manner) by Helmholtz[20]: a patient with toothache usually can't indicate whether the troubling spot is in the upper or lower jaw! Helmholtz explained this by noting that

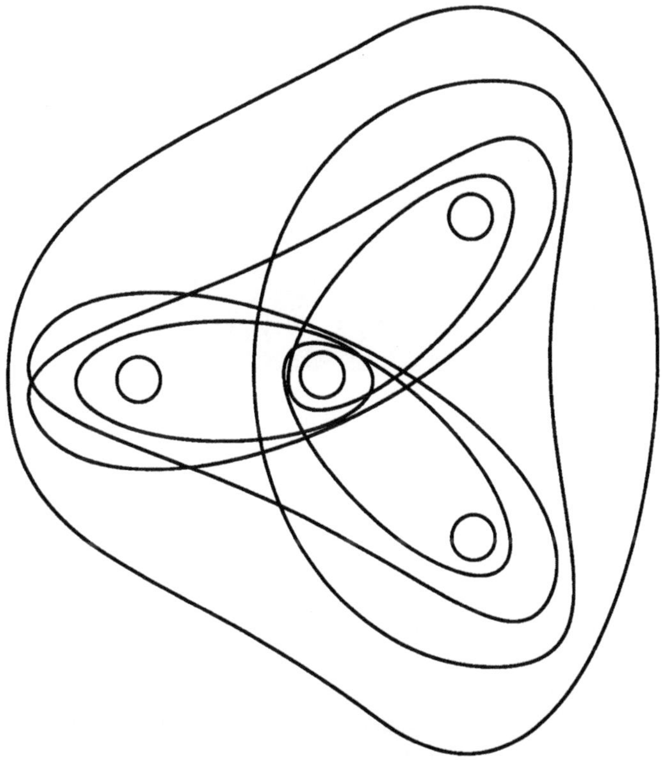

Figure 2: A simple configuration of eleven heavily overlapping receptive fields.

in chewing our food the neurons at the upper and lower jaw are always stimulated in synchrony. For the brain they are "coincident" in our definition, though physically they are well separated.

A somewhat less trivial example would be the case in which the responses where highly correlated, but not identical. In such a case the points are *related* without being coincident, I will call them "overlapping" (or "incident"). I will use the notation $\mathcal{A} \sim \mathcal{B}$ for the relation "$\mathcal{A}$ is incident with $\mathcal{B}$". Notice that this information is in principle available to the brain itself, whereas the position of my probe isn't. The relation is obviously symmetrical, that is, the relations $\mathcal{A} \sim \mathcal{B}$ and $\mathcal{B} \sim \mathcal{A}$ imply each other. Moreover, you (trivially) have $\mathcal{A} \sim \mathcal{A}$. From the relation of incidence I can derive many other useful relations via logical operations, for instance, I define "$\mathcal{A}$ is included in $\mathcal{B}$ (or $\mathcal{A} \subset \mathcal{B}$)"

Figure 3: The correlation matrix corresponding to the receptive field layout shown in figure 1. This sums up the brain's knowledge of the receptive field layout.

as follows (see figure 1): "$\mathcal{A} \subset \mathcal{B}$ if and only if it is true that for any point $\mathcal{C}$ the condition $\mathcal{C} \sim \mathcal{A}$ implies the truth of the condition $\mathcal{C} \sim \mathcal{B}$. (Many colleagues will prefer: "$\mathcal{A} \subset \mathcal{B}$: $\forall_{\mathcal{C}}$ $(\mathcal{C} \sim \mathcal{A}) \Rightarrow (\mathcal{C} \sim \mathcal{B})$".) This definition is reasonable if you draw simple pictures in which the points are indicated as circular discs in the plane (see figure 1).

In a similar vein I can define a notion of "convex hull". A point $\mathcal{Q}$ is defined to be a member of the convex hull of a set of points ("polytope") $\mathcal{P}_1, \mathcal{P}_2, \ldots, \mathcal{P}_n$ if and only if it is establised that for any point $\mathcal{S}$ that includes each of the $\mathcal{P}_i$, it can also be established that $\mathcal{Q} \subset \mathcal{S}$. This notion of convex hull is important because it opens the way to build higher dimensional entities. For instance the "line $\mathcal{L}$" defined by the points $\mathcal{P}, \mathcal{Q}$ is the collection of points that contains $\mathcal{S}$ if and only if of the triple $\mathcal{P}, \mathcal{Q}, \mathcal{S}$ one lies in the convex hull of two others. The prototypical convex hulls are based on polytopes of $(n+1)$ points, where n denotes the *dimension* of the space. For instance, in the plane we consider the *triangles*, which are convex hulls based upon point triples. A (nondegenerated) planar triangle contains more than the union of its sides. (A "side" is the convex hull of any subset of two points of the triple.) On the other hand any fourtuple of points leads to a convex hull that is the mere union of its component triangles. This means that we may empirically deduce the dimension of the space (*e.g.*, the planar nature of the retina) from the correlation structure of the set of points (in the example the correlation structure of the optic nerve fibers).

The relation of overlap is fundamental. Indeed, if the points happen to conform to the pattern of circular discs on the plane, you may "boot up" the topology (even Euclidean geometry!) from the overlap relations alone, much in the spirit of the example of the "line". The figures 2, 3 and 4 illustrate a simple example. The example is instructive because you can easily check it by hand in only a few minutes. First construct the correlation matrix by checking the overlap relations in the geometrical layout of receptive fields. Then find the inclusion relations through application of the logical rule discussed above. This yields

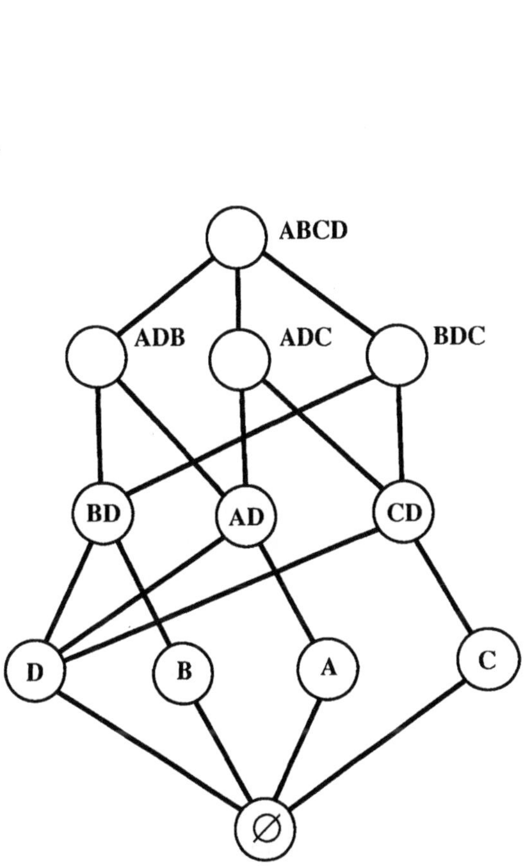

Figure 4: The Hasse diagram of the partial order of inclusion of the receptive field layout shown in figure 1. The partial order can be derived from the correlation matrix shown in figure 2 using the logical rule mentioned in the text. In this diagram the lowermost node is an additional formal receptive field "the empty receptive field" that is conventionally included in *any* of the real receptive fields.

the partial order of units that is illustrated in the Hasse diagram: in this figure a node located above another node and connected via edges includes the lower node. This Hasse diagram again admits of simple geometrical interpretations, the simplest one being the "Y-branch" shown in the final figure of the series. (See figure 5.) This Y-branch neatly sums up the topology of the original receptive fields layout. Note that you can construct it from the correlation matrix alone though: monitoring multi unit activity allows you to reconstruct the geometry. Because the correlations are objectively present in the brain (no homunculus needed), we conclude that the brain may indeed boot up the topology from the signals alone!

This was very nicely shown by Huntington[21], who constructed an Euclidean geometry with "points of finite size". Of course, there's no guarantee that the actual relations will conform to this pattern of overlapping circular discs. It is easy enough

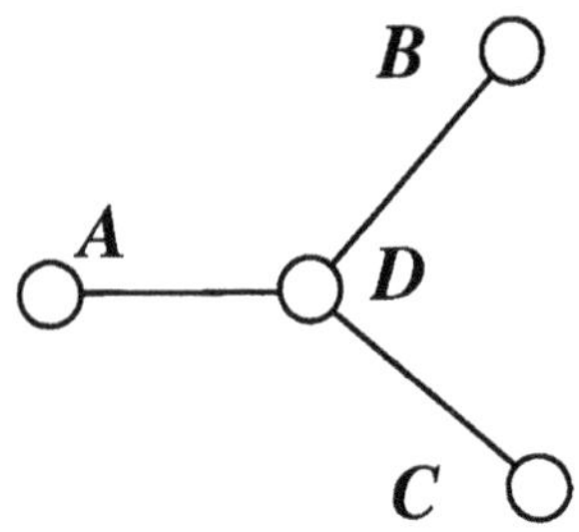

Figure 5: The geometrical configuration defined by the Hasse diagram shown in figure 4 is a one dimensional space. However, it is not a line, as its points cannot be linearly ordered. It contains a bifurcation point. Apart from the points A, B, C and D (which appear as "points" in the figure), one has to conceive of various "larger points" such as [AB] (which contains A and B, but not any other point) and [ABD] (which contains A, B, D, [AD] and [BD]), *etc.* The eleven receptive fields are represented through the points { A, B, C, D, [AD], [BD], [CD], [ADB], [ADC], [BDC], [ABCD]}.

to construct counterexamples, for instance consider the collection of annular shapes (the apertures defined by unequal, concentric circles) in the plane. A somewhat less weird example is the collection of all convex sets of squares of a chess board. (An instructive—because counterintuitive—example that one easily studies in the margin: the space is one-dimensional(!), though not a line, for if I delete an internal point it doesn't fall apart into two pieces!) For some reason (diffusion processes acting through morphogenesis may have to do with this) actual visual receptive fields tend to be convex regions, and one is in good shape: the overlap relations (correlations between receptive field responses) encode the topology and no homunculus is required to impose Lotze's local sign.

For the present purpose it suffices to have seen the *possibility* of a fully operational definition of topology.

4 THE INFINITESIMAL DOMAIN

4.1 The n–jet

"Infinitesimal" has to be taken *cum grano salis*: I really mean "local". However, the idea of an "infinitesimal neighborhood" has such important connotations[22] that I prefer its use here. If the retinal illuminance distribution would be differentiable as many times as needed (this is easily enforced as I will show later), you may compute all spatial derivatives up to order n say. Then you can perform the following calculations on the basis of these $(n+1)(n+2)/2$ (in dimension 2) numbers, without further reference to the illuminance distribution itself:

- you can approximate the illuminance distribution through its nth order Taylor expansion in a neighborhood of the point;

- you can approximate various transformations of the illuminance distribution (*e.g.*, shifted, rotated, expanded, sheared versions) in that neighborhood;
- you can compute various geometrical properties of the illuminance distribution, such as the illuminance gradient, the isophote curvature, *etc.* The number of independent measures you can obtain depends on n of course;
- you can compute various summary descriptions of the local illuminance distribution (*i.e.*, various textural properties).

No geometrical expertise is needed to do these computations: all the necessary data are already contained in the values of the derivatives. One may envisage such neighborhoods to exist as "columns" in some somatotopic projection of the retina, each column containing the operators necessary to compute the spatial derivatives. Figure 6 illustrates the receptive fields that could together make up the two jet at some location for some particular scale. This is a minimum basis: there could be many more receptive fields, but they would not add to the completeness of the two jet (although they might help gain additional noise immunity, *etc.*). The illustration shows just one particular implementation. Receptive fields of different structure (for instance with polar–azimuthal symmetry) might have been used as well[23]. This makes no difference since all functions in the two jet can be constructed from any basis set. That's why the nature of the receptive fields themselves reveals *very little* about the power of the assembly (the two jet), a point that has important consequences for electrophysiology.

Of course it would be somewhat of a blemish if you had to find the derivatives in any *particular* coordinate system. Every estimate of illuminance structure for the undedicated visual system should be independent of any such arbitrary specialization.

For the formal description[24] we disregard such specializations by way of a simple trick: we don't speak of derivatives at all, but only of the equivalence classes of all illuminance distributions that differ at most in some mth ($m > n$) spatial derivative. Such equivalence classes are known as "n–jets" in modern mathematics and are indeed geometrical objects (*i.e.*, exist independent of any particular coordinatization). The n–jets can be *represented* in many different ways, for instance through specification of the (mixed, partial) derivatives in some Cartesian system, but this is not necessary. Many different representations (Cartesian, polar, Gabor–type wavelets, *etc.*) can be constructed and are useful in various contexts, although they all specify the same datum: the n–jet of the illuminance distribution.

4.2 Linearity or nonlinearity?

The derivatives, as a basis for the n–jet, are *linear* functionals of the stimulus. This is often regarded as a drawback, because linear systems are supposedly trivial. Various answers may be given to such objections, *e.g.*, being trivial doesn't implicate being useless: it may actually be an advantage to have a representation with simple and transparent properties. But the more correct answer to the objection is that the true entities represented in the n–jet are the coordinate independent geometrical invariants, and these are

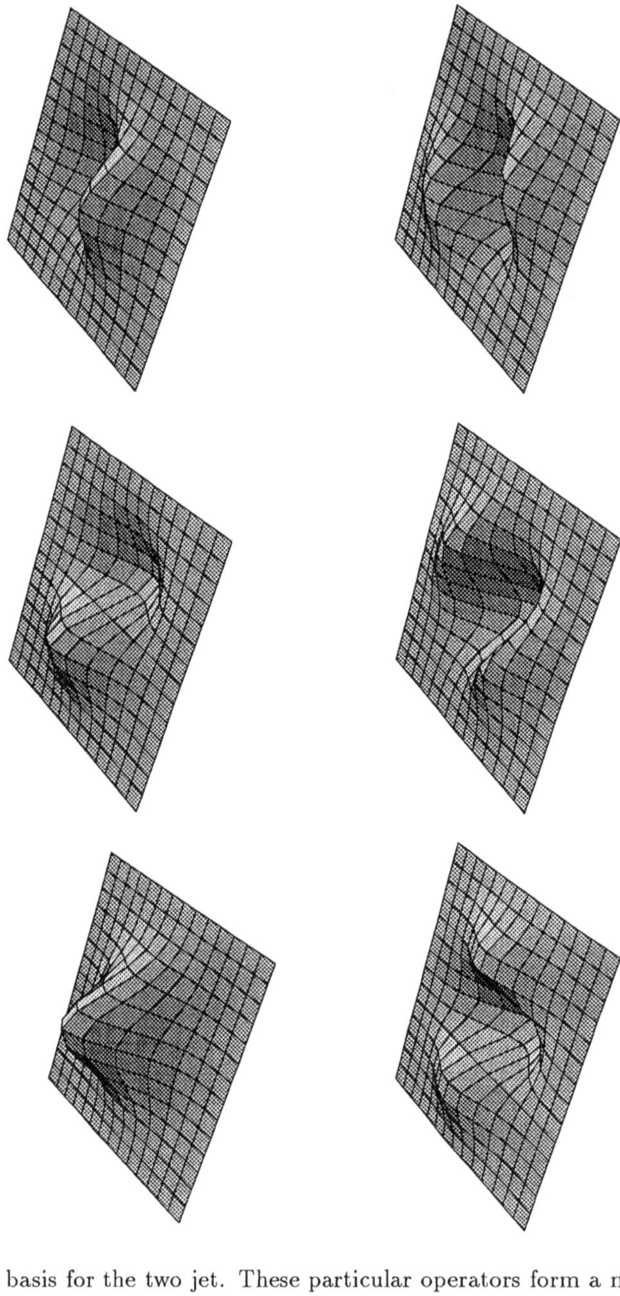

Figure 6: One basis for the two jet. These particular operators form a minimal basis of directional derivatives that resemble Gabor type wavelets.

algebraic, typically *nonlinear*, combinations of the derivatives. (Just think of the definition of isophote curvature which contains such combinations of first and second order derivatives as $\frac{\partial I}{\partial x_i}\frac{\partial I}{\partial x_j}\frac{\partial^2 I}{\partial x_i \partial x_j}$.)

The dichotomy linear versus nonlinear is actually not very useful and may even depend on your description of the system.

4.3 Multiresolution representations

In the discussion of the n–jet I have deviously glossed over the problems of continuity, noise sensitivity and support area, and have not at all specified what the size of the "neighborhood" might be in the first place. The fact is that the undedicated system knows no preferred size. It has to function on *all* sizes. For any visual routine one has to distinguish four characteristic lengths:

grainsize that is the smallest length over which significant change can be established. For the human visual system the grainsize depends on optical factors and the receptor diameters;

scope that is the greatest distance spanned by sampling units. For the human visual system this is the diameter of the retina;

inner scale that is the "pixel size" used for the computation. It has to be smaller than the scope and larger than the grainsize;

outer scale that is the "region of interest" used for the computation. It must be smaller than the scope, and larger than the inner scale.

If the inner scale is close to the grainsize you suffer from "insufficient resolution", if the outer scale is close to the scope you run into "boundary problems". The visual agent will strife to get rid of these problems through exploratory behavior (getting closer for detail, stepping back to gain an overview). Typically the outer scale is much smaller than the scope whereas the inner scale far exceeds the grainsize. In such cases you may forget about grainsize and scope: only inner and outer scale matter in typical local computations.

What scale should one pick[25,26]? That depends on the task and on the optical input. It's not so good if you don't see the wood for the trees, but the reverse is equally bad. The uncommitted agent has to be able to process optical structure over a range of scales, as indeed humans seem to be able to[27,28]. This brings up the problem that the same stimulus will need to have multiple representations, and that these representations had better always agree on what they hold in common. Specifically, fine detail should be "summarized" or "generalized" in a coarse representation and placed in "context" there, whereas coarse detail should find its "cause" on the finer scale. This causal dependence, or directedness, is familiar from the structure of geographical atlases. The requirement that the multiresolution representation should be causal is a very strong one, and together with the general symmetries one imposes on the uncommitted system it fully determines the structure of the possible receptive fields: no choice is left.

Figure 7 shows a very simple example. It depicts a segment of an "edge". Actually, this is a thresholded grayscale picture. The threshold is identical for all pictures. The illuminance pattern is shown as detected by sample arrays of various inner scales. The receptive field profiles were isotropic Gaussian functions: this is the *only* way to be certain of a "causal behavior" over scale changes. As you see in the example small details are progressively removed as one moves towards coarser scales. In noncausal systems it could happen that details were *created* on moving to coarser scales, a phenomenon that certainly should be avoided. (Imagine a geographical atlas that showed increasing detail on going from fine to coarse scale!) Only receptive fields based upon the Gaussian profile (and its spatial derivatives) can respect the various front end symmetries and causality. Thus the common remark in much neural modelling that such profiles are convenient descriptions because they look like realistic profiles and are mathematically simple completely misses the mark—such profiles are the *only* choice and not a mere convenience. Other profiles would violate some of the symmetries or causal behavior, in many cases with objectionable results. At best, some different profiles might provide useful specializations of the universal structure, such as systems exploiting the violation of certain symmetries (size invariance, preferred directions, ...) in the agent's biotope.

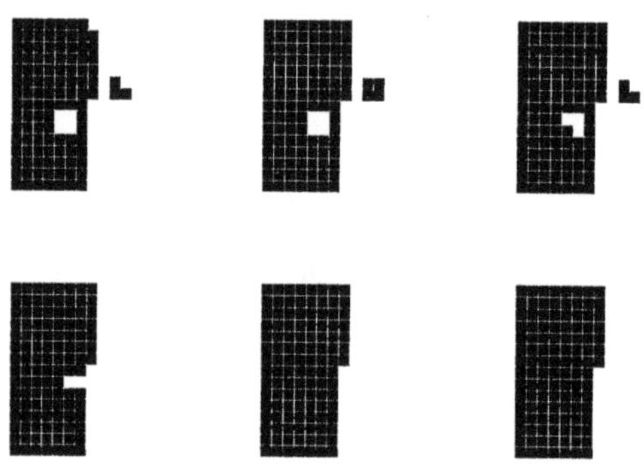

Figure 7: A perturbed edge at various levels of resolution (steps of square root of two of inner scale). The direction of increase of inner scale should be visually obvious: blurring "generalizes", whereas deblurring "articulates".

In the multiresolution representation the "points" *must* have Gaussian profiles if causality and symmetries are to be respected, and one possible basis for the n–jet at any given scale is the set of all directional derivatives up to the n–th order of such a Gaussian profile. This turns out to yield the right "regularization" for the derivatives automatically. The receptive fields with n–th order directional derivatives of a Gaussian profile compute *exact* n–th order directional derivatives of the retinal illuminance pattern *as seen by a point of the corresponding size*, that is of a suitably blurred pattern. As a result the derivatives always exist and the process is robust against noise.

As is well known the derivatives of Gaussian profiles are quite similar to what one encounters in the primate visual system[29,30,31,32].

4.4 Equivalent inputs

The local n–jet can be represented as a $m = (n+1)(n+2)/2$–tuple of numbers (receptive field activities say). Such a m–tuple of numbers comprises a local *observation* of the retinal illuminance distribution. The brain has only these numbers to work with: they are all of the optical input available to the brain—locally at least. Given that you may easily dream up infinitely many illuminance distributions (given sufficient time), we meet with a problem. The stimulus space is an infinite dimensional Hilbert space, whereas the space of all observations is only (part of) a m–dimensional vector space. It must be the case that an observation typically signifies an *infinite number* of possible stimuli. Thus the brain can never find out what the stimulus really is like, but can only classify stimuli in terms of equivalence classes. The equivalence class of all optical inputs that give rise to a single observation is known as a "metamery class", and the various illuminance patterns are known as "metamers". For ease of reference you may pick one metamer (the "simplest" in some sense) and use this "canonical stimulus" as representative for the full metamery class. Such a canonical member sums up all knowledge available and is the "image" the brain may entertain of the "true" optical input (which exists only for the external observer).

Canonical images may be picked in various ways. However, in a great many natural circumstances certain constraints apply that greatly limit the freedom of choice. Suppose you look at a collection of objects under some illuminant. Except from occasional specularities (mirror reflections of the sources) the luminances in the scene will be strictly limited from above. Of course the luminances are also limited from below, since they are always positive. In such cases it is *a priori* clear that not all observations are physically possible. Thus the observation space (which is $\mathbf{R}^m$) can be divided into two regions, a certain region $\mathcal{A}$ (say) of observations that admit of a physical interpretation, and its complement $\mathcal{A}^\dagger$ that contains the "impossible" observations. Of course the impossible observations may well occur as brain states, it is just that they have no causal explanation in terms of physical stimuli. If such an impossible observation is encountered we may consider it as an "error message", or just ignore it. If the observation is in the interior of $\mathcal{A}$ there will in general exist infinitely many equally valid physical interpretations (the metamers). If an observation is a member of the *boundary* $\partial\mathcal{A}$ of $\mathcal{A}$ we meet with a very singular case: such observations turn out to admit only unique physical interpretations. Such an observation thus *fully specifies* the actual stimulus (locally that is). I will refer to these observations as "boundary observations". Of course such observations are not likely to occur in realistic situations though.

One may easily prove several theorems that characterize the geometrical nature of the region $\mathcal{A}$ as a subset of $\mathbf{R}^m$, that is observation space. The proofs are easy modifications of similar proofs by Erwin Schrödinger[6,7] relating to the case of object colors under various illuminants. One proves that:

- the region $\mathcal{A}$ is simply connected, in fact, it is *convex*;
- the boundary $\partial\mathcal{A}$ of the region $\mathcal{A}$ is smooth, except at two points, namely the origin and the point that corresponds to the observation due to a maximum uniform retinal illuminance ("white" picture);
- all "gray pictures" (uniform illuminance) map to the straight line segment that connects the origin with the observation due to the white picture;
- the midpoint of this gray axis is a symmetry center of the region.

In addition to this we note the (really trivial) fact that the map from the stimulus domain to observation space is a *linear* one. It follows that any possible observation can be "explained" as due to an *attenuated* boundary observation. This provides a simple procedure to define canonical pictures in a systematical manner.

So what are those canonical pictures like? The answer depends on the dimension of observation space, or—equivalently—on the order n of the n–jet. For $n = 0$ (only illuminance samples) the canonical pictures are just the gray pictures. Indeed, for any picture we may find a uniform gray picture that would yield the identical observation, and this is clearly the simplest (and thus the preferred in the absence of further prior knowledge) interpretation of that observation. For $n = 1$ each observation admits an interpretation as a suitably offset and oriented edge. Of course one may only trust the observation in a sufficiently small neighborhood: if the edge is far away the canonical picture is effectively a gray picture. Thus the first order jet allows local interpretations in terms of uniform pictures and straight, offset step edges.

The case $n = 2$ is somewhat more interesting. Note that the zeroth order receptive fields are mere Gaussian weighted integrators of luminance, whereas the first order ones are like "edge detector"–like simple cells and the second order ones as Hubel and Wiesel's "line detectors". Thus the second order jet is like a very simplified model of a column in area V1. The canonical pictures turn out to be uniform blobs on a uniform background, with general quadrics (ellipses, hyperbolas) for blob boundaries. If one limits the attention to restricted neighborhoods, then this gives rise to the following qualitatively different pictures:

- uniform gray pictures;
- offset curved step edges of various orientations;
- offset curved and tapered light or dark bars of various orientations;
- elongated and offset light or dark blobs of various orientations.

Again, *any* observation admits one of these interpretations (or is an "error message"). For the second order jet only uniform pictures, edges, bars and blobs exist, the system is totally "blind" for anything else.

For higher order jets the number of qualitatively different pictures of the n–jets increases enormously. The available evidence suggests that the columns in area V1 have to be modeled with jets of an order as high as four: then the number of qualitatively different pictures already runs into the hundreds. It doesn't seem far fetched to interprete these canonical pictures as reminiscent of the "textons" revealed by psychophysical procedures[33].

4.5 Features and feature detectors

The notion of a "feature" and the related notion of "feature detector" frequently turn up in psychophysics[34,33], neurophysiology[32], image processing[35] and computer vision[16]. These notions are curiously ill defined. In some cases one has some operational definition of "feature" and then defines feature detectors as mechanisms fit to respond selectively to these features. In other cases one defines or implements a "feature detector" and then defines features in terms of what this mechanism reponds best to. In almost no case does an author escape the vicious circle lurking here. Common expressions like "edge detector", or "line detector" are subject to this criticism.

The construct of the canonical pictures of a general n–jet provides a natural definition of "feature": features are the qualitatively different canonical pictures of the n–jet. Thus blobs, edges and bars are features of the two–jet.

There is a problem here though: it is not possible to define "feature *detectors*" at all. For instance, the "bar" is one of the features of the two–jet, but there doesn't exist a corresponding "bar *detector*". It is true that the two–jet typically contains receptive fields with profiles that look exactly like those of the "bar detectors" one finds in the literature. However, such units cannot be said to detect bars: only the two–jet as a whole detects bars. The output of a given "bar detector" may be identical in cases where the two–jet detects a bar, a blob, or an edge. This is to say that the interpretation of the response of any given operator depends on the context of all other responses in the jet.

5 CONNECTIONS

The possible computations in a n-jet for rather small n are quite limited and will rarely suffice to meet the requirements of an optically guided agent capable of sophisticated behavior. The great advantages of the jet construction are obvious enough:

- no geometrical expertise is needed except for what has been put into the wiring in the first place. The need for continuous (re–)calibration and sophisticated higher order visual routines is obviated;
- since everything is prewired the processing can be very fast, in parallel, and "pre–attentive";
- universal representation need not be sacrificed, yet the structure is very simple.

The drawbacks are equally obvious though:

- only a fixed repertoire of simple computations is supported;

- it is not obvious how to increase the region of interest (outer scale) without decreasing the resolution (inner scale) if the order of the jet has to be kept within bounds. Indeed, although the primate visual system may well go to as high as order four, image processing practitioners find it prudent not to exceed order one or two. (But CCD cameras are not yet as versatile as the eye).

It is merely good designers horse sense to invest in a system that starts off by building a parallel representation of jets of as high an order as is reasonable. Then one has to build layers on top of this that somehow combine the representations of noncoincident jets by "gluing them together" as it were. In figure 8 you have a simple example of an obviously relevant configuration of local jet activities. Of course such a pattern is actually buried in a field of local jet activities that doesn't fit this pattern so well. Moreover, there will be perturbations, gaps, and so forth. How to find such multilocal geometry is a highly nontrivial task that has not yet been solved in a satisfactory fashion by engineers even after years of activity in the field of image processing. The human visual system seems to do remarkably well on the whole, although it is easy enough to reveal obvious shortcomings in the laboratory[15,36]. Such shortcomings, or blindnesses to geometrical relations in the visual field, are to be considered *mental deficiencies*, since all the necessary data certainly enter the brain via the optic nerve.

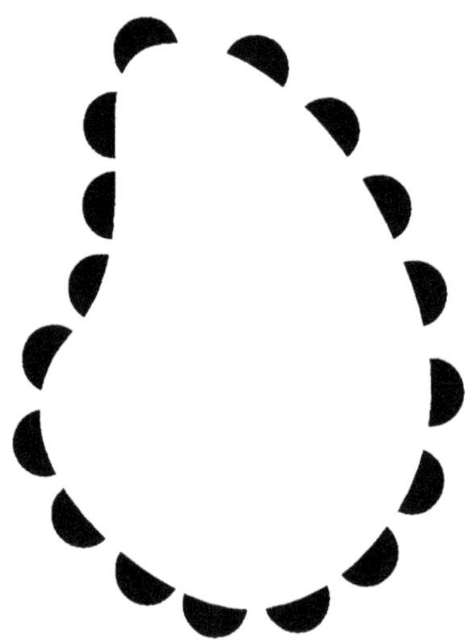

Figure 8: A number of two jet observations at different locations may well indicate some larger configuration of considerable interest. However, it is no easy task to find such a multilocal order!

This same problem arises in differential geometry, and we may well look at the way geometers have solved the problem before we start to speculate on visual systems. The geometer considers the Cartesian product of a base space and a fiber space. (Known as a (trivial) fiber bundle.) For the base space we may take the visual field, for the fiber space the jets. The "fiber over a point" is just the jet at that point, and the "natural projection" reveals for each element of the bundle "where the jet is at". The problem is to put additional structure on this "jet bundle". The standard way to do this is to define a "connection", that is basically a way to compare jets at different (but close) locations. Once a connection has been established, jets at different positions can be compared via an integration of the connection along a curve that connects the locations. (Of course it may well turn out that the result depends on the choice of path.)

In the simplest nontrivial case (first order jet) the connection boils down to the definition of a way of "parallel transport" of a vector. In the brain this may assume the form of a set of fibers connecting neighboring columns in such a manner as to associate simple cells with like, and/or dissociate those with unlike orientations. It seems *a priori* likely that such a system would have to be calibrated via exposure of the system to the external world. One way to do this would be to exploit the results of voluntory eye movements. Rotations of the eye-ball about an axis in the equitorial (or frontoparallel) plane would put the receptive fields to be associated via parallel transport into successive correspondence. Such eye-movements satisfy Listing's law and thus will be the rule. The idea that the geometrical structure of the visual field might well be due to eye movements goes back to Helmholtz[37] and (especially) Poincaré[38].

The use of the connection is best explained through a simple example. Suppose the system wants to extract "edge curvature". In the two-jet this can be interpreted as the curvature of the local isophote (locus of constant retinal illuminance). Of course this only makes sense if the gradient of the illuminance turns out to be high. The isophote curvature is a particular algebraic combination of first and second order partial spatial derivatives. It can be implemented as a simple pre-wired circuit. No further geometrical expertise is required. The algorithm is simple: either you use a fairly complicated algebraic expression that involves all first and second order derivatives ("edge and line finders", this could be pre-wired into a simple neural network), or you find the direction of the illuminance gradient first. Then the curvature is just the ratio of the two receptive fields illustrated in figure 9 (the upper profile is like an "end stopped cell"). Of course the latter method assumes some expertise in the column. Such expertise might take the form of a system of connections ("pointers") that would allow you to find a "line finder" parallel or orthogonal to any "edge finder", and so forth. Again, this implicates a local neural network.

It appears likely that this type of algorithm runs in the human front end visual system[40].

In the one-jet such a simple computation is not possible. However, the computation may be done in the one-jet *bundle*. First one finds two fibers at different locations on the same isophote. This can be done if the system is able to compare the local illuminance estimates. (This is not trivial if the system uses independent local automatic volume controls to keep the responses within bounds. Then the local amplification factors must be available.) Next one uses first order operators ("edge detectors") to find the gradient direction at the location of each fiber. These are compared via the connection (parallel transport of vectors!) and the angle of turn is measured. That the gradient turns when one moves from one place to the next means that the isophote must be curved. Indeed,

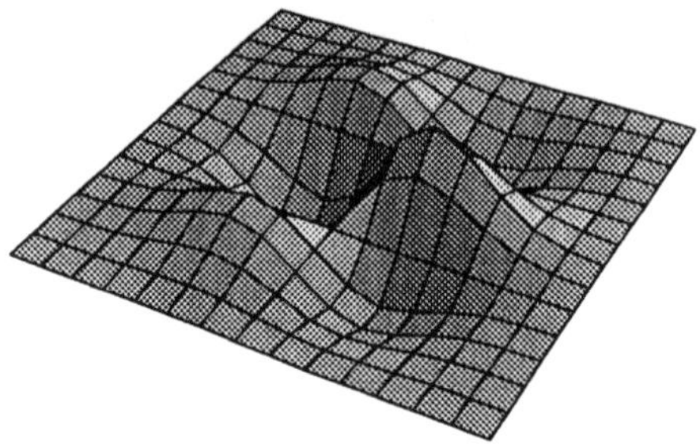

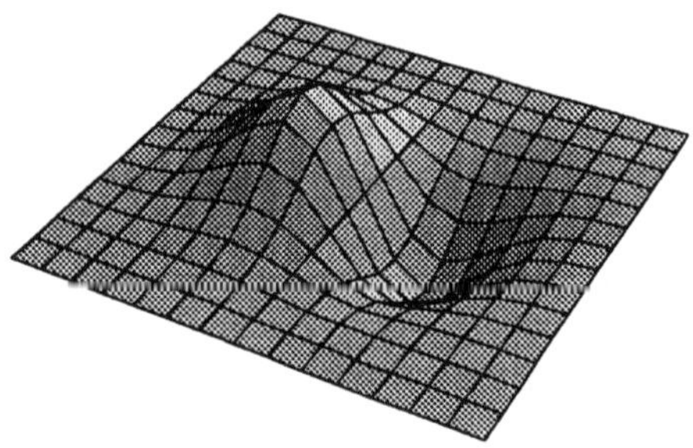

Figure 9: Boundary curvature detection in the two jet: the boundary curvature is just the ratio of responses of these two receptive fields. (Of course the boundary direction has to be determined first.) Notice that this is a purely local operation that requires no explicit geometrical expertise.

the curvature equals the angle of turn divided by the distance of the locations where the fibers are at. Figure 10 illustrates this case. The curvature is simply $\kappa = \delta\Phi/ds$. Clearly one also needs to be able to measure this distance. Thus the curvature can indeed be computed in the first–order jet bundle even though it can't be computed in any individual one–jet. But rather high demands on various types of geometrical expertise are made. Such expertise has to be provided by systems on top of the jet bundle somehow.

In the zeroth order jet bundle the situation is even worse. It is again possible to compute isophote curvature, but that at the cost of much complicated geometry. One first finds three (different) locations where the retinal illuminance is equal. By testing other locations one verifies that a reasonable illuminance gradient exists. (The input might be a uniform field, then the compution would fail.) Next one has to assess the metrical relations of these points, the three distances for instance. Then one may fit a unique circle through these points and compute its radius (the case is illustrated in figure

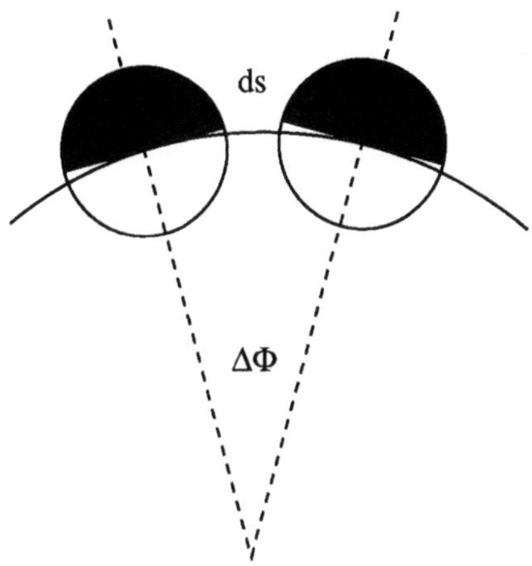

Figure 10: Two first order fields from different fibers at the same illuminance. If the two directions and the distance are known, then the curvature can be estimated simply as the angular turn per unit distance.

11). The curvature is the reciprocal of the radius of this circle.

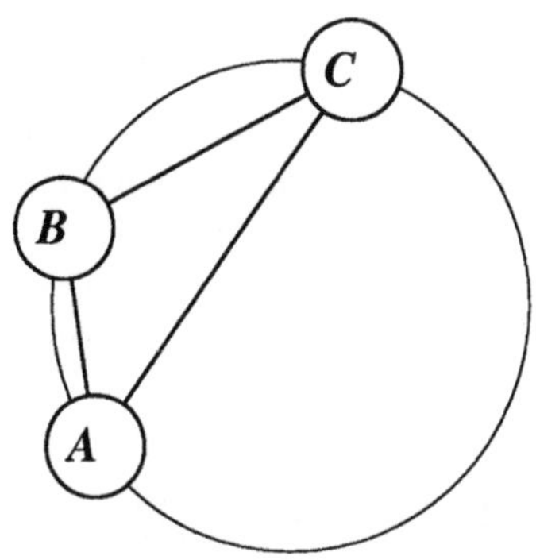

Figure 11: Three zeroth order fields from different fibers at the same illuminance. If the three distances are known, then the curvature can be estimated via the simple algebraic expression for the radius of the circumscribed circle of the triangle.

It is quite clear from such examples that one should perform the required calculations locally if at all possible. It would be nice to establish connections to deal with demands that can't be met in this way. The examples are simple ones, it is easy to conceive of much more complicated—and potentially important—cases[40].

It seems that the capabilities of doing multilocal calculations is quite limited in human observers. In many cases the computations can only be done when the attention is focussed on the region of interest and definitely not "effortlessly" in parallel. Doubtless various forms of connection have been established, but the higher order routines that operate on the jet bundle are scarce and not very sophisticated. In many cases the human visual system doesn't even give it a try.

6 PROJECTIVE STRUCTURE

The jets represent the geometrical structure of infinitesimal neighborhoods. We also need to establish the global geometrical structure of the visual field, the base space of the fiber bundle. Until here we have only elaborated how a topology of neighborhoods might be booted up (the local sign problem). Apart from that one needs a *projective structure*, basically the nexus of the "straight lines" of the visual field.

It is well known that the human observer judges objective straight lines in the physical environment as subjectively curved[41,42], at least if the lines are of great extension and don't pass through the center of the visual field. However, this fact will be of little interest to the present discussion. When I use the term "straight lines" I mean the ability to judge whether a triple of points is *perceptually collinear* or not. If this leads to consistent results I will say that a projective structure of the visual field has been established. The "veridicality" or conformity to the projective structure of the external physical world is an entirely different issue.

By "consistency" I mean roughly that the collinearity and incidence (point on line, line through point, intersection point of two lines, line defined by two points) judgements conform within reasonable tolerances to one of the axiomatic systems of the projective plane.

It is not known how well human visual judgements approximate projective structure. Certainly the rudiments are present, and over restricted regions it is even rather precise. As far as neurophysiology is concerned we know even less. (*E.g.*, are there "straight line assemblies", sensitive to the fact that certain subsets of receptive fields happen to be collinear?) The way the projective structure is acquired is unclear, although eye movements are almost certainly implicated[43].

7 METRICAL STRUCTURE

The metrical structure can be regarded as a further specialization of the projective structure. Again, it is well documented that human observers make characteristic mistakes (often called "illusions") when the task is to judge metrical relations in the physical environment. Veridicality is not the issue here. If a human observer can reliable equate a vertical stretch with a horizontal one this is a metrical feat, even if the stretches turn out (as they do) to be physically different.

People can indeed compare vertical to horizontal stretches with good reproducibility. This feat is even harder to acquire than the projective structure. Clearly no eye movement will ever bring horizontal and vertical stretches into superposition, thus eye movements do not suffice to acquire this ability. The only way to calibrate this type of comparison is via body movements of the agent. By turning the head you may bring stretches of different orientation into coincidence and thus compare them. Somewhat more comfortably you may manipulate objects while looking at them. Turning objects about the line of regard will enable you to get calibrated.

Thus projective structure may be "visual", but metrical structure almost certainly must be calibrated via visuomotor processes.

8 VISUAL FIELD AND VISUAL WORLD

In this paper I have almost singularly discussed the geometry of the visual field. It is clear that the survival of the optically guided agent will primarily depend on the visuomotor competences tuned to the visual *world*.

For some "depth cues" such as binocular stereo, motion parallax, or accommodation it is rather straightforward to base visual routines tuned to the visual world on the jet bundle. For other cues this is far less straightforward. In any case the step from local "germs" of visual world structure to a global representation and its interrelations with motor actions is a major research problem.

9 THE BRAIN A GEOMETRY ENGINE?

How viable is the notion of the brain as a "geometry engine"? Of course the notion has to be restricted to the "front end" visual system, that is the mainly data driven, bottom up, syntactical part. I think that the prospects for theories of visual front ends as geometry engines are rather good.

Geometry seems very fit as a language in which to describe both the neuroanatomical structure (for instance the somatotopic columnar structure of area V1 as a fourth order jet bundle) as well as the structure of elementary visual competences.

This is important, because the existence of a great body of mathematical knowledge of the structure of geometries then enables us to proceed much faster in the development of theories in visual neurophysiology and psychophysics. That such theories are powerful is increasingly clear from the many succesful applications in image processing and computer vision. Applications to physiology and psychology are understandably harder, due to the experimental constraints and enormous complexities involved. It appears to me that this even more stresses the need for principled theoretical approaches to guide our research.

References

[1] L.M.Blumenthal and K. Menger: Studies in geometry, San Francisco: W.H.Freeman (1970).

[2] F.Klein: Vergleichende Betrachtungen über neuere geometrische Forschungen (so called "Erlanger Programm"), Math.Ann. **43** (1893), pp. 63–100.

[3] L.M.Blumenthal: A modern view of geometry, New York: Dover (1961).

[4] F.Klein: Elementarmathematik vom höheren Standpunkte aus, 3rd. ed., Springer, Berlin (1928); Dritter Teil: Von der Versinnlichung idealer Gebilde durch Zeichnungen und Modelle.

[5] G.Berkeley: A Treatise Concerning the Principles of Human Knowledge, Dublin (1710).

[6] E.Schrödinger: Theorie der Pigmente von grösster Leuchtkraft, Ann. Physik **62** (1920), p. 603.

[7] E.Schrödinger: Grundlinien einer Theorie der Farbenmetrik im Tagessehen I, II, Ann. Physik **63** (1920), p. 397 and p. 427.

[8] R.K.Luneburg: Mathematical analysis of binocular vision, Princeton NJ: Princeton University Press (1947).

[9] J.J.Koenderink: The concept of local sign, in: Limits in Perception, Eds.: A.J. van Doorn, W.A. van de Grind and J.J. Koenderink, Utrecht: VNU Science Press (1984).

[10] W.C.Hoffman: The Lie algebra of visual perception. J.Math.Psychol. **3** (1966), pp. 65-98.

[11] J.J.Koenderink: Operational significance of receptive field assemblies, Biol. Cybern. **58** (1988), pp. 163–171.

[12] J.J.Koenderink: The brain a geometry engine, Psychological Research **52** (1990), pp. 122–127.

[13] H.Lotze: Mikrokosmos, Leipzig: Hirzel (1884).

[14] E.Mach: Analyse der Empfindungen, Jena: Fischer (1886).

[15] W.Metzger: Gesetze des Sehens, Frankfurt a.M.: Verlag Waldemar Kramer (1975).

[16] B.K.P.Horn: Robot Vision, Cambridge Mass.: The MIT Press (1986).

[17] J.J.Koenderink: The structure of images, Biol. Cybern. **50** (1984), pp. 363–370.

[18] M.Spivak: Differential Geometry, 5 Vols., Berkeley Ca.: Publish or perish, Inc. (1970, 1970, 1975, 1975, 1975).

[19] J.J Koenderink and A. J. van Doorn: Representation of local geometry in the visual system, Biol. Cybern. **55** (1987), pp. 367–375.

[20] H.von Helmholtz: Vorträge und Reden, 2 Vols., Brunswick: Fr Vieweg und Sohn (1884).

[21] E.V.Huntington: A set of postulates for abstract geometry expressed in terms of the single relation of inclusion, Math.Ann. **73** (1913), pp. 522–559.

[22] H.Cartan: Differential Calculus, Paris: Hermann (1983).

[23] J.J.Koenderink and A.J. van Doorn: Receptive Field Families, Biol. Cybern. **63** (1990), pp. 291–297.

[24] R.Thom: Stabilité structurelle et morphogénèse, Paris: W.A.Benjamin, Inc. (1972).

[25] J.Ruskin: Modern painters, London: George Routledge and sons (1843).

[26] A.P.Witkin: Scale–Space Filtering, Proc. IJCAI, Karlsruhe (1983), pp. 1019–1021.

[27] J.J.Koenderink and A. J. van Doorn: Visual detection of spatial contrast; Influence of location in the visual field, target extent and illuminance level, Biol. Cybern. **30** (1978), pp. 157–167.

[28] P.Bijl and J. J. Koenderink: Visibility of blobs with a gaussian luminance profile, Vision Res. **29** (1989), pp. 447–456.

[29] J.P.Jones and L.A. Palmer: The two–dimensional spatial structure of simple receptive fields in the cat striate cortex, J.Neurophysiol. **58** (1987), pp. 1187–1211.

[30] J.P.Jones and L. A. Palmer: An evaluation of the two-dimensional Gabor filter model of simple receptive fields in cat striate cortex, J. Neurophysiol. **58** (1987), pp. 1233–1258.

[31] R.A.Young: The gaussian derivative theory of spatial vision: Analysis of cortical cell receptive field line-weighting profiles, General Motors Res. Tech. Rep.**GMR-4920**, May (1985).

[32] M.Livingstone and D. Hubel: Segregation of form, color, movement and depth: Anatomy, physiology, and perception, Science **240** (1988), pp. 740–749.

[33] B.Julesz and J. Bergen: Textons, the fundamental elements in preattentive vision and perception of textures, Bell Systems Technical Journal **62**(6) (1983), pp. 1619–1645.

[34] C.B.Blakemore and F.W. Campbell: On the existence of neurones in the human visual system selectively sensitive to the orientation and size of retinal images, J.Physiol. **203** (1969), pp.237–260.

[35] J.Canny: Finding Edges and Lines in Images, MIT AI Laboratory Technical Report 720, June (1983).

[36] I.Rock: The logic of perception, Cambridge Mass.: MIT Press (1983).

[37] H.von Helmholtz: Handbuch der physiologischen Optik, 2nd. ed., Hamburg and Leipzig: Leopold Voss (1896).

[38] H.Poincaré: La valeur de la science, Paris: Flammarion (1905).

[39] J.J.Koenderink and W. Richards: Two-dimensional curvature operators, J. Opt. Soc. Am. **A5** (1988), pp. 1136–1141.

[40] J.J.Koenderink and A.J.van Doorn: Second-order optic flow, J.Opt.Soc.Am. **A9** (1992), pp. 530-538.

[41] M.H.Pirenne: Optics, painting and photography, Cambridge: Cambridge University Press (1970).

[42] G.Hauck: Die subjektive Perspektive und die horizontalen Curvaturen des Dorischen Styls, Stuttgart (1875).

[43] J.R.Platt: How we see straight lines, Scient.Am., June (1960), pp. 121–129.

Acknowledgement This work was done as part of the ESPRIT Basic Research Action "InSight" of the European Commission.

Brain Theory / A. Aertsen (Editor)

Learning Algorithms and Network Architectures

T. Poggio and F. Girosi[a]

[a]Artificial Intelligence Laboratory and Center for Biological and Computational Learning, Department of Brain and Cognitive Sciences, Massachusetts Institute of Technology, Cambridge, Massachusetts 02139, USA

Abstract

The thesis of this essay is that there is nothing special or particularly new about neural networks, since they can be regarded as a graphical notation for specific regression and classification techniques. There are several classical techniques in estimation and function approximation. We will review some of the most interesting schemes and compare them with the best neural network based techniques. In particular, we will discuss a new scheme that seems to have broad theoretical foundations and even a plausible neurobiological implementation.

1. THE NEURAL NETWORKS EPIDEMIC

Let us first briefly define Neural Networks. Neural Networks, also called sometime Connectionist Systems, are systems with many "simple" processors (e.g. threshold units), highly connected. Learning - by changing weights of connections - is their most interesting aspect.

Of course, Neural Network are far from providing magic performance, despite the hype in the popular press. Despite their name, they have very little to do with real neurons. Several of the proposed schemes do not work particularly well. Most are a rehash of older techniques in pattern recognition and statistics. The enthusiasm around Neural Networks has taken the form of a real intellectual epidemics with too much hype and exagerations. It is of course not too strange for epidemics of ideas to develop, often with a period of about 20-25 years, for reason that should be obvious. In the case of Neural Networks the previous peaks of the epidemics had names such as Gestalt in the 20's, Cybernetics in the 40's, Perceptrons in the 60's. It is natural to ask what is infectious in Neural Networks today. The answer contains both bad and good aspects:

- the fascination of the magic box (*no need to sweat about specific problems*)
- emphasis on learning
- the term *neural* (more or less the only thing in common with the brain)

---

[1]This paper has previously appeared in **Exploring Brain Functions: Models in Neuroscience**, Proceedings of the Dahlem Conference, T. Poggio and D. Glaser, eds., John Wiley & Sons Ltd, 1992.

- "low-level" learning (i.e. good emphasis on "boring" but useful statistical methods)
- fine-grained parallelism

In summary, the most interesting contribution of Neural Networks is the emphasis on simple forms of learning from examples and the demonstration that interesting results can be obtained in several cases. Until recently, however, even the best techniques lacked in solid theoretical basis and a good understanding of their working principles. Progress however is possible and actually ongoing. We are in fact witnessing a revival of interest in statistical and estimation techniques. It is possible that, years from now, the main legacy of the Neural Networks epidemics of today will be a renewed interest in a variety of statistical and function approximation techniques and the recognition that the latter may be relevant to the problem of intelligence and of the brain.

2. LEARNING AS FUNCTION APPROXIMATION

Most neural networks attempt to synthesize modules that transduce inputs into desired outputs from a set of correct input-output pairs, called examples. Some of the best known applications are a network that maps english spelling into its phonetic pronounciation (Sejnowski and Rosenberg, 1987) and a network that learns the mapping corresponding to a chaotic dynamical system, thereby predicting the future from its past (Lapedes and Farber, 1987). The main thesis of this paper is that problems of learning from examples of this type can be reformulated and formalized as problems of approximating a multivariate function from a set of sparse data points.

In order to see how the problem of learning is related to approximation theory let us consider the typical structure of examples of learning problems. In all the cases an instance of the problem is given by a set D of input-output pairs, $D \equiv \{(x_i, y_i) \in X \times Y\}_{i=1}^{N}$, belonging to some input and output space, X and Y (see fig. 1).

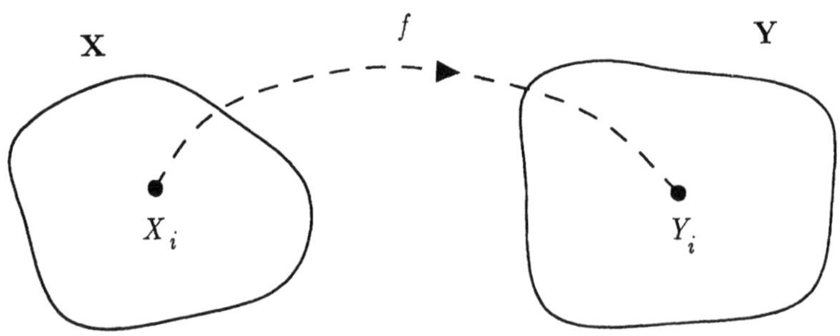

Figure 1 The examples (x_i, y_i) are points belonging to the graph of a function $f : X \to Y$.

We assume that there is some relationship between the input and the output, and consider the pairs (x_i, y_i) as *examples* of it. Therefore we assume that it exists a map

$$f : X \rightarrow Y$$

with the property:

$$f(x_i) = y_i \quad i = 1, \ldots, N \; .$$

Learning, or generalizing, means being able to estimate the function at points of its domain X where no data are available. From a mathematical point of view this means estimating the function f from the knowledge of a subset D ot its graph, that is from a set of sparse data. Therefore from this point of view the problem of learning is equivalent to the problem of approximating a multivariate function from sparse data. An appropriate mathematical domain of this problem is *approximation theory*, a well consolidated branch of mathematics. A large body of results in this field can be exploited to analyze the problem of learning from examples in a rigorous mathematical framework.

3. APPROXIMATION SCHEMES AND NETWORKS

Suppose that the set $S = \{(\mathbf{x}_i, y_i) \in R^d \times R\}_{i=1}^N$ of data has been obtained by random sampling, possibly in presence of noise, a function f belonging to some set G of functions defined on R^d. We are interested in recovering the function f, or an estimate of it, from the set of data S. Several techniques have been developed, in the fields of statistics and approximation theory, in order to find solutions to this problem (Wahba, 1990; Rice, 1964, 1969; Schumaker, 1981; Eubank, 1988). A common technique consists in approximating the function $f(\mathbf{x})$ by a parametric function $F(\mathbf{p}, \mathbf{x})$ having a fixed number of parameters $\mathbf{p}$ belonging to some set P. For a choice of a specific F, the problem is then to find the set of parameters $\mathbf{p}$ that provides the best possible approximation of f on the set of "examples" (the data set S). This is the *learning* step, and it is usually done by finding those parameters that minimize the least square error on the data set, that is solving the following minimization problem:

$$\min_{\mathbf{p} \in P} \sum_{i=1}^{N} (F(\mathbf{p}, \mathbf{x}_i) - y_i)^2$$

Needless to say, it is very important to choose an approximating function F that can represent f as well as possible. There would be little point in trying to learn, if the chosen approximation function $F(\mathbf{p}, \mathbf{x})$ could only give a very poor representation of $f(\mathbf{x})$, even with optimal parameter values.

3.1. Networks are a Notation for Approximation Schemes

Almost all approximation schemes can be mapped into some kind of network that has come in recent years to be dubbed as a "neural network" [2]. Networks, after all, can be regarded as a graphic notation for a large class of algorithms. In the context of

[2] Although the relation to biological neurons is in most cases weak at best.

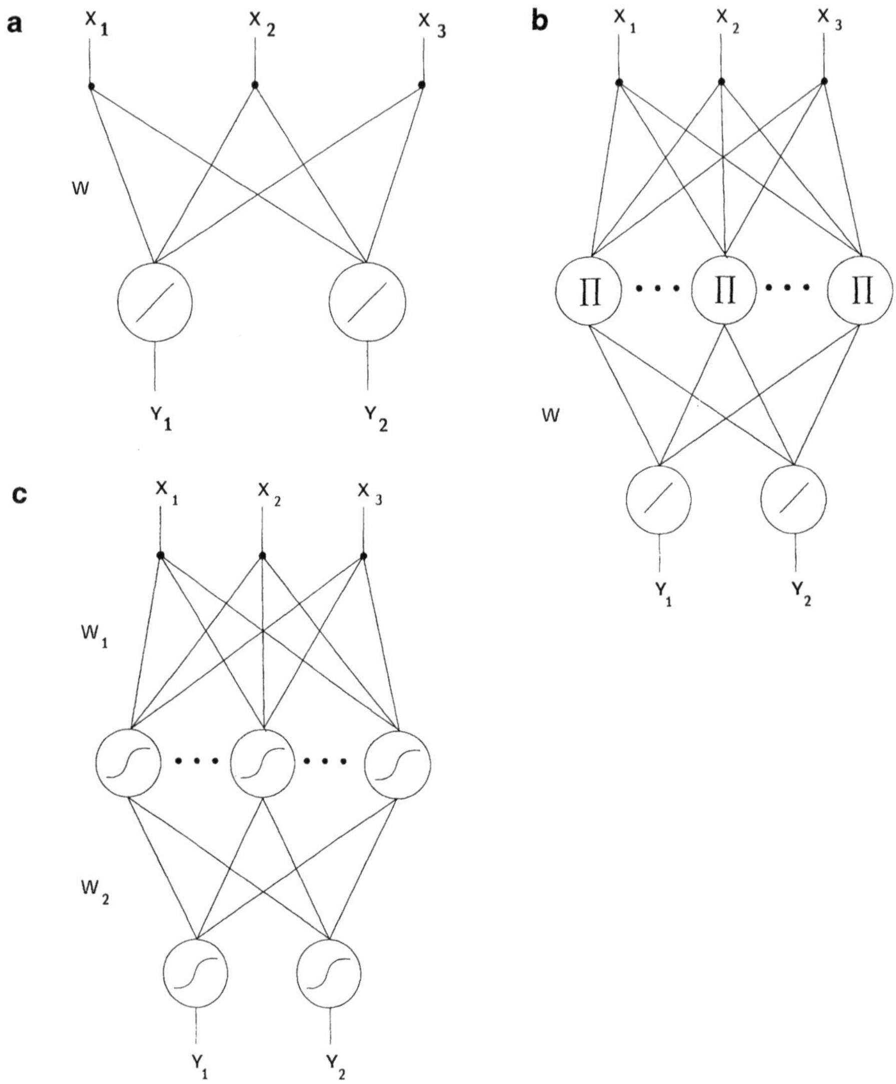

Figure 2 (a) A linear approximating function maps into a network without a hidden layer with linear units. The "weights" of the connections correspond to the matrix P – the linear estimator. (b) Polynomial estimators and other linear combinations of nonlinear "features" of the input correspond to networks with one hidden layer. In the case of polynomials the hidden units correspond to products (Π) and powers of the components of the input vector. The first layer of connections is fixed; the second is modified during "training". (c) A Multilayer perceptron with one layer of hidden sigmoid units. Both sets of connections – from the input layer to the hidden units (w_1) and from there to the output layer (w_2) – are modified during training.

our discussion, a network is a function represented by the composition of many basic functions.

We now consider a few examples of $F(\mathbf{p}, \mathbf{x})$, shown in figure 2:

- the classical linear approximation scheme corresponds to the following choice:

$$F(\mathbf{p}, \mathbf{x}) = P\ \mathbf{x}$$

  where P is an $m \times d$ matrix and $\mathbf{x}$ is an d-dimensional vector. It corresponds to a network without hidden units;

- the classical quadratic approximation scheme corresponds to the choice:

$$F(\mathbf{p}, \mathbf{x}) = a + \mathbf{b} \cdot \mathbf{x} + \mathbf{x} \cdot C\mathbf{x}$$

  where a is a number, $\mathbf{b}$ is a d-dimensional vector and C is a $d \times d$ matrix. Higher order terms can be added, and the corresponding parametric function can be represented by the network of fig. 2b;

- the nested sigmoids scheme can be written as

$$F(\mathbf{p}, \mathbf{x}) = \sigma(\sum_n c_n \sigma(\mathbf{w}_n \cdot \mathbf{x}))$$

  where σ is a sigmoidal function, and corresponds to a multilayer network of units that compute weighted sums of their inputs, and then perform a sigmoidal transformation of this sum. This scheme, known as Multilayer Perceptron, MLP in short, is unusual in the classical theory of the approximation of continuous functions and will be discussed later.

In general, the problem of which approximation scheme to use – that is which network – is independent from the problem of which algorithm should be applied for finding the optimal set of parameters $\mathbf{p}$, that is for learning. An approach that works in general, though it may not be the most efficient in any specific case, is some relaxation method, such as gradient descent or conjugate gradient or simulated annealing, in parameter space, attempting to minimize the least square error over the set of examples.

In the case of multilayer perceptrons, a version of gradient descent, called BackPropagation (BP), is by far the most popular technique. However, as other deterministic techniques, it suffers of the problem of local minima, and in many cases non deterministic techniques, as simulated annealing (Kirkpatrick, Gelatt and Vecchi, 1983) or genetic algorithms (Goldberg, 1989), may be more efficient. For instance, a simple method that does not require calculation of derivatives (that have been used and found surprisingly efficient by Caprile and Girosi, 1990) consists in looking for random changes (controlled in appropriate ways) in the parameter values that reduce the error over the set of examples. The size of the random changes is controlled by a simple adaptive scheme.

4. THERE ARE MANY APPROXIMATION SCHEMES AND CORRESPONDING NETWORKS

A direct consequence of the point of view we have outlined above is that there are several approximation schemes and associated networks that could be used for the problem of learning from examples. A non-exhaustive list includes CART, MARS, (Breiman et al., 1984), Projection Pursuit Regression (PPR) (Friedman and Stuetzle, 1981; Huber, 1985) , Parzen Windows and Kernel Methods , K-nearest Neighbors (Duda and Hart, 1973) , Linear Weighted Regression, Vector Quantization. In the following, we will consider just two general classes of approximation schemes – Regularization Networks and Ridge Functions – and a specific approximation scheme for each one: Hyper Basis Functions (HBF) and Multilayer Perceptrons (MLP).

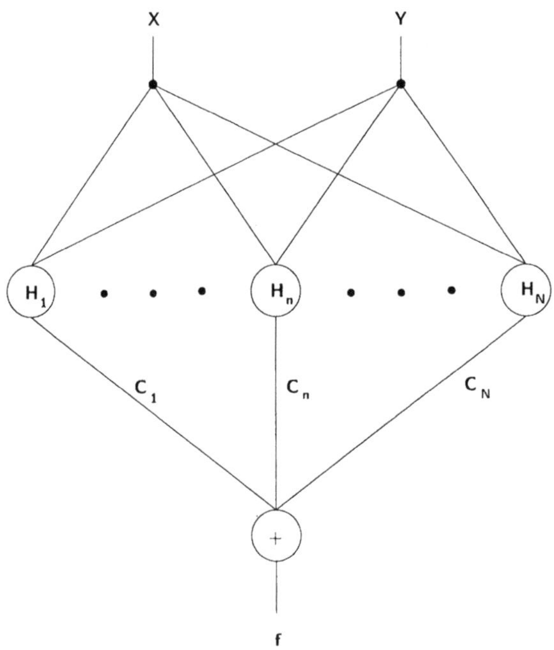

Figure 3 The most general network with one layer of hidden units. Here we show the two-dimensional case, in which $\mathbf{x} = (x, y)$. Each function H_i can depend on a set of unknown parameters, that are computed during the learning phase, as well as the coefficients c_i.

5. TWO CLASSES OF APPROXIMATION SCHEMES

In this section we will discuss two classes of function approximation schemes, the first one being represented by hyperbasis functions (Hbf) and the second one by multilayer perceptrons (MLP).

Consider the network of fig. 3 as a notation for approximation schemes of the type

$$f(\mathbf{x}) = \sum_{i=1}^{n} c_i H_i(\mathbf{x}) \tag{1}$$

where H_i are functions that depend on some parameters to be estimated together with the coefficients c_i.

Depending on the function H_i different approximation schemes can be obtained. Two possible choices are:

1. **Regularization networks**, that correspond to the choice:

$$H_i(\mathbf{x}) = h(\|\mathbf{x} - \mathbf{t}_i\|_{\mathbf{W}})$$

   where h is a (conditionally) positive definite function, the $\mathbf{t}_i$ are d-dimensional vectors called "centers", $\mathbf{W}$ is a $d \times d$ matrix, and we have defined the *weighted norm* $\|\cdot\|_{\mathbf{W}}$ as

$$\|\mathbf{x}\|^2_{\mathbf{W}} = \mathbf{x}^T \mathbf{W}^T \mathbf{W} \mathbf{x} .$$

   This class of approximation schemes can be formally derived from regularization theory, (Tikhonov and Arsenin, 1977; Bertero, 1986). which has an elegant bayesian interpretation as optimal estimation. It includes as special or limiting cases:

   - splines
   - kernel estimation techniques and
   - Parzen windows
   - Potential methods
   - Hyperbf and RBF

2. **Ridge Functions**, that correspond to the choice:

$$H_i(\mathbf{x}) = h_i(\mathbf{x} \cdot \mathbf{w}_i + \theta_i)$$

   where the $\mathbf{w}_i$ are d-dimensional vectors called "weights", and the parameters θ_i are bias terms. This form of approximation does not seem to derive from a variational principle and does not have a Bayesan interpretation. Several techniques are included in this class:

- **Projection Pursuit Regression (PPR).** It corresponds to an expansion of the type eq. (1) in which all the coefficients c_i are set to 1, and

$$H_i(\mathbf{x}) = h_i(\mathbf{x} \cdot \mathbf{w}_i)$$

where the vectors $\mathbf{w}_i$ are normalized ($\|\mathbf{w}_i\| = 1$). The functions h_i are determined by means of some non parametric estimation technique, in an iterative way, together with the directions $\mathbf{w}_i$.

- **Flexible Fourier Series.** The function h_i are all equal to the cosine (or sine) function, and therefore:

$$H_i(\mathbf{x}) = \cos(\mathbf{x} \cdot \mathbf{w}_i + \theta_i) \ . \qquad (2)$$

The interpretation of the "weights" $\mathbf{w}_i$ is the following: they represent the fundamental frequency content of the underlying function.

- **Multilayer Perceptrons.** The functions h_i are all equal to a sigmoidal function:

$$H_i(\mathbf{x}) = \sigma(\mathbf{x} \cdot \mathbf{w}_i + \theta_i) \ . \qquad (3)$$

The function σ is usually defined as

$$\sigma(x) = \frac{1}{1 + e^{-x}}$$

but other choices, (as the hyperbolic tangent), are equivalent, as long as the sigmoidal shape is mantained.
While in the approximation by trigonometric functions the parameters $\mathbf{w}_i$ have a simple interpretation in terms of frequencies, in this case the meaning of the $\mathbf{w}_i$ is less clear. If the expansion (3) is considered as an instance of projection pursuit regression, (Friedman and Stuetzle, 1981; Huber, 1985) the $\mathbf{w}_i$ are the "relevant directions" that are supposed to encode most of the information about the function.

The main difference between these two classes of approximation schemes seems to be related to the use of the scalar product $\mathbf{x} \cdot \mathbf{w}_i$ instead of the weighted Euclidean distance $\|\mathbf{x} - \mathbf{t}_i\|_\mathbf{W}$, as argument of the parametric functions.

6. THE HYPERBF NETWORKS

We do not derive here the Hyperbf formulation from Bayes principles and regularization (a review is in Poggio and Girosi, 1990b). The main result is the derivation from regularization of the scheme

$$f(\mathbf{x}) = \sum_{i=1}^{n} c_i h(\|\mathbf{x} - \mathbf{t}_i\|_{\mathbf{W}}) + p(\mathbf{x}) , \tag{4}$$

where $p(\mathbf{x})$ is a polynomial of low degree, which is equivalent to a network with one hidden layer (see fig. 4). The formulation of the learning problem in terms of regularization is satisfying from a theoretical point of view, since it establishes connections with a large body of results in the area of Bayes estimation and in the theory of approximation of multivariate functions. In particular, eq. 4 can be used to define generalized splines in any dimension. A special case (for $\mathbf{W} = \mathbf{I}$ and fixed $\mathbf{t}_i$) corresponds to an interpolation technique, called Radial Basis Functions, that has been extensively studied (Powell, 1987; Franke, 1982; Micchelli, 1986; Broomhead and Lowe, 1988).

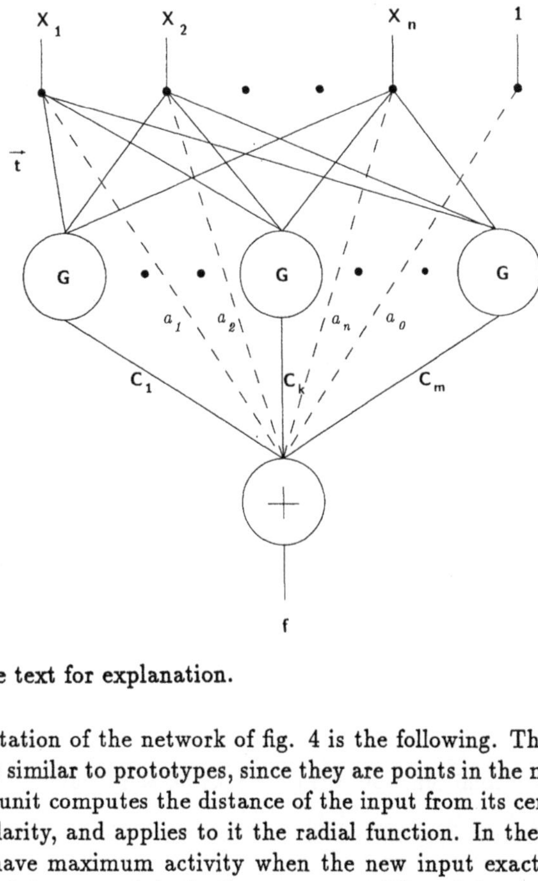

Figure 4 See text for explanation.

The interpretation of the network of fig. 4 is the following. The centers of the radial functions are similar to prototypes, since they are points in the multidimensional input space. Each unit computes the distance of the input from its center, that is a measure of their similarity, and applies to it the radial function. In the case of the Gaussian, a unit will have maximum activity when the new input exactly matches its center. The output of the network is the linear superposition of the activities of all the radial functions in the network. The corresponding weights are found during learning by minimizing a measure of the error between the network's prediction and each of the examples. At the same time, the centers of the radial functions are also updated during learning. Moving the centers is equivalent to modifying the corresponding prototypes.

7. CONNECTIONS BETWEEN MLP AND HBF

The HyperBF network is similar to the architecture used for backpropagation, being a multilayer network with one hidden layer and two sets of adjustable parameters (a HyperBF network with diagonal $\mathbf{W}$ has the same number of parameters of a MLP network with the same number of inputs and of hidden units). Its boolean limiting version carves the input space into hyperspheres, each corresponding to a center: a radial unit is active if the input vector is within a certain radius of its center and is otherwise silent. The boolean limit of backpropagation carves the space with hyperplanes. With an arbitrary number of units each network can approximate the other, since each network can approximate arbitrarily well continuous functions on a limited interval.

Both multilayer perceptrons and Hyper Basis Functions have good approximation properties, theoretically and experimentally. Are they related? Maruyama, Girosi and Poggio (1991) have observed that for normalized inputs, multilayer perceptron networks *are* radial function networks (albeit with a non-standard radial function). This provides an interpretation of the weights $\mathbf{W}$ as centers $\mathbf{t}_\alpha$ of the radial function network, and therefore as equivalent to *templates.*

7.1. Normalized Inputs: Ridge Functions are Radial Functions

We summarize here the observation of Maruyama et al. on the relation between MLP and RBF. The main difference between these two schemes is that MLP is based on scalar products $(\mathbf{x}\cdot\mathbf{w})$ and RBF is based on distances $(\|\mathbf{x}-\mathbf{t}\|)$. However we notice that if the inputs are normalized, the following equality holds:

$$\mathbf{x}\cdot\mathbf{w} = \frac{1}{2}[1+\mathbf{w}^2 - \|\mathbf{x}-\mathbf{w}\|^2] \tag{5}$$

therefore showing that scalar product and distances are closely related on the sphere. Under the hypothesis of normalized inputs, and using identity (5) the output of a Multilayer Perceptron can always be written as:

$$f(\mathbf{x}) = \sum_{i=1}^{n} c_i S_i(\|\mathbf{x}-\mathbf{w_i}\|)\ . \tag{6}$$

where the shape of the radial function $S_i(r)$ depends on some parameter C_i (see fig. 5):

$$S_i(r) = \frac{1}{1+C_i e^{r^2}}\ .$$

We notice that if the value of C_i is large $(C_i \gg 1)$, the basis function S_i is very similar to a Gaussian. Moreover, with a proper choice of the bias parameters, it is always possible to make the C_i equal to the same given number. Therefore it is also possible to obtain a network in which all the hidden units have the same activation function, very similar to a Gaussian.

This implies that a MLP network can always simulate any Gaussian RBF network, if both networks consist of the same number of units. It is less clear what the relation is between RBF and MLP networks.

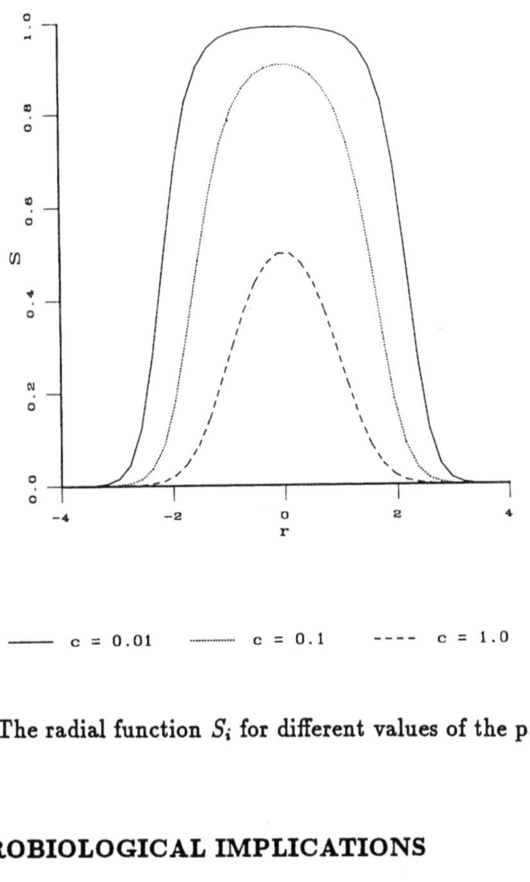

Figure 5 The radial function S_i for different values of the parameter C_i.

8. NEUROBIOLOGICAL IMPLICATIONS

How do MLP and HBF networks compare from the point of view of neurobiological plausibility? We summarize a few observations from Poggio and Girosi (1989).

8.1. Implementation Issues

8.1.1. MLP

MLP networks have a straighforward implementation in terms of linear excitation and inhibition and of the threshold mechanism of the spike for the sigmoidal nonlinearity. The latter could also be implemented in terms of the pre-postsynaptic relationship between presynaptic voltage and postsynaptic voltage. In either case this implementation requires one neuron per sigmoidal unit in the network.

8.1.2. HBF

The synthesis of (weighted) radial basis functions in high dimensions can be achieved in an interesting way (with diagonal $\mathbf{W}$) for Gaussian Radial functions, since a multi-dimensional Gaussian function can be represented as the product of lower dimensional Gaussians. For instance, a 2D Gaussian radial function centered in $\mathbf{t}$ can be written as:

$$G(\|\mathbf{x}-\mathbf{t}\|_W^2) \equiv e^{-\|\mathbf{x}-\mathbf{t}\|_W^2} = e^{-\frac{(x_1-t_1)^2}{2\sigma_1^2}} e^{-\frac{(x_2-t_2)^2}{2\sigma_2^2}} , \tag{7}$$

with $\sigma_1 = 1/w_1$ and $\sigma_2 = 1/w_2$, where w_1 and w_2 are the elements of the matrix $\mathbf{W}$ assumed, in this section, to be diagonal.

This dimensionality factorization is especially attractive from the physiological point of view, since it is difficult to imagine how neurons could compute $G(\|\mathbf{x}-\mathbf{t}_\alpha\|^2)$. The scheme of figure 6, on the other hand, is physiologically plausible. Gaussian radial functions in one, two and possibly three dimensions can be implemented as *receptive fields* by weighted connections from the sensor arrays (or some retinotopic array of

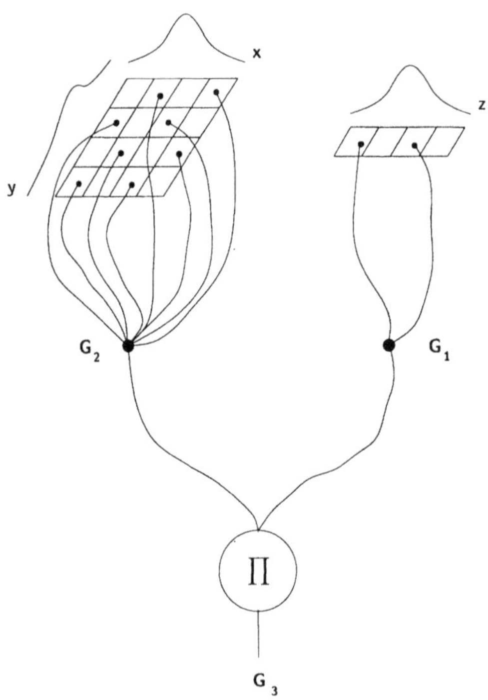

Figure 6 A three-dimensional radial gaussian implemented by multiplying two-dimensional gaussian and one-dimensional gaussian receptive fields. 2D gaussians acting on a retinotopic map can be regarded as representing 2D "features", while the radial basis function represents the "template" resulting from the conjunction of those lower-dimensional features.

units representing with their activity the position of features). Gaussians in higher dimensions can then be synthesized as *products of one and two dimensional receptive fields.*

This scheme has three additional interesting features:

1. the multidimensional radial functions are synthesized directly by appropriately weighted connections from the sensor arrays, without any need of an explicit computation of the norm and the exponential.

2. 2D Gaussians operating on the sensor array or on a retinotopic array of features extracted by some preprocessing transduce the implicit position of features in the array into a number (the activity of the unit).

3. 2D Gaussians acting on a retinotopic map can be regarded each as representing one 2D "feature," i.e., a component of the input vector, while each center represents the "template," resulting from the conjunction of those lower-dimensional features. Notice that in this analogy the radial basis function is the AND of several features and could also include the negation of certain features, that is the AND NOT of them. $\mathbf{W}$ weights the importance of the different features.

The multiplication operation required by the previous interpretation of Gaussian GRBFs to perform the "conjunction" of Gaussian receptive fields is not too implausible from a biophysical point of view. It could be performed by several biophysical mechanisms (see Koch and Poggio, 1987), such as

1. inhibition of the silent type and related circuitry (see Torre and Poggio, 1978; Poggio and Torre, 1978)

2. the AND-like mechanism of NMDA receptors

3. a logarithmic transformation, followed by summation, followed by exponentiation. The logarithmic and exponential characteristic could be implemented in appropriate ranges by the sigmoid-like pre-to-postsynaptic voltage transduction of many synapses.

If the first or the second mechanism is used, the product of figure 3 can be performed directly on the dendritic tree of the neuron representing the corresponding radial function (alternatively, each dendritic tree may perform pairwise products only, in which case a logarithmic number of cells would be required). The scheme also requires a certain amount of memory per basis unit, in order to store the center vector. In the case of Gaussian receptive fields used to synthesize Gaussian radial basis functions, the center vector is effectively stored in the position of the 2D (or 1D) receptive fields and in their connections to the product unit(s).

The linear terms (the direct connections from the inputs to the output in figure 1) can be realized directly as inputs to the output neuron that summates linearly its synaptic inputs (an output nonlinearity is allowed and will not change the basic form of the model, see Poggio and Girosi, 1989). They may also be realized through intermediate linear units.

Another possible implementation of GRBF (but not HBF) is by exploiting the equivalence with MLP for normalized inputs. If the inputs are normalized (like it is the case for unary input representations of the type used in NetTalk) a GRBF network could be implemented as a MLP network by using threshold units. There is the problem, however, of normalizing appropriately the inputs in a biologically plausible way.

8.1.3. MLP and HBF: Mechanisms for Learning

Do the update schemes have a physiologically plausible implementation? Consider first the steepest descent methods, which require derivatives. The update of the weights from the hidden units to the output in terms of steepest descent scheme is not too unlikely, since the resulting equations are, for both MLPs and HBFs, a quasi-hebbian scheme which requires only a small amount of neural circuitry. Updates of the other layer (in MLP) or of the centers (in GRBFs) seems much more difficult to implement for a network of real neurons in terms of steepest sescent schemes.

Methods such as the random-step method (Caprile and Girosi, 1990), which do not require calculation of derivatives are biologically much more plausible. In the Gaussian case, with basis functions synthesized through the product of Gaussian receptive fields, moving the centers means establishing or erasing connections to the product unit. A similar argument can be made also about the learning of the matrix $\mathbf{W}$. Notice that in the diagonal Gaussian case the parameters to be changed are exactly the σ of the Gaussians, i.e., the spread of the associated receptive fields. Notice also that the σ for all centers on one particular dimension is the same, suggesting that the learning of w_i may involve the modification of the scale factor in the input arrays rather than a change in the dendritic spread of the postsynaptic neurons.

In all these schemes the real problem consists in how to provide the "teacher" input.

9. MLPS AND HBFS AS THEORIES OF THE BRAIN

As theories of the brain (or of parts of it) both MLPs and HBFs networks replace computation with memory. They are equivalent to modules that work as *interpolating look-up tables.* In a previous paper one of us has discussed how theories of this type can be regarded as a modern version of ideas of the "grandmother cell" type (Poggio, 1990).

The proposal that much information processing in the brain is performed through modules that are similar to *enhanced look-up tables* is attractive for many reasons. It also promises to bring closer apparently orthogonal views, such as the *immediate perception* of Gibson (1950) and the *representational theory* of Marr (1982), since almost iconic "snapshots" of the world may allow the synthesis of computational mechanisms completely equivalent to vision algorithms such as, say, structure-from-motion. The idea seems to change significantly the computational perspective on several vision tasks. As a simple example, consider the different specific tasks of hyperacuity, as considered by the psychophysicists. The proposal would suggest that an appropriate module for the task, somewhat similar to a new "routine," may be synthesized by learning in the brain (see Poggio, Fahle and Edelman, 1991).

The claim common to both theories (MLP and HBF) is that the brain can be explained in part in terms of approximation modules. In the case of HBF networks these modules can be considered as a direct extension of look-up tables. MLP networks cannot be interpreted as direct extensions of look-up tables (they are more similar to an extension of multidimensional Fourier series), though the case of normalized inputs shows that they are similar to using templates.

The HBF theory predicts that population coding (broadly tuned neurons combined linearly) is a consequence of extending a look-up table scheme – corresponding to interval coding – to yield interpolation (or more precisely approximation, since the examples may be noisy), that is generalization. In other words, population coding is a direct and strong prediction of HBF schemes. The HBF model seem to bear suggestive similarities with observations on the cortical cells apparently involved in visual face recognition. MLP networks have been recently claimed to be consistent with a number of physiological data. Before looking for critical experiments capable of distinguishing among them, it should be emphasized that they are not the only possible schemes for learning from examples. One of the main points of this paper is indeed that several different approximation schemes are theoretically possible.

10. A GENERAL LIMITATION

The most critical problem for all schemes for learning from examples that rely on very weak a priori information such as HyperBf and MLPs has to do with the number of examples needed for learning any specific task or, in other words, with the problem of *sample complexity: how many samples are needed to achieve a given degree of accuracy?*

It is well known, and rather intuitive, that if we do not impose any constraint on the class of functions we approximate, the sample complexity will increase exponentially with the dimensionality of the function. This phenomenon seems to be very discouraging, because indicates that the complexity of the approximation increases incredibily fast as the dimension grows, and for this reason it has been called "the curse of dimensionality". However it is also intuitive that the smoother a function is, the less number of data points will be required to approximate it within some given accuracy. Very general results from functional analysis (Vitushkin, 1977, 1954; Lorentz, 1986), statistics (Stone, 1980, 1982) and approximation theory (Cox, 1984; Wahba, 1990) indicate that the complexity of classes of functions, and therefore the sample complexity, seems to depend on the ratio between the dimensionality d and a smoothness index p of the function, where by smoothness index we mean a measure of the number of derivatives of the function. Typical results assert that the number of examples N needed to approximate a function with error ϵ is

$$N \propto \left(\frac{1}{\epsilon}\right)^{\frac{d}{p}}$$

where d and p have been defined above.

So we might see the way to escape the curse of dimensionality: if we let smoothness increase with the dimension, the sample complexity will not grow when the dimensions grows, because the effect of the dimension will be canceled by the smoothness. These results are asymptotic and should be taken with some caution. They show both the severity of the problem and also ways around it, such as trading smoothness of the mapping with incresing dimensionality.

It is very likely, also in light of very recent results (Barron, 1991; Girosi and Anzellotti, 1991) that all networks, including MLPs and HyperBF, will have the same fundamental limitations from the point of view of sample complexity. This is probably a very deep limitation of neural networks of any type and in fact of all approaches to learning that start with very little prior information, in other words, that really start from a kind of *tabula rasa.* These results may be considered the mathematical "proof" that learning from a *tabula rasa* is out of the question, and that in general, the scientific problem lies in characterizing the right mixture of nature and nurture, that is, in characterizing how much pre-existing representation is needed in order to be able to learn effectively.

But even if the individual does require specific structures and representations to learn perceptual and motor tasks, evolution has actually developed both representations and learning schemes. Thus both ontogenetic and epigenetic aspects of learning are critical for any theory that attempts in the long run to understand the brain and the nature of intelligence.

REFERENCES

1. A.R. Barron. Universal approximation bounds for superpositions of a sigmoidal function. Technical Report 58, Department of Statistics, University of Illinois at Urbana-Champaign, Champaign, IL, March 1991.
2. M. Bertero. Regularization methods for linear inverse problems. In C. G. Talenti, editor, *Inverse Problems.* Springer-Verlag, Berlin, 1986.
3. L. Breiman et al. *Classification and Regression Trees.* The Wadsworth statistics/probability series. Wadsworth, Belmont, CA, 1984.
4. D.S. Broomhead and D. Lowe. Multivariable functional interpolation and adaptive networks. *Complex Systems,* 2:321–355, 1988.
5. B. Caprile and F. Girosi. A nondeterministic minimization algorithm. A.I. Memo 1254, Artificial Intelligence Laboratory, Massachusetts Institute of Technology, Cambridge, MA, September 1990.
6. D.D. Cox. Multivariate smoothing spline functions. *SIAM J. Numer. Anal.,* 21:789–813, 1984.
7. G. Cybenko. Approximation by superposition of a sigmoidal function. *Math. Control Systems Signals,* 2(4):303–314, 1989.
8. R. O. Duda and P. E. Hart. *Pattern Classification and Scene Analysis.* Wiley, New York, 1973.
9. R.L. Eubank. *Spline Smoothing and Nonparametric Regression,* volume 90 of *Statistics, textbooks and monographs.* Marcel Dekker, Basel, 1988.
10. R. Franke. Scattered data interpolation: tests of some method. *Math. Comp.,* 38(5):181–200, 1982.

11. J.H. Friedman and W. Stuetzle. Projection pursuit regression. *Journal of the American Statistical Association*, 76(376):817–823, 1981.
12. K. Funahashi. On the approximate realization of continuous mappings by neural networks. *Neural Networks*, 2:183–192, 1989.
13. J.J. Gibson. *The Perception of the Visual World.* Houghton Mifflin, Boston, 1950.
14. F. Girosi and T. Poggio. Networks and the best approximation property. *Biological Cybernetics*, 63:169–176, 1990a.
15. D.E. Goldberg. *Genetic algorithms in search, optimization, and machine learning.* Addison-Wesley, Reading, MA, 1989.
16. P.J. Huber. Projection pursuit. *The Annals of Statistics*, 13(2):435–475, 1985.
17. S. Kirkpatrick, C.D. Gelatt, and M.P. Vecchi. Optimization by simulated annealing. *Science*, 220:219–227, 1983.
18. C. Koch and T. Poggio. Biophysics of computational systems: Neurons, synapses, and membranes. In G.M. Edelman, W.E. Gall, and W.M. Cowan, editors, *Synaptic Function*, pages 637–697. John Wiley and Sons, New York, 1987.
19. A. Lapedes and R. Farber. How neural nets work. In Dana Z. Anderson, editor, *Neural Information Processing Systems*, pages 442–456. Am. Inst. Physics, NY, 1988. Proceedings of the Denver, 1987 Conference.
20. G. G. Lorentz. *Approximation of Functions.* Chelsea Publishing Co., New York, 1986.
21. David Marr. *Vision: A Computational Investigation into the Human Representation and Processing of Visual Information.* W.H. Freeman and Company, San Francisco, 1982.
22. M. Maruyama, F. Girosi, and T. Poggio. A connection between HBF and MLP. A.I. Memo No. 1291, Artificial Intelligence Laboratory, Massachusetts Institute of Technology, 1991.
23. C. A. Micchelli. Interpolation of scattered data: distance matrices and conditionally positive definite functions. *Constr. Approx.*, 2:11–22, 1986.
24. T. Poggio. A theory of how the brain might work. In *Cold Spring Harbor Symposia on Quantitative Biology*, pages 899–910. Cold Spring Harbor Laboratory Press, 1990.
25. T. Poggio, M. Fahle, and S. Edelman. Synthesis of visual modules from examples: learning hyperacuity. A.I. Memo No. 1271, Artificial Intelligence Laboratory, Massachusetts Institute of Technology, 1991.
26. T. Poggio and F. Girosi. A theory of networks for approximation and learning. A.I. Memo No. 1140, Artificial Intelligence Laboratory, Massachusetts Institute of Technology, 1989.
27. T. Poggio and F. Girosi. Regularization algorithms for learning that are equivalent to multilayer networks. *Science*, 247:978–982, 1990a.
28. T. Poggio and F. Girosi. Networks for approximation and learning. *Proceedings of the IEEE*, 78(9), September 1990b.
29. T. Poggio and V. Torre. A theory of synaptic interactions. In W.E. Reichardt and T. Poggio, editors, *Theoretical approaches in neurobiology*, pages 28–38. The M.I.T Press, Cambridge, MA, 1978.
30. M. J. D. Powell. Radial basis functions for multivariable interpolation: a review.

In J. C. Mason and M. G. Cox, editors, *Algorithms for Approximation.* Clarendon Press, Oxford, 1987.

31. J.R. Rice. *The approximation of functions, Vol. 1.* Addison-Wesley, Reading, MA, 1964.
32. J.R. Rice. *The approximation of functions, Vol. 2.* Addison-Wesley, Reading, MA, 1969.
33. L.L. Schumaker. *Spline functions: basic theory.* John Wiley and Sons, New York, 1981.
34. T. J. Sejnowski and C. R. Rosenberg. Parallel networks that learn to pronounce english text. *Complex Systems*, 1:145–168, 1987.
35. C. J. Stone. Optimal rates of convergence for nonparametric estimators. *Ann. Stat.*, 8(6):1348–1360, 1980.
36. C. J. Stone. Optimal global rates of convergence for nonparametric regression. *Ann. Stat.*, 10:1040–1053, 1982.
37. A. N. Tikhonov and V. Y. Arsenin. *Solutions of Ill-posed Problems.* W. H. Winston, Washington, D.C., 1977.
38. V. Torre and T. Poggio. An application: a synaptic mechanism possibly underlying motion detection. In W.E. Reichardt and T. Poggio, editors, *Theoretical approaches in neurobiology*, pages 39–46. The M.I.T Press, Cambridge, MA, 1978.
39. A. G. Vitushkin. On Hilbert's thirteenth problem. *Dokl. Akad. Nauk SSSR*, 95:701–704, 1954.
40. A. G. Vitushkin. *On representation of functions by means of superpositions and related topics.* L' Enseignement Mathematique, 1977.
41. G. Wahba. *Splines Models for Observational Data.* Series in Applied Mathematics, Vol. 59, SIAM, Philadelphia, 1990.

Brain Theory / A. Aertsen (Editor)

Processing of synaptic signals in fly visual interneurons selectively responsive to small moving objects

Alexander Borst and Martin Egelhaaf

Max-Planck-Institut für biologische Kybernetik, Spemannstraße 38, D-7400 Tübingen

Abstract
In the fly identified visual interneurons are found which respond selectively to small moving objects. Recently, some of the neural elements could be identified which constitute the underlying circuit. We review the available experimental data and compare them with some model circuitries.

1. INTRODUCTION

A moving observer generates on the retina a continuous flow of the images of the environment. The distribution of motion signals on the retina, the so-called optic flow-field, thereby depends both on the trajectory of the observer through the world as well as on the structure of the environment. If the observer passes a nearby object the image of the object moves faster on the retina than the background. Therefore, the flow-field contains discontinuities with a small patch containing larger motion vectors than the rest of the flow-field which indicates the existence of nearby objects. If evaluated appropriately, motion information can, therefore, be used to segregate the environment into salient objects and a more distant background ('figure-ground discrimination').

The importance of relative motion as a visual cue for figure-ground discrimination is illustrated by the fact that many animal species are able to detect objects which differ from their surroundings only by the velocity at which they move (e.g. flies: Egelhaaf et al. 1988; bees: Srinivasan et al. 1990; locusts: Collett and Paterson, 1991; primates: Miles and Kawano, 1987; Regan and Beverley, 1984). Accordingly, neurons that respond best to small objects moving relative to their background are found in various phylogenetically divergent animal groups (e.g. hawkmoths: Collett 1971 1972; dragonflies: Olberg 1981, 1986; hoverflies: Collett and King 1975; blowflies: Egelhaaf 1985b; toads: Tsai 1990; pigeons: Frost and Nakayama 1983, Frost et al. 1988; cats: Grünau and Frost 1983; monkeys: Allman et al. 1985, Tanaka et al. 1986, Davidson and Bender, 1991). Despite the widespread significance of object detection by relative motion, not much is known so far about the underlying cellular mechanisms in most species. As will be shown in this article the situation is different in flies.

To study figure-ground discrimination flies are well suited as experimental animals: (i) They track moving objects in free flight (Land and Collett, 1974; Wehrhahn, 1979; Wagner, 1986). They also detect and fixate objects even when they differ from their background only by motion (Virsik and Reichardt, 1976; Reichardt and Poggio, 1979; Bülthoff, 1981; Reichardt et al. 1983; 1989; Egelhaaf, 1985a). (ii) The major neurons in their main center of motion computation are amenable to electrophysiological analysis and can be identified individually on the basis of their invariant anatomical structure and response characteristics (Hausen, 1981; 1984; Hausen and Egelhaaf, 1989). (iii) Moreover, by comparing behavioral and electrophysiological data obtained under the same stimulus conditions some of these neurons, the so-called figure-detection or FD-cells, have been concluded to play a decisive role in mediating figure-ground discrimination behavior (Egelhaaf, 1985a-c; Egelhaaf, 1987; Egelhaaf et al. 1988; Reichardt et al. 1989).

Here we review recent experiments that led to the elucidation of the neural circuit that tunes a particular interneuron in the fly's visual system to the motion of small objects relative to the background. Knowing the major elements of a circuit, however, is not sufficient to understand how the circuit operates. More information is required on the synaptic relationships, the synaptic transfer characteristics as well as on the passive and active membrane properties of the neurons involved. Since this knowledge is only beginning to be available in the case of the circuit responsible for small-field tuning of the particular fly interneuron discussed here, we try to constrain the possibilities by model simulations of various variants of the basic experimentally established wiring scheme.

2. FIGURE-GROUND DISCRIMINATION BY THE FLY

In a series of experiments (Reichardt et al., 1983, Egelhaaf 1985a, Reichardt et al., 1989), the reaction of a fly flying tethered to a torque compensator has been determined in response to a figure moving relative to a background (Fig.1, left). These experiments undoubtedly demonstrate that flies are able to discriminate between an object and the background solely by means of relative motion between these two: As long as figure and background move in synchrony, the fly's torque follows the motion of the drum. This compensatory optomotor responses would stabilize the visual world on its eyes if it were allowed to turn freely. During this period of synchronous motion no response towards the object is displayed by the fly. Only when figure and background move relative to each other the mean torque of the fly shifts towards the side of the figure and characteristic torque peaks appear in the trace indicating that the animal tries to turn towards the figure (Fig.1, right).

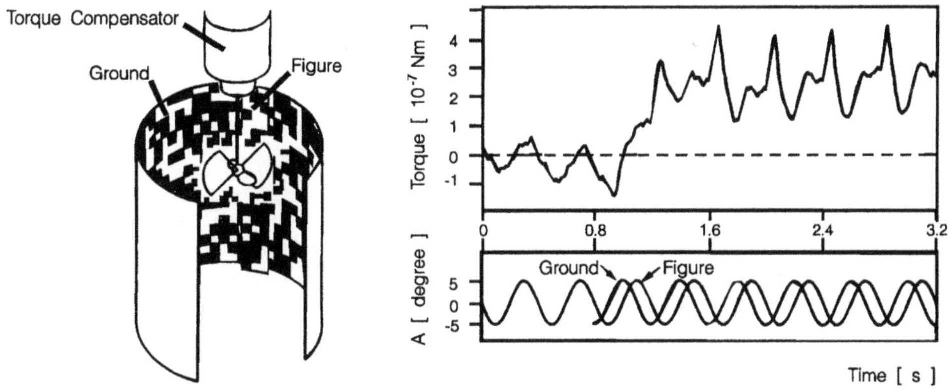

Fig.1 (Left) Experimental set-up of a behavioral figure-ground experiment. A fly is suspended to a torque meter and placed in the middle of a textured drum (background) with a stripe (figure) covered with the same texture placed in front of the fly's right eye. Drum and stripe can be controlled independently. (Right) Torque response profile of the fly *Calliphora erythrocephala.* A textured stripe of 7.2° width was oscillated sinusoidally about an angular position of 30° to the right of the fly (2.5 Hz oscillation frequency, 5° oscillation amplitude). The bottom trace shows the deviation of figure and ground from their mean position. During the first two cycles, figure and ground move in synchrony. At time 0.8s they are set to a phase angle of 90° . The plotted curve represents the average of 50 sweeps of a single fly. (Modified after Egelhaaf, 1985a).

3. RESPONSE PROPERTIES OF FIGURE-DETECTION CELLS

The main center of motion computation of the fly visual system is the posterior part of the third visual neuropil, the lobula plate. There reside about 50 individually identifiable nerve cells with large dendritic trees, the so-called tangential cells (Hausen, 1981; Hausen and Egelhaaf, 1989). They are specifically tuned to various kinds of retinal motion patterns as are encountered by the animal when moving around in its environment. Part of these neurons have been shown to play important roles in visual orientation behavior (Hausen, 1981; Hausen and Egelhaaf, 1989; Hausen and Wehrhahn, 1983; 1990; Egelhaaf et al. 1988; Egelhaaf and Borst, 1993). Most of the tangential cells acquire their characteristic motion selectivity by two basic processing steps. (i) With their extended dendritic trees they spatially pool the output of large numbers of retinotopically organized local motion sensitive elements (Borst and

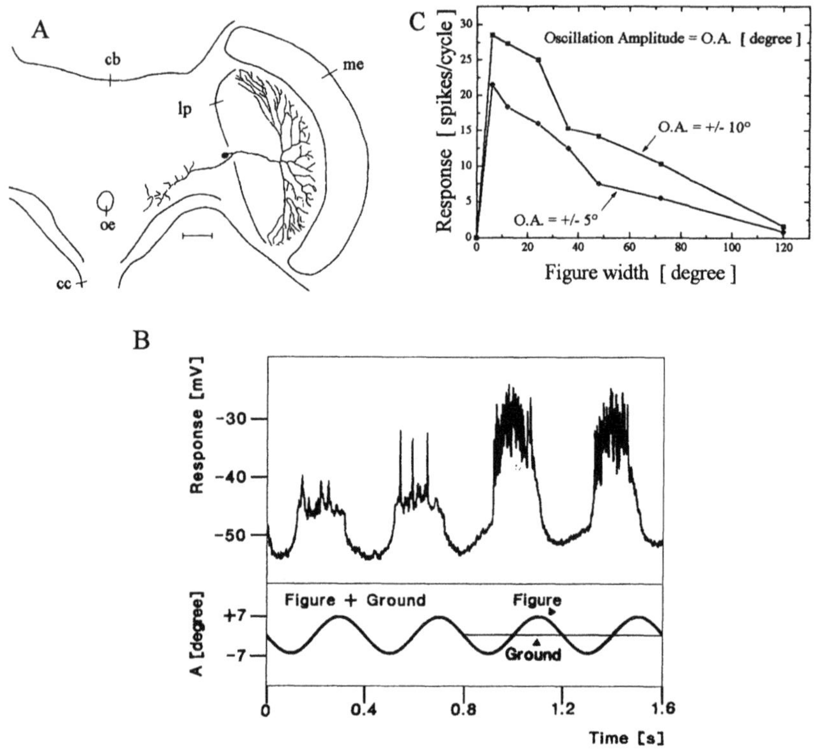

Fig.2 (A) Structure of an FD1-cell as reconstructed from an intracellular LY injection. The cell is drawn into a schematic frontal view of the brain. Abbreviations: cb central brain, cc cervical connective, me medulla, lp lobula plate, oe oesophagus. Scale bar = 100 μm. (B) Intracellular recording of an FD1-cell. After two cycles of synchronous oscillation of a 12°-wide textured figure and a binocular background (from -120° to +120° horizontal extent) the ground stopped moving, while the figure continued oscillating for another two cycles (2.5 Hz oscillations frequency, 7° oscillation amplitude). The figure was positioned in the cells excitatory receptive field at an angular horizontal position of 10°. The FD1-cell is much more excited when the figure oscillates in front of a stationary background than during synchronous motion. (C) Dependence of the FD1-cell response on the angular width of a pattern moving within the ipsilateral visual field. The pattern was oscillating in front of a stationary background at a frequency of 2.5 Hz with two different oscillation amplitudes ('O.A.'). As a distinguishing feature of FD-cells, the cell's response is maximum for small patterns and decreases with increasing pattern size. Note that for the smaller oscillation amplitude the peak response remains at small pattern sizes while only the overall response amplitude is reduced (A,B: Modified after Egelhaaf, 1985b; C: Egelhaaf, unpublished).

Egelhaaf, 1989; Egelhaaf et al., 1989; Borst and Egelhaaf, 1990; Haag et al., 1992). By this retinotopic input, the tangential cells become directionally selective to motion in large parts of the ipsilateral visual field. (ii) The response properties of part of the tangential cells are further shaped by excitatory or inhibitory input from other tangential cells of the ipsi- or contralateral lobula plate.

These network interactions have been unravelled in detail for a particular cell type, the FD1-cell that is most sensitive to the motion of small objects. Fig.2 shows the anatomy of an FD1-cell together with its response characteristics. The cell has an extensive dendritic arborization within the lateral lobula plate and, accordingly, its excitatory receptive field in the fronto-lateral part of the eye. It sends its axon into the central brain (Fig.2A). As is a common property of most invertebrate neurons, the cell body is connected with the axon through a thin cell body fibre. When the cell is stimulated in the same way as in the behavioral experiments (see Fig.1), it displays a pronounced sensitivity for motion when the figure moves relative to the background. The cell responds only weakly when both move in synchrony (Fig.2B). When the response of the FD1-cell is measured as a function of pattern size, its response is maximum for small pattern widths and decreases when the pattern becomes more extended in space ('small-field tuning'). This is true when the pattern is increased within the ipsilateral visual field (Fig.2C). However, the response of the FD1-cell to a small moving object is also reduced when there is large-field motion in front of the contralateral eye (Egelhaaf, 1985b). Consequently, binocular rotatory large-field motion reduces its response to small-field motion most strongly.

4. INPUT CIRCUITRY OF FIGURE-DETECTION CELLS

The specific tuning of the FD1-cell in the right half of the brain to small-field and relative motion has been concluded to be mediated by the inhibitory action of one or several elements that respond best to binocular clockwise large-field motion (Egelhaaf, 1985b,c). There are several lobula plate tangential cells that have the appropriate preferred direction to act as the elements that inhibit the FD1-cell during large-field motion. Two of the HS-cells (HSN-, HSE-cell) as well as the two CH-cells (DCH- and VCH-cell) (Hausen, 1976a,b; 1982a,b; Eckert and Dvorak, 1983) respond maximally to rotatory large-field motion and, hence, under the conditions when the FD1-cell's response is reduced maximally (Egelhaaf, 1985b). From these the CH-cells were the most plausible candidates because they have been suggested to be GABAergic (Meyer et al. 1986) and, thus, most likely inhibitory elements.

From the above mentioned cells only the VCH-cell represents the large-field-inhibitor of the FD1-cell and, thus, mediates its small-field selectivity. This was recently demonstrated by two inactivation techniques: (i) The GABA-antagonist picrotoxinin (Benson, 1988; Rauh et al. 1990) was injected into the haemolymph. As a consequence the sensitivity of the FD1-cell to small moving objects is eliminated (Egelhaaf, 1990;

Warzecha et al., 1993). (ii) Using the photoinactivation technique (Miller and Selverston, 1979; Selverston and Miller, 1980) various candidates of the large field inhibitor were individually ablated and the response of the FD1-cell was recorded before and after ablation. After photoinactivation of the DCH-cell or HSE-cell the response of the FD1-cell did not change significantly. Only when the VCH-cell was ablated the response of the FD1-cell to large-field motion became about three times as large as before, whereas the response to small-field motion did not change considerably. Hence, out of all the candidates, the VCH-cell is the one that appears to be responsible for tuning the FD1-cell to the motion of small objects (Warzecha et al., 1992, 1993).

Knowing the identity of the circuit elements and knowing that one is inhibiting the other is, however, not sufficient to understand how the circuit works at a cellular level. Since the effectiveness of an inhibitory signal depends on the site of inhibition (e.g. Vu and Krasne, 1992), one also needs to have information about the contact sites and the transmission characteristics between these elements. In the present context, it is important to know where the FD1-cell is contacted by the VCH-cell and where the VCH-cell receives its input signals from the contralateral and the ipsilateral eye. The VCH-cell has two arborizations, one in the lobula plate and the other in the ventrolateral brain (Hausen, 1976b; Eckert and Dvorak, 1983). In principle, both arborizations could be input sites either of the ipsi- or contralateral input elements or of both of them. One method which allows to map the synaptically induced spatial activity distribution within single cells is the optical recording technique. In the semi-intact fly where the cells can still be activated and inhibited by their natural synaptic input this technique can be employed because the arborizations of the tangential cells in the lobula plate are more or less two-dimensional lying less than 50 μm below and almost parallel to the surface of the brain. Using the calcium-sensitive dye fura-2 (Grynkiewicz et al., 1985) we could show that, in addition to the presynaptic terminal and the cell body, calcium accumulates also in the dendritic tree. In particular, by motion stimuli restricted to small areas of the visual field, calcium accumulates only in selected parts of the dendrite (Borst and Egelhaaf, 1992).

We applied this technique to the VCH-cell (Egelhaaf et al., 1993). In each of these experiments a single cell was filled with the fluorescent calcium indicator fura-2. Visually induced intracellular calcium accumulation served exclusively as a marker of the sites that are activated by synaptic input. The resulting images represent the relative fluorescence changes of the cell induced by motion stimulation (for details, see Borst and Egelhaaf, 1992). In a first series of experiments a VCH-cell was activated either by front-to-back motion in the ipsilateral visual field or by back-to-front motion presented to the contralateral eye. Ipsilateral motion led to calcium accumulation only in the main arborization in the lobula plate and not in the arborization in the ventrolateral brain. In contrast, contralateral motion induced an increase in calcium concentration simultaneously in both arborizations. Two conclusions can be drawn on the basis of these data: (i) Since the small arborization in the ventrolateral brain is only

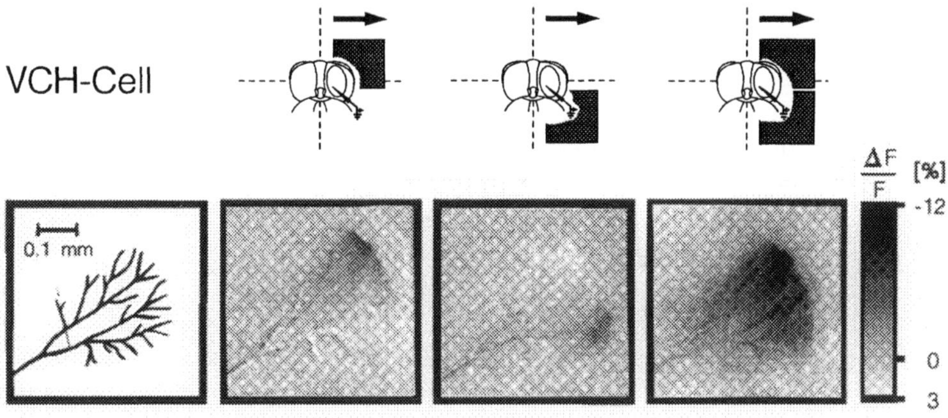

Fig.3 Retinotopic input organization of the VCH-cell. Motion in the preferred direction in the ipsilateral visual field of the VCH-cell leads to an increase in free cytosolic calcium in locally restricted dendritic branches. Two stimulus patterns (diameter about 20°) were placed in the visual field in such a way that each of them, when presented alone, induced membrane potential changes. The stimulus conditions are shown schematically on top of each image with the shaded area indicating the place where the grating was positioned in the visual field of the fly. Left diagram: Reconstruction from the fluorescence intensity image (100x100 pixels; 100 ms exposure time) of a VCH-cell filled with fura-2 taken at 380 nm excitation in the living animal. Other diagrams: Changes in the raw fluorescence ($\Delta F/F$, without background subtraction) of the same cell after 9 s of preferred direction motion in the dorsal, ventral or simultaneously in both parts of the receptive field. Negative $\Delta F/F$ coded by darker grey levels correspond to an accumulation of calcium. Motion in the ventral and dorsal part of the receptive field leads to localized calcium accumulation in ventral and dorsal dendritic branches, respectively, revealing the retinotopic input organization of the VCH-cell. Note the overproportional increase in calcium accumulation when the cell is stimulated by motion with both patterns (Modified after Egelhaaf et al., 1993).

labelled during contralateral motion, although the cell is depolarized under both stimulus conditions, it is suggested that it represents a postsynaptic site receiving its input exclusively from the contralateral eye. (ii) The labelling of the main arborization in the lobula plate during motion in front of either eye may indicate that it is postsynaptic to both ipsi- and contralateral input elements.

These conclusions are further substantiated in experiments where the cell injected with fura-2 was stimulated by horizontal motion of a grating pattern in either the dorsal or ventral part of its ipsilateral receptive field, respectively, or in both regions simultaneously (see schematic diagrams above the plots in Fig.3). When the cell is

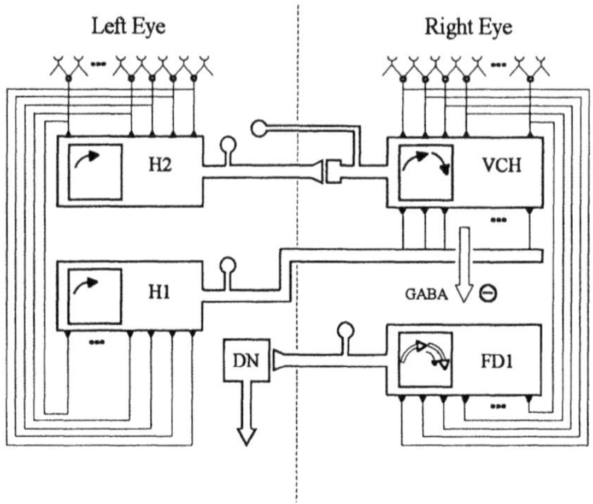

Fig.4 Working hypothesis of the neural circuit responsible for tuning the FD1-cell to small moving objects. The VCH-cell is assumed to receive its ipsilateral input from retinotopically organized local motion sensitive elements and its contralateral input from two tangential cells of the other optic lobe (H1- and H2-cell). The VCH-cell inhibits the FD1-cell that, in turn, is assumed to be involved, via descending neurons (DN), in the detection and fixation of small moving objects. The inhibitory connection between the VCH- and FD1-cell is indicated by an arrow. The arrows in the boxes within the schematic dendrites of the various cells indicate the stimulus conditions that lead to either maximum excitation (filled arrows) or, in the case of the FD1-cell, to maximum inhibition (open arrows); large arrows indicate large-field motion, the small arrow in the schematic FD1-cell indicates small-field motion. The arrows in the left and right halves of the boxes denote motion in the left and the right visual field, respectively. The dashed vertical line indicates the midline of the brain. The thin branches of the axons with their circular endings represent the cell body fibres and the cell bodies, respectively. (Modified after Egelhaaf et al., 1993).

activated by motion stimulation in the upper part of the ipsilateral receptive field, only the dorsal dendrites are labelled. Only the ventral dendritic branches are labelled when the motion stimulus is presented in the lower part of the ipsiilateral visual field. When motion is presented in both regions the entire dendritic tree appears labelled. In conclusion, the activity labelling induced by motion in different parts of the visual field clearly shows that the VCH-cell receives its ipsilateral motion input from retinotopically organized input elements. It should be noted that the labelling induced by motion in both stimulus areas is much larger than the summated activity as induced during motion in each area alone. This feature was observed in most VCH-cells investigated so far and will be further discussed later on.

From these and other experimental data, the following circuit can be derived (Fig.4) which is thought to underly the small-field tuning of the FD1-cell. It consists of at least four identified tangential neurons which are all located in the lobula plate: The H1- and H2-cell, and the VCH- and the FD1-cell. They all receive ipsilateral input from a retinotopic array of local, motion-sensitive elements which they spatially integrate with their large dendrites. By this, the H1- and H2-cell are excited most strongly by back-to-front motion whereas the VCH- and FD1-cell have front-to-back motion as their preferred direction within their ipsilateral receptive field (see arrows indicating preferred directions in Fig.4). The VCH-cell receives additional excitatory input from the contralateral visual field via the H1- and H2-cell at two different locations (Hausen, 1976a,b; Eckert and Dvorak, 1983; Egelhaaf et al., 1993). The VCH-cell, in turn, inhibits the FD1-cell. Interestingly, the lobula plate arborization of the VCH-cell seems to be an input and output-site of the neuron at the same time indicating a dendro-dendritic interaction between VCH- and FD1-cell. The functional consequences of this type of neuronal architecture will be discussed below.

5. MODELING THE CIRCUIT

The knowledge about the identity of the various circuit elements, their contact sites and their response characteristics should, in principle, allow to reconstitute the FD1-cell properties in a model simulation. Closer inspection, however, reveals that all this knowledge is by no means detailed enough to sufficiently constrain the mechanism that may account for the specific selectivity of the FD1-cell for small objects. For instance, it is not clear, so far, how the synaptic input of the various cells is transformed into the corresponding postsynaptic signals. However, as will be shown below, the synaptic transmission characteristics are critical for the performance of the circuit. Since the mechanism underlying small-field tuning of the FD1-cell cannot yet be specified unambiguously, various model alternatives will be compared in the following and related to our experimental findings. Since the optimal object size of the FD1-cell was found to be smaller than the size of its excitatory receptive field, the small-field tuning within the confines of the receptive field is the most challenging aspect of the circuit sketched in Fig.4. Therefore, we restrict our theoretical considerations to this point. All the models we will consider have several properties in common: (i) They consist of two cells which represent the VCH- and the FD1-cell, respectively, one inhibiting the other, and (ii) both cells receive retinotopic synaptic input from an identical array of local motion-sensitive elements. At first glance, these simple assumptions may be expected to lead easily to an element that responds best to small moving objects. However, careful considerations suggest that matters may not be as simple. For instance, as soon as one allows other stimulus parameters to vary in addition to stimulus size and requires the FD1-cell to exhibit small-field tuning not only at a particular stimulus strength but for a wide range of stimulus parameters, the constraints imposed on the mechansim become much more demanding. These problems arise because the VCH-cell, as well as the other lobula plate tangential cells are ambiguous

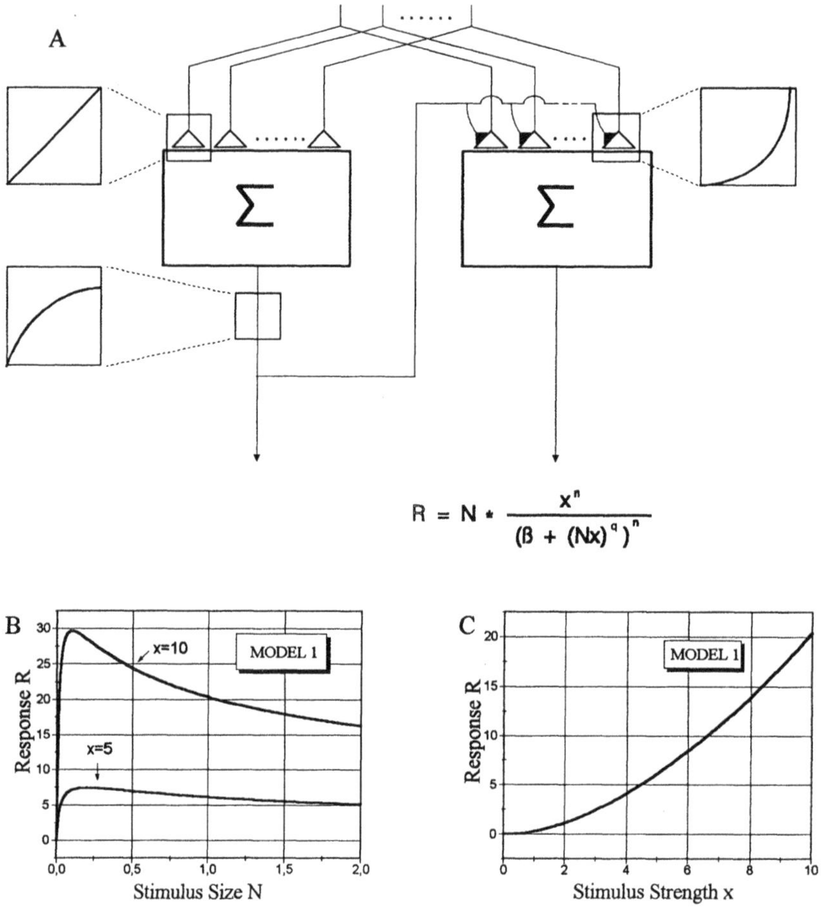

Fig.5 Original model proposed to account for small-field tuning of a nerve cell. (A) Structure of the model: Two cells symbolized by the large rectangles receive input from an array of motion detectors (open synaptic terminals). They are assumed to summate these input signals in a linear way (Σ). The output signal of the left element, the 'pool-cell', inhibits presynaptically (filled synaptic terminals) the input lines to the right element, the FD1-cell. The transmission characteristics of the various synapses shown as input-output functions (input on the x-axis, output on the y-axis) are zoomed in. (B) Output signal of the small-field element as a function of stimulus size for two different stimulus strengths (x=5, x=10). (C) Output signal of the small-field element as a function of stimulus strength at a given stimulus size (N=1). For further explanation see text.

with respect to different stimulus parameters. For instance, they may respond with the same overall amplitude and thus inhibit the FD1-cell in essentially the same way when a small object moves at a high velocity compared with an extended object moving at a smaller velocity. However, the experimental data available so far indicate that the FD1-cell shows small-field tuning at various pattern velocities (see Fig.2C). As will be shown below there are, in principle, two ways to overcome this ambiguity problem at least to some extent. The first model we shall consider was proposed by Reichardt and colleagues to explain the phenomenon of gain-control found in the optomotor behavior and lobula plate tangential cells of flies (Poggio et al., 1981; Reichardt et al., 1983; Fig.5). As it turned out, the property of small-field tuning could also be mimicked by this model by an appropriate choice of parameters (Reichardt et al., 1983; Egelhaaf, 1985c).

Reichardt and colleagues introduced the idea of a pool cell sharing its input with another cell which becomes in turn inhibited by the pool cell. Both neurons are assumed to integrate the output signals of local motion detectors. According to their terminology, the size of the stimulus will be denoted in the following as N and the amplitude of the local motion detector output as x. If everything is linear, the signal of an integrating element is Nx. The different inhibitory effect of stimulus size and local stimulus strength onto the small-field element was modelled by assuming that the two neurons handle their retinotopic input in a different way: Whereas the synapses onto the small-field element posses a pronounced expansive transmission characteristic (modelled by a power function with an exponent of 3), the synapses onto the inhibitory element are assumed to behave linearly. Moreover, the inhibitory neuron is assumed to saturate and its response is approximated by $(Nx)^q$ with $0<q<1$. As a specific type of inhibitory interaction between two neurons the authors assumed a shunting inhibition. In this case, the equilibrium potential of the inhibitory ion is close to the resting potential of the postsynaptic element. To be effective, the conductance change exerted by the inhibitory signal has to be large with respect to the other conductances (leak and excitatory conductance). As follows immediately from the equation for an electrical equivalent circuit for a passive patch of membrane, the excitatory input then becomes essentially divided by the inhibitory input. Thus, given a presynaptic shunting inhibition with β as the shunting coefficient (equivalent to the leak conductance of the cell) the input signals of the small-field element turn into $x/(\beta+(Nx)^q)$, which after expansive transmission (power of n) become $(x/(\beta+(Nx)^q))^n$. Assuming linear spatial integration within the small-field element, the response R becomes

$$R = N*(x/(\beta+(Nx)^q))^n \qquad (1)$$

The model together with its response as a function of stimulus size and amplitude is shown in Fig.5. As one can see, the model displays small-field tuning for various local stimulus strengths (Fig.5B). Furthermore, its response increases monotonically with

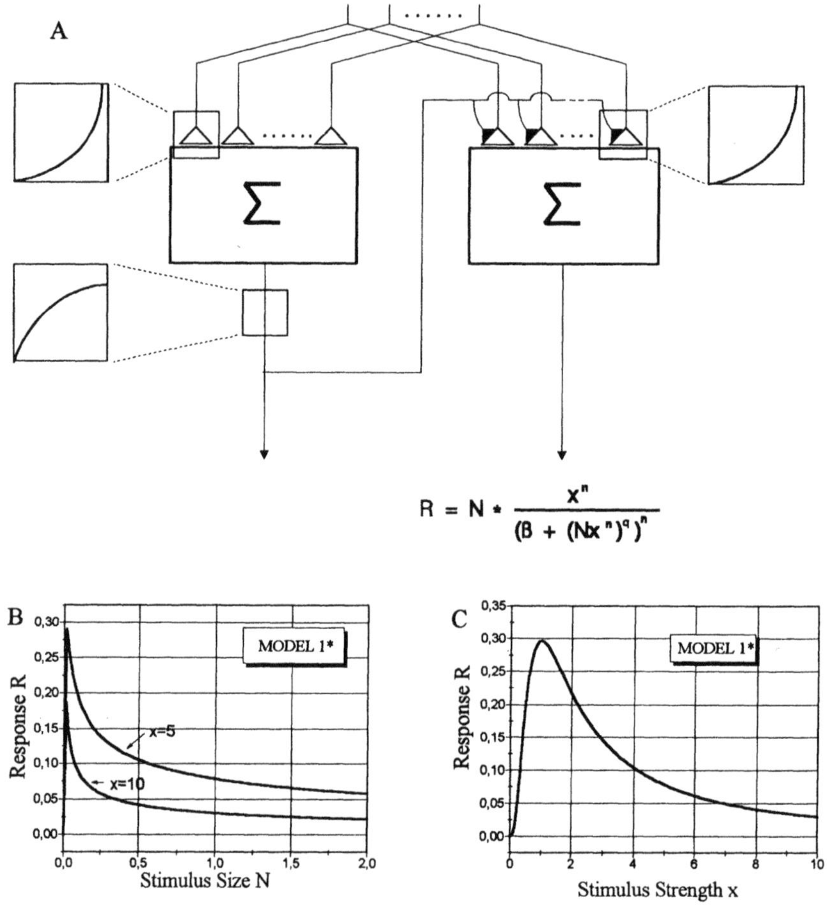

Fig.6 Same model as shown in Fig.5 with the only difference that here the synapses of the motion-sensitive input elements are assumed to have identical expansive transmission characteristics (power of 3) onto both integrating neurons. (A) Structure of the model. (B) Output signal of the small-field element as a function of stimulus size for two different stimulus strengths (x=5, x=10). (C) Output signal of the small-field element as a function of stimulus strength at a given stimulus size (N=1). Note that in contrast to the model shown in Fig.5 the function is no longer monotonically increasing with stimulus strength.

stimulus strength such as velocity (Fig.5C). Hence its behavior quite nicely mimicks the response of the FD1-cell. We can take a second look at the model parameters and ask for which numerical range of the various parameters the model displays these desired properties. To do so, we may simplify equation (1) by setting the shunting coefficient β equal to zero. Then R is proportional to N^{1-nq} and to $x^{n*(1-q)}$. If we ask that R is decreasing with N and increasing with x, n and q have to be chosen such that $1-nq < 0$, and $n(1-q) > 0$. Obviously, this is the case e.g. for $q=0.5$ and $n=3$. These values are chosen in the graphs shown in Fig.5B and C.

The crucial assumption of this model is that both neurons have input synapses with fundamentally different transmission characteristics. To understand the impact of this assumption, the same circuit as is shown in Fig.5 is modelled with the only difference that both neurons treat their retinotopic input in the same manner, i.e. in the example shown in Fig.6 through an expansive nonlinearity (power of 3). The response then becomes

$$R = N*(x/(\beta+(Nx^n)^q))^n \qquad (2)$$

As one can see (Fig.6), this has the following consequences: The neuron still displays small-field tuning for different local stimulus parameters (Fig.6B). However, for a certain range of stimulus parameters the response amplitude becomes the larger the smaller the local stimulus signal is. Hence, if the response is plotted as a function of stimulus strength for a given stimulus size, it shows a peaked dependence with decreasing amplitude for large stimulus amplitudes (Fig.6C). Although such an astonishing response behavior has not been observed in the FD1-cell, so far, the model could still account for the experimental data given that the stimulus parameters are restricted to the operating range of the model where the response is monotonically increasing with local stimulus strength (left part of the graph in Fig.6C). Interestingly, one can demonstrate that the particular features of this model is independent of the specific values of the parameters n and q: If we again ask that R is monotonically decreasing with N and increasing with x, 1-nq has to be smaller than zero, and n(1-nq) has to be larger than zero. Obviously, this can never be the case for any positive values of n and q so ever.

The next model we shall consider takes into account experimentally established facts that were not known when the original model for figure-ground discrimination has been proposed. (i) The dendro-dendritic nature of of the VCH-FD1-cell interaction. Such a construction allows, at least in principle, to have different inhibitory signals for a small pattern moving at a large velocity and a large pattern moving at a small velocity, because both stimuli may lead to a different activity distribution in the VCH-cell dendrite. (ii) The calcium signal observed within the dendrite of the VCH-cell seems to increase overproportionally with stimulus size (see Fig.4). Based on light-microscopical anatomical criteria these arborizations show presynaptic specializations ('blebs') and, therfore, are likely to be also an output region of the cell (Hausen,

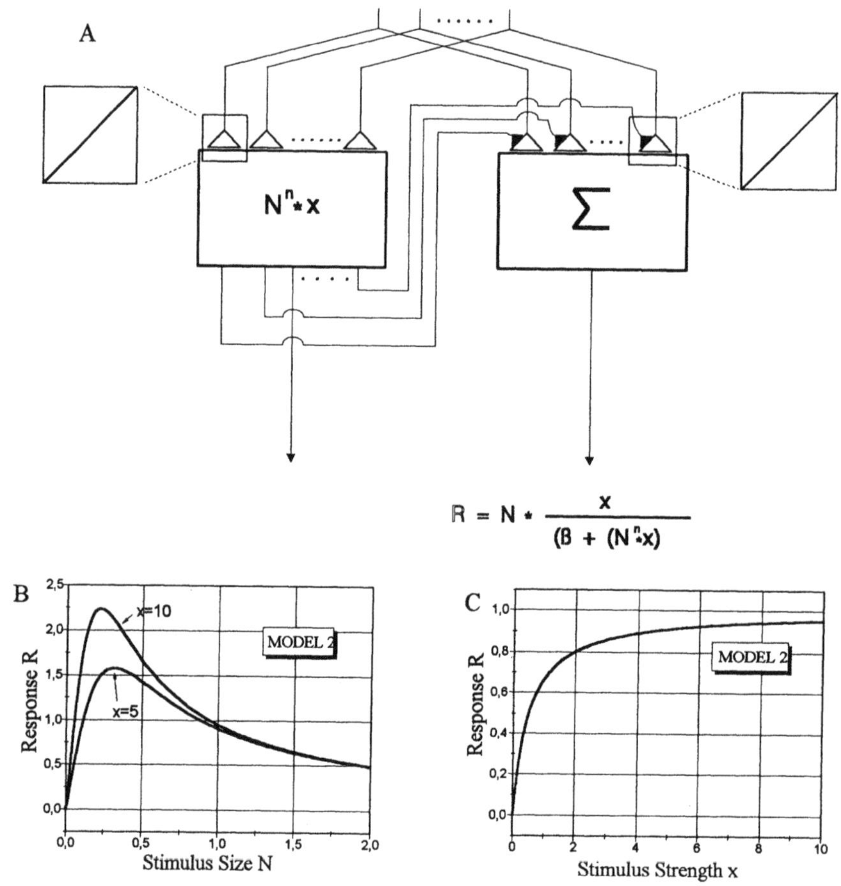

Fig.7 Alternative model to account for small-field tuning of a nerve cell. (A) Structure of the model. In contrast to the model shown in Fig.5 the input synapses to both elements are assumed to possess identical linear transmission characteristics. Here, the crucial assumption is that the inhibitory signals exerted from the left element increases overproportionally with stimulus size but not with stimulus strength. (B) Output signal of the small-field element as a function of stimulus size for two different stimulus strengths (x=5, x=10). (C) Output signal of the small-field element as a function of stimulus strength at a given stimulus size (N=1).

1976a). Since calcium is known to regulate transmitter release, the images shown in Fig.4 are presumed to represent the amplitude and distribution of the inhibitory signal the VCH-cell exerts onto the FD1-cell. Given this, we assume that the inhibitory signal, unlike the electrical response of the VCH-cell, increases overproportionally with stimulus size but not with local stimulus strength. We account for this in formal terms by an expansive nonlinearity which acts only on N but not on x: $N^n x$. Assuming either pre- or postsynaptic inhibition, the response of the small-field neuron becomes

$$R = Nx / (\beta + N^n x) \quad (3)$$

This circuit together with its resulting behavior is shown in Fig.7. The circuit displays small-field tuning for both local stimulus parameters (Fig.7B). Furthermore, its response is a monotonic function of local stimulus strength. Again, as the model shown in Fig.5, the response is in accordance with the response characteristics of the FD1-cell as recorded from the fly brain. Considering the specific values of the parameter n, one immediately sees that as long as n is larger than 1, the circuit will display small-field tuning, since R is decreasing with N. Independent from that the response is always monotonically increasing with x and shows a saturating behavior.

6. CONCLUSIONS AND PROSPECTS

We have provided experimental evidence for a neural circuit that is responsible for the small-field tuning of a particular nerve cell, the so-called FD1-cell, in the fly visual system. However, the knowledge about the identity and connectivity between different identified neurons is not sufficient to understand at a detailed biophysical level how this circuit provides small-field tuning. Two alternative models are presented. The core of the first one has been proposed by Reichardt and colleagues (Poggio et al., 1981; Reichardt et al., 1983) at a time before the large-field inhibitor of the FD1-cell was identified. This model assumes presynaptic inhibition of the small-field cell through a so-called pool-cell. By the experiments of Warzecha et al. (1992, 1993), this pool-cell has been identified as the VCH-cell.

The anatomical characteristics of the VCH-cell, however, prompted us to propose a slightly different wiring scheme. Since the arborization of this cell in the third visual ganglion is suggested to be postsynaptic to retinotopic motion sensitive elements and, in addition shows presynaptic specializations, it is suggestive to assume that it inhibits the FD1-cell *via* a dendro-dendritic interaction. Furthermore, optical recording experiments revealed an over-proportional increase of calcium concentration in this arborization with increasing pattern size. If calcium is assumed to control transmitter release, this finding would imply that the inhibitory strength increases overproportionally when the stimulus pattern increases.

Both alternative models critically rely on specific assumptions to disentangle the consequences of an increasing pattern size on the one hand and velocity or contrast on the other hand and, thus, to guarantee small-field selectivity of the FD1-cell for a wide range of contrasts and velocities. The original model has to assume different transmission characteristics of the local motion detector output synapses onto the VCH- and FD1-cell. In contrast the new model is less restrictive in this respect, but has to assume that the strength of the inhibitory signal increases overproportionally with increasing pattern size but linearly or less than linearly with variations in contrast or velocity.

Is there experimental evidence to distinguish between these two mechanisms? There are several arguments against assuming different synaptic transmission characteristics between the local motion detectors and the VCH- and FD1-cell, respectively. The most compelling one comes from the recent photoinactivation experiments of Warzecha et al. (1992, 1993). After selectively ablating the VCH-cell, the small-field selectivity of the FD1-cell disappears, obviously because the inhibitor of the FD1-cell has been killed. If the transmission characteristic between the movement detectors and the FD1-cell corresponds to an expansive nonlinearity as proposed by the original model, removal of the inhibitory input should lead to a dramatic increase in the overall response amplitude of the FD1-cell. This, however, is in contrast to the experimental findings where the response amplitude during small-field motion is not much altered after the VCH-cell has been killed (Warzecha et al., 1992, 1993).

The only other way we see at present to achieve small-field tuning irrespective of contrast and velocity is to assume that the inhibitory signal exerted by the VCH-cell depends in a different way on pattern size than on local stimulus strength, such as has been implemented in the model shown in Fig.7. Although this can be easily done formally, it is by no means a trivial problem how real neurons achieve that the inhibitory signal increases overproportionally with pattern size but linearly or according to a compressive nonlinearity with increasing velocity and contrast. The anatomy of the VCH-cell and our recent optical recording results provide a hint how this may be possible. Since the interaction between VCH- and FD1-cell is dendro-dendritic, the activity distribution in the VCH-cell and, thus, the inhibitory signal is likely to differ considerably when it is excited, for instance, by a small pattern moving with a larger velocity or an extended pattern moving with a smaller velocity. The over-proportional increase in calcium concentration with increasing pattern size may well correspond to the required expansive nonlinear increase in inhibitory strength. However, before the model shown in Fig.7 can be accepted to account for the mechanism tuning the FD1-cell to small moving objects, it has to be determined whether the calcium signal increases less than proportionally with increasing pattern velocity or contrast.

Finally one further point should be mentioned that has not yet been taken into account in any model explaining small-field tuning of the FD1-cell. So far the model cells are fed by detectors that faithfully encode the time course of pattern velocity. However, as is known from many experiments, this assumption is not correct. For instance, the output of each individual local movement detector is not constant even when the pattern moves with a constant velocity but modulated over time with the temporal frequency of the pattern and its second harmonic (Egelhaaf et al., 1989, 1990). Only if also these complexities are taken into account, we will be able to understand the circuit underlying the small-field tuning of the FD1-cell at a really deep level.

7. ACKNOWLEDGEMENTS

We are grateful to T.Brotz, A.-K.Warzecha and J.M.Zanker for critically reading the manuscript and to M.Bauch for technical assistance.

8. REFERENCES

ALLMAN, J., MIEZIN, F. and MCGUINNESS, E. Direction- and velocity-specific responses from beyond the classical receptive field in the middle temporal visual area (MT). *Perception* 14: 105-126, 1985.

BENSON, J. A. Transmitter receptors on insect neuronal somata: GABAergic and cholinergic pharmacology. In: *Neurotox '88: Molecular basis of drug and pesticide action,* edited by G. G. Lunt. Elsevier, 1988, p. 193-205.

BORST, A. and EGELHAAF, M. Principles of visual motion detection. *Trends Neurosci.* 12: 297-306, 1989.

BORST, A. and EGELHAAF, M. Direction selectivity of fly motion-sensitive neurons is computed in a two-stage process. *Proc. Natl. Acad. Sci. USA* 87: 9363-9367, 1990.

BORST, A. and EGELHAAF, M. In vivo imaging of calcium accumulation in fly interneurons as elicited by visual motion stimulation. *Proc. Natl. Acad. Sci. USA* 89: 4139-4143, 1992.

BÜLTHOFF, H. Figure-ground discrimination in the visual system of Drosophila melanogaster. *Biol. Cybern.* 41: 139-145, 1981.

COLLETT, T. S. Visual neurones for tracking moving targets. *Nature* 232: 127-130, 1971.

COLLETT, T. S. Visual Neurones in the Anterior Optic Tract of the Privet Hawk Moth. *J. Comp. Physiol. A* 78: 396-433, 1972.

COLLETT, T. S. AND KING, A. J. Vision during flight. In: *The compound eye and vision of insects,* edited by G. A. Horridge. Oxford: Clarendon Press, 1975, p. 437-466.

COLLETT, T. S. and PATERSON, C. J. Relative motion parallax and target localization in the locust, *Schistocerca gregaria. J. Comp. Physiol. A* 169: 615-621, 1991.

DAVIDSON, R. M. and BENDER, D. B. Selectivity for relative motion in the monkey superior colliculus. *J. Neurophysiol.* 65: 1115-1133, 1991.

ECKERT, H. and DVORAK, D. R. The centrifugal horizontal cells in the lobula plate of the blowfly, *Phaenicia sericata. J. Insect. Physiol.* 29: 547-560, 1983.

EGELHAAF, M. On the neuronal basis of figure-ground discrimination by relative motion in the visual system of the fly. I. Behavioural constraints imposed on the neuronal network and the role of the optomotor system. *Biol. Cybern.* 52: 123-140, 1985a.

EGELHAAF, M. On the neuronal basis of figure-ground discrimination by relative motion in the visual system of the fly. II. Figure-Detection Cells, a new class of visual interneurones. *Biol. Cybern.* 52: 195-209, 1985b.

EGELHAAF, M. On the neuronal basis of figure-ground discrimination by relative motion in the visual system of the fly. III. Possible input circuitries and behavioural significance of the FD-Cells. *Biol. Cybern.* 52: 267-280, 1985c.

EGELHAAF, M. Dynamic properties of two control systems underlying visually guided turning in house-flies. *J. Comp. Physiol. A* 161: 777-783, 1987.

EGELHAAF, M. How do fly FD-cells acquire their sensitivity to small-field motion. *Naturwissenschaften* 77: 182-185, 1990.

EGELHAAF, M. AND BORST, A. Motion computation and visual orientation in flies. *Comp.Biochem.Physiol*, 1993 (in press).

EGELHAAF, M., BORST, A. and REICHARDT, W. Computational structure of a biological motion detection system as revealed by local detector analysis in the fly's nervous system. *J. Opt. Soc. Am. A* 6: 1070-1087, 1989.

EGELHAAF, M., BORST, A. and PILZ, B. The role of GABA in detecting visual motion. *Brain Res.* 509: 156-160, 1990.

EGELHAAF, M., HAUSEN, K., REICHARDT, W. and WEHRHAHN, C. Visual course control in flies relies on neuronal computation of object and background motion. *Trends Neurosci.* 11: 351-358, 1988.

EGELHAAF, M., BORST, A., WARZECHA, A. -K., FLECKS, S. and WILDEMANN, A. Neural circuit tuning fly visual neurons to motion of small objects. II. Input organization of inhibitory circuit elements by electrophysiological and optical recording techniques. *J. Neurophysiol.* 1993 (in press).

FROST, B. J., CAVANAGH, P. and MORGAN, B. Deep tectal cells in pigeons respond to kinematograms. *J. Comp. Physiol. A* 162: 639-647, 1988.

FROST, B. J. and NAKAYAMA, K. Single visual neurons code opposing motion independent of direction. *Science* 220: 744-745, 1983.

GRÜNAU, M. VON and FROST, B. J. Double-opponent-process mechanism underlying RF-structure of directionally specific cells of cat lateral suprasylvian visual area. *Exp. Brain Res.* 49: 84-92, 1983.

GRYNKIEWICZ, G., POENIE, M. and TSIEN, R. Y. A new generation of Ca2+ Indicators with greatly improved fluorescence properties. *The Journal of Biological Chemists* 260,6: 3440-3450, 1985.

HAAG, J., EGELHAAF, M. and BORST, A. Dendritic integration of visual motion information in the fly. *Neurosci. Lett.* 140: 173-176, 1992.

HAUSEN, K. Functional charaterization and anatomical identification of motion sensitive neurons in the lobula plate of the blowfly *Calliphora erythrocephala. Z. Naturforsch.* 31 c: 629-633, 1976a.

HAUSEN, K. Struktur, Funktion und Konnektivität bewegungsempfindlicher Interneurone im dritten optischen Neuropil der Schmeissfliege *Calliphora erythrocephala*. University of Tübingen: Doctoral Dissertation, 1976b.

HAUSEN, K. Monocular and binocular computation of motion in the lobula plate of the fly. *Verh. Dtsch. Zool. Ges.* 74: 49-70, 1981.

HAUSEN, K. Motion sensitive interneurons in the optomotor system of the fly. I. The Horizontal Cells: Structure and signals. *Biol. Cybern.* 45: 143-156, 1982a.

HAUSEN, K. Motion sensitive interneurons in the optomotor system of the fly. II. The Horizontal Cells: Receptive field organization and response characteristics. *Biol. Cybern.* 46: 67-79, 1982b.

HAUSEN, K. The lobula-complex of the fly: Structure, function and significance in visual behaviour. In: *Photoreception and vision in invertebrates*, edited by M. A. Ali. Plenum Press, 1984, p. 523-559.

HAUSEN, K. AND EGELHAAF, M. Neural mechanisms of visual course control in insects. In: *Facets of vision*, edited by D. Stavenga and R. Hardie. Berlin,Heidelberg,New York,London,Paris,Tokyo: Springer-Verlag, 1989, p. 391-424.

HAUSEN, K. and WEHRHAHN, C. Microsurgical lesion of horizontal cells changes optomotor yaw responses in the blowfly Calliphora erythrocephala. *Proc. R. Soc. Lond. B.* 219: 211-216, 1983.

HAUSEN, K. and WEHRHAHN, C. Neural circuits mediating visual flight in flies. II. Separation of two control systems by microsurgical brain lesions. *J. Neurosci.* 10: 351-360, 1990.

LAND, M. F. and COLLETT, T. S. Chasing behaviour of houseflies (*Fannia canicularis*). A description and analysis. *J. Comp. Physiol.* 89: 331-357, 1974.

MEYER, E. P., MATUTE, C., STREIT, P. and NÄSSEL, D. R. Insect optic lobe neurons identifiable with monoclonal antibodies to GABA. *Histochemistry* 84: 207-216, 1986.

MILES, F. A. and KAWANO, K. Visual stabilization of the eyes. *Trends Neurosci.* 10: 153-158, 1987.

MILLER, J. P. and SELVERSTON, A. I. Rapid killing of single neurons by irradiation of intracellularly injected dye. *Science* 206: 702-704, 1979.

OLBERG, R. M. Object- and self-movement detectors in the ventral nerve cord of the dragonfly. *J. Comp. Physiol.* 141: 327-334, 1981.

OLBERG, R. M. Identified target-selective visual interneurons descending from the dragonfly brain. *J. Comp. Physiol.* 159: 827-840, 1986.

POGGIO, T., REICHARDT, W. and HAUSEN, K. A neuronal circuitry for relative movement discrimination by the visual system of the fly. *Naturwissenschaften* 68: 443-446, 1981.

RAUH, J. J., LUMMIS, S. C. R. and SATTELLE, D. B. Pharmacological and biochemical properties of insect GABA receptors. *Trends Pharmacol. Sci.* 11: 325-329, 1990.

REGAN, D. and BEVERLEY, K. I. Figure-ground segregation by motion contrast and by luminance contrast. *J. Opt. Soc. Am. A* 1: 433-442, 1984.

REICHARDT, W., EGELHAAF, M. and GUO, A. Processing of figure and background motion in the visual system of the fly. *Biol. Cybern.* 61: 327-345, 1989.

REICHARDT, W. and POGGIO, T. Figure-ground discrimination by relative movement in the visual system of the fly. Part I: Experimental results. *Biol. Cybern.* 35: 81-100, 1979.

REICHARDT, W., POGGIO, T. and HAUSEN, K. Figure-ground discrimination by relative movement in the visual system of the fly. Part II: Towards the neural circuitry. *Biol. Cybern.* 46 (Suppl.): 1-30, 1983.

SELVERSTON, A. I. and MILLER, J. P. Mechanisms underlying pattern generation in lobster stomatogastric ganglion as determined by selective inactivation of identified neurons. I. Pyloric System. *J. Neurophysiol.* 44,6: 1102-1120, 1980.

SRINIVASAN, M. V., LEHRER, M. and HORRIDGE, G. A. Visual figure-ground discrimination in the honeybee: the role of motion parallax at boundaries. *Proc. R. Soc. Lond. B.* 238: 331-350, 1990.

TANAKA, K., HIKOSAKA, K., SAITO, H., YUKIE, M., FUKADA, Y. and IWAI, E. Analysis of local and wide-field movements in the superior temporal visual areas of the macaque monkey. *J. Neurosci.* 6: 134-144, 1986.

TSAI, H. -J. Response of toad's tectal neurons to in-phase and anti-phase movements of object and textured background. *J. Comp. Physiol. A* 167: 857-863, 1990.

VU, E.T. and KRASNE, F.B. Evidence for a computational distinction between proximal and distal neuronal inhibition. *Science* 255: 1710-1712, 1992.

VIRSIK, R. and REICHARDT, W. Detection and tracking of moving objects by the fly *Musca domestica. Biol. Cybern.* 23: 83-98, 1976.

WAGNER, H. Flight performance and visual control of the flight of the free-flying housefly (*Musca domestica*). II. Pursuit of targets. *Phil. Trans. R. Soc. Lond. B* 312: 553-579, 1986.

WARZECHA, A. -K., BORST, A. and EGELHAAF, M. Photo-ablation of single neurons in the fly visual system reveals neural circuit for the detection of small moving objects. *Neurosci. Let.* 141: 119-122, 1992.

WARZECHA, A. -K., EGELHAAF, M. and BORST, A. Neural circuit tuning fly visual neurons to motion of small objects. I. Dissection of the circuit by pharmacological and photoinactivation techniques. *J. Neurophysiol.* 1993 (in press).

WEHRHAHN, C. Sex-specific differences in the chasing behaviour of houseflies (*Musca*). *Biol. Cybern.* 32: 239-241, 1979.

Brain Theory / A. Aertsen (Editor)

Encoding of 'nothing' in the peripheral auditory pathway: Nervous activity associated with a spectral gap

E. de Boer and A.J. Breed

Auditory Physics, Rm D2-210,
Academic Medical Centre, Meibergdreef 9,
1105 AZ Amsterdam, The Netherlands.

Abstract

Because the cochlea performs a frequency analysis, the spectrum of the incoming acoustical signal is split up into many parts and each part is encoded by a group of auditory-nerve fibres. As far as the rate of firing is concerned, the resolution limit is set by the bandwidth of the cochlear filters. In terms of the distribution of timing instants of neural firings, resolution may well be finer, however. This forms the subject treated in this paper. Experiments were done on responses of single auditory-nerve fibres recorded in Mongolian gerbils. Stimuli were bands of noise in which a gap was made centred at the Characteristic Frequency of the fibre. Results are, in short, that one fibre cannot encode such a gap. As a rough estimate, responses of several hundreds of fibres must be combined to obtain sufficient sharpness of analysis. Convergence of this nature may be possible in higher centres of the auditory pathway.

1. INTRODUCTION

The inner ear is never silent. The nerve fibres that go from the cochlea (inner ear) to the brain always carry irregular trains of spikes. With acoustical stimulation of the ear the average activity of some groups of fibres increases (it never decreases), and the firing instants regroup themselves into another pattern, this constitutes all the information about the sound that goes to the central nervous system (CNS). An essential aspect is that the firing of auditory-nerve fibres is a stochastic process, yet many acoustical stimuli sound as if they were perfectly stationary.

It is well-known that the cochlea carries out a kind of Fourier analysis which means that the activity in every nerve fibre of the auditory nerve reflects the spectral content of the stimulus in a certain (narrow) frequency region that is specific to that fibre. In this sense, a fibre-specific frequency response can be associated with a fibre, this frequency response is of the band-pass type and the relative bandwidth is of the order of 20 percent - it has an extension towards the lower frequencies but that is unimportant for the present work.

The first and simplest-to-measure parameter of fibre activity is the rate of firing. If the stimulus contains important frequency components in the pass band of a fibre, that fibre is likely to respond with an increase of the rate of firing over the spontaneous rate. Sounds that are important for communication (especially, human communication) generally are rich in spectral detail. For a vowel in human speech much of the coding goes into the shape of the power spectrum: the latter shows two or more peaks (formants) that are characteristic for that vowel. Accordingly, for vowel sounds it can be expected that two or more groups of nerve fibres are more excited than others, namely, those that are tuned to the formant frequency regions. It has been found that the representation of spectral content of a vowel in terms of firing rates distributed over the fibres of the auditory nerve is insufficient for explaining auditory recognition of vowels (Sachs and Young, 1979). Accordingly, it has been hypothesized that the fine detail of the temporal distribution of firing instants is the supplementary carrier of information (Sachs and Young, 1980). And this is the view we will entertain in what follows. We will be concerned with the temporal information that is (actually and theoretically) present in the information going to the brain. The actual mechanism by which temporal information would be decoded by the CNS will be left out of consideration here. However, we will estimate to which extent the information coming from different fibres has to be combined in order to make optimal analysis possible.

It should be clear that the ensemble of auditory-nerve fibres can faithfully represent spectral details only inasmuch as the associated frequency response curves and the synaptic mechanisms allow it. Any detail that is finer than the width of the frequency response function, for example, would be greatly lost, and certainly in terms of the rate of firing. What information would still be present in the detailed temporal firing pattern constitutes the main theme of the present paper. It will be shown that information about finer details than frequency resolution would admit is really present in the temporal firing pattern - a kind of 'hyper-acuity' in the ear. It will be shown also that there is a fundamental limit to the recovery of that information: this is only possible when the output of many fibres is combined. This might constitute one of the tasks of a higher-order centre in the auditory pathway.

As an extreme example of spectral shape, the case of a spectral 'zero', or a *spectral gap*, will be treated. Assume that the stimulus spectrum contains a frequency band with zero intensity density surrounded by bands with nonzero intensity density. If the width of the spectral gap is smaller than the width of the band-pass response function associated with a fibre, there will be little evidence of the gap in the firing rate, i.e., the rate will be substantially the same whether the gap is present or not. However, in the details of the firing instants, information about the gap would certainly be present but we will need a special analysis method to recover it. Actually, such a method exists, it is called the correlation-function method, and Appendix A describes the theory underlying this assertion. The central question of this paper is: to what extent do the nerve fibres faithfully represent spectral gaps (and edges) in their timing patterns? Or, stated differently: how is 'nothing', the absence of 'something', coded in the auditory nerve? In our analysis we will use the correlation-function method and we leave aside the question as to whether the CNS would be able to apply an analysis method similar to it. The answer to the question posed is important for understanding processing of speech (nasal semi-vowels are often characterized by spectral gaps) but also for understanding how nonlinear distortion which is inherent in the functioning of the inner ear makes itself known in the discharges of the nerve fibres. As a by-product of the work we obtain information about stochastic aspects of encoding and decoding, and this is pretty important for understanding an organ that works close to its natural limits.

2. COCHLEAR ENCODING - GENERAL PROPERTIES

This section briefly describes the elements of the transduction in the cochlea and its mathematical-physical analysis. Input to the cochlea is the sound impinging on the ear-drum membrane; the sound is converted into vibration which is communicated to the fluid with which the channels of the cochlea are filled. As is known, the cochlea has a shape resembling that of a snail-shell, it contains long and narrow channels filled with fluid and separated by membrane-like cell aggregates. As the result of stimulation a wave is set up which travels along the length of the main partition between the cochlear channels, the Basilar Membrane (BM). The wave causes the partition to vibrate and, consequently, to excite the nerve fibres of the auditory nerve. Consider the case where the stimulus is a pure tone, in that case the wave will produce a vibration maximum in a specific region of the cochlea. A high-frequency tone produces the maximum in the basal, and a low-frequency tone in the apical region of the cochlea.

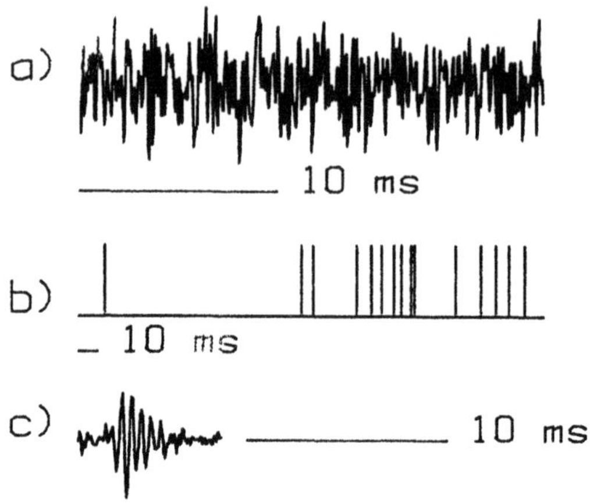

Figure 1-a: stimulus waveform, b: response of one fibre of the auditory nerve, c: revcor function (note the different time scales).

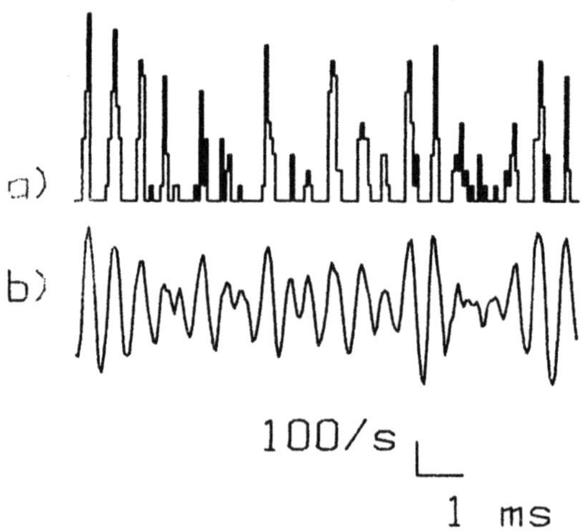

Figure 2-a: firing probability. b: Linear Excitatory Signal (LES)

In a more precise form, the excitation of one of the fibres of the auditory nerve can be described as follows. Associated with each location we define a location-specific *filter*. The filter can be thought of as a linear filter. The filter has a frequency response which mainly shows band-pass character. We can consider the stimulus signal, to be called x(t), as being filtered by this filter, and to give rise to spikes in the filtered form. Let us call the filtered stimulus signal for the i-th fibre the Linear Excitatory Signal (LES) - linear because it is obtained from a linear filtering procedure - and let us designate it by $y_i(t)$. It may be expected that $y_i(t)$ can serve as a predictor of the probability density $p_i(t)$ of firing of the fibre. In fact, it turns out that $p_i(t)$ is close to a nonlinear memoryless function of the LES $y_i(t)$. In particular, when $y_i(t)$ is negative, $p_i(t)$ is zero, and when $y_i(t)$ is positive and small, $p_i(t)$ is approximately proportional to $y_i(t)$. That $p_i(t)$ shows saturation for large positive excursions of $y_i(t)$ is unimportant. This simple description of the relation between $y_i(t)$ and $p_i(t)$ is a good approximation, there are definitely nonlinear effects in the filtering, and adaptation effects in the process of neural triggering, but in the context of the present section they are small and can be neglected.

Figures 1 and 2 illustrate the actual relation between stimulus and response, and between $y_i(t)$ and $p_i(t)$ as it can be deduced from recordings made in an animal experiment. We will first briefly outline the method of measurement (more details will be given later). In the experimental animal the auditory nerve is made accessible and a recording is made of the response of a single fibre of the nerve. The sound stimulus - produced by an earphone - is either a pure tone with adjustable frequency and intensity or a wide band of random noise (actually, the noise signal is pseudo-random noise: its waveform is periodic). With the pure-tone stimulus the frequency to which the fibre is most sensitive (generally known as the Characteristic Frequency, abbreviated: CF) is determined, next the centre of the noise-bands's spectrum is adjusted to it and the noise stimulus is switched on. The noise signal is presented for a few minutes and the detailed firing instants of the fibre's response are recorded and processed in their relation to the stimulus.

Figure 1 shows, in part (a), a segment of the noise waveform x(t) and, in part (b), the firing instants of the fibre under study, marked by vertical bars. In the course of the 233 msec depicted, very few of such events (usually called: spikes) occur. An estimate of the filtering involved in transduction is obtained by applying the 'reverse-correlation' ('revcor') technique (de Boer and de Jongh, 1978) to a long recording of the type shown. This technique is essentially a direct application of the cross-correlation procedure described in Appendix A, a few specific details are given in Appendix B. The result of the procedure is the *reverse-correlation function* (abbreviated revcor function) ; this can, actually, be interpreted as the impulse response of the filter that best approximates the transduction from the stimulus x(t) to the series of firing instants. Figure 1-c shows the revcor function $h_1(t)$ derived from a large number (12000) of spikes recorded from the nerve fibre. Because of the stochasticity of firing, $h_1(t)$ is a statistical estimate and will be partially corrupted by fluctuations. Processing of more spikes reduces this effect; if needed, the waveform can be further improved by smoothing.

Figure 2 shows the relation between the LES $y_1(t)$ and $p_1(t)$. The stimulus signal x(t) is presented continuously, over many periods - its waveform is periodic - and the firing instants are accumulated in a period histogram (a PST histogram of which the zero point corresponds to the starting point of one period of the stimulus). Figure 2-a shows a segment of the PSTH, this can be interpreted as an estimate of the probability density function $p_1(t)$. The histogram shows a number of isolated lobes separated by gaps. In Figure 2-b the corresponding segment of the computed LES $y_1(t)$ is drawn, this function is computed from the stimulus x(t) by convolution with the (previously determined) revcor function $h_1(t)$. The LES shows a number of oscillations and, on cursory inspection, the lobes of the PSTH correspond to the positive excursions of the LES. From closer inspection of the two functions it is clear that *the firing probability density $p_1(t)$ is approximately proportional to the rectified $y_1(t)$ signal*. Like $h_1(t)$, $y_1(t)$ shows small irregularities due to fluctuations but these are small compared to the main oscillations. The fluctuations in $p_1(t)$ are much greater because only few spikes are accumulated in each bin of the PSTH.

It should be evident that a search for finer details in the $y_1(t)$-$p_1(t)$ relation will be severely contaminated by the large fluctuations in $p_1(t)$. Suffice it to say that from many extended recordings of the type as shown by Figure 2, the general conclusion has been drawn that, on the average, nonlinear and time-dependent effects in the stimulus-response relation are negligible (de Boer and de Jongh, 1978). With a more specialized analysis technique it has proven possible to extract information about the nature of the (small) nonlinearities involved from a record like the one presented here - but from a much longer one. The results show that one of the familiar types of frequency-selective nonlinearity of the cochlea, namely, amplitude compression for frequencies above the CF, also exists in the case of noise stimulation (Breed et al., 1992) but, in general, the effect is found to be weak. Whether or not the same is true for other distortion products is an open question but we can certainly expect nonlinear effects to confound spectral gaps. For this reason we will describe some aspects of nonlinear distortion in general, and the presence of distortion products in a spectral gap, in more detail. Later we will return to the extraction of properties of cochlear distortion from neural recordings. In due course we will also use simulated neural signals.

3. NONLINEAR DISTORTION - STATIONARY AND RANDOM SIGNALS

A sinusoidally varying signal with frequency f_o produces, when it is distorted in its waveform, a periodic signal with the same frequency. The distortion is thus evident in the production of 'harmonics', i.e., components with frequencies that are integral multiples of f_o ($2\ f_o$, $3\ f_o$, etc.). When a waveform consisting of two sinusoidal signals (two primary components), with frequencies f_1 and f_2, is distorted, the spectrum of distortion products is richer. In particular, there will arise distortion components called combination tones (CTs), these have frequencies $n_1\ f_1 \pm n_2\ f_2$ - where n_1 and n_2 are integers. Of these CTs, the cubic difference tone (CDT) with frequency $2\ f_1 - f_2$ is particularly important. We can hear the CDT generated in our ear when the ratio f_2/f_1 is between 1.1 and 1.4. This CDT, as well as other audible distortion components, will lie below the spectral region from f_1 to f_2 occupied by the primary tones. Other distortion products will lie above those tones in frequency, these tones we don't hear but they will be generated all the same. One example: the CDT with frequency $2\ f_2 - f_1$. When, now, the number of primary components is increased, and all components occupy a relatively narrow band of frequencies, many of the generated distortion components will have frequencies below that band and, therefore, they can be isolated and analyzed. This is the ratio behind the use of stimulus signals with sharp spectral edges, in particular, signals with a gap in their spectrum. The gap should be centred on the CF of the fibre studied.

A certain amount of nonlinear distortion occurs in the inner ear and from the foregoing it is evident that the production of extra frequency components limits the spectral analysis that the inner ear (cochlea) can carry out. This constitutes an interesting field for study: for once, it gives insight into the mechanism of the cochlea, and, on the other hand, it helps us to assess the potentiality and limitation of cochlear encoding of information. From the foregoing it is clear that it is worthwhile to use stimulus signals that have one or more sharp edges in their spectrum. In relation to what has been described about the reverse-correlation function, we have done our simulations and experiments with stimuli belonging to the class of pseudo-random noise signals, i.,e., signals that possess most of the characteristics of random signals yet have a periodic waveform. Their spectra are not continuous, they consist of discrete and equidistant components. All component frequencies - including those of the distortion products - are integral multiples $n\ f_o$ (where n is an integer) of the fundamental frequency f_o with which the waveform repeats itself, and will consequently be labelled by their rank number n.

When the stimulus consists of noise with a narrow gap in the spectrum we may expect that the gap will be filled by distortion products.

When the nonlinearity is thought to occur *after* the filtering of the stimulus by the cochlea, the levels of the distortion components occurring in the spectral gap can be predicted, and this can be done for any type of nonlinearity. As already mentioned in section 2, the response probability density $p_1(t)$ resembles a rectified version of a linearly filtered stimulus $y_1(t)$ so it is logical to consider *rectification* as the first form of nonlinear distortion to study - note that it occurs after the filtering of the stimulus. We will do this first with synthetic signals and computed responses, and later apply the results to actual recordings from physiological experiments.

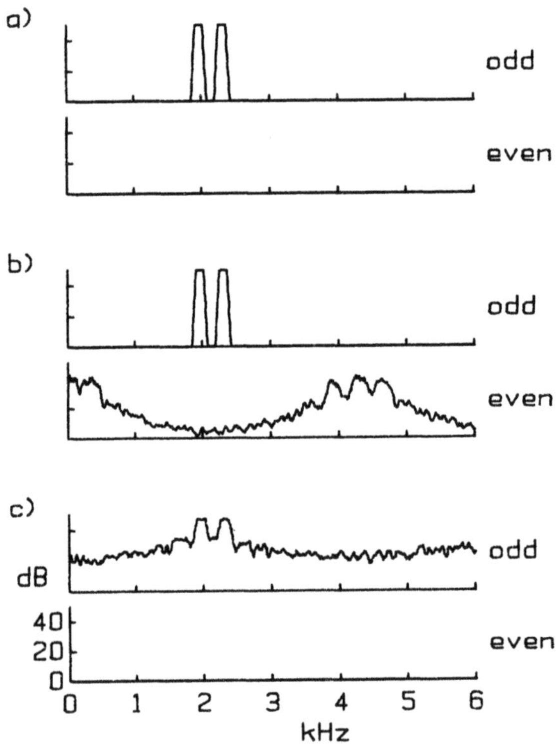

Figure 3. Spectra of stylized LES functions; components of even and odd order separated. (a): spectrum of primary signal, (b): spectrum of rectified signal, (c): spectrum of clipped signal.

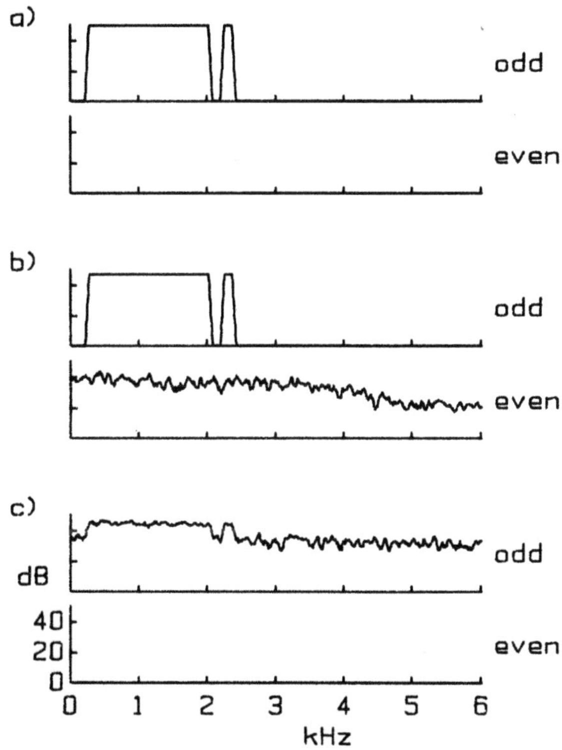

Figure 4. Spectra of stylized LES functions; primary signal has a larger bandwidth. Lay-out as in Figure 3.

In the first phase we will use stylized signals that resemble LES signals. Figure 3-a shows the spectrum of a band-limited pseudo-random signal (with a relative bandwidth about equal to that of a typical revcor spectrum) which has a narrow gap in it. To enable us to isolate the effect of the rectifier the signal is made to consist entirely of *odd* multiples of the frequency f_o. New frequency components that are generated by the rectification process can only be even combinations of the original component frequencies and thus will be *even* multiples of f_o. Therefore, the figure is made to show odd- and even-numbered spectral components separately. Figure 3-b illustrates the - moderately smoothed - spectrum of the signal after it has been rectified. Within the primary band of frequencies the distortion components are pretty weak; stronger distortion is seen in the region of the second harmonics of the primary components. The relative amplitudes of the distortion components in the spectral gap are pretty small, that they depend on the bandwidth of the signal is shown by Figure 4 where the bandwidth is four times larger. A provisional conclusion to draw is that distortion by the rectifying process would not be harmful to our intended analysis of (odd-order) combination-tone distortion.

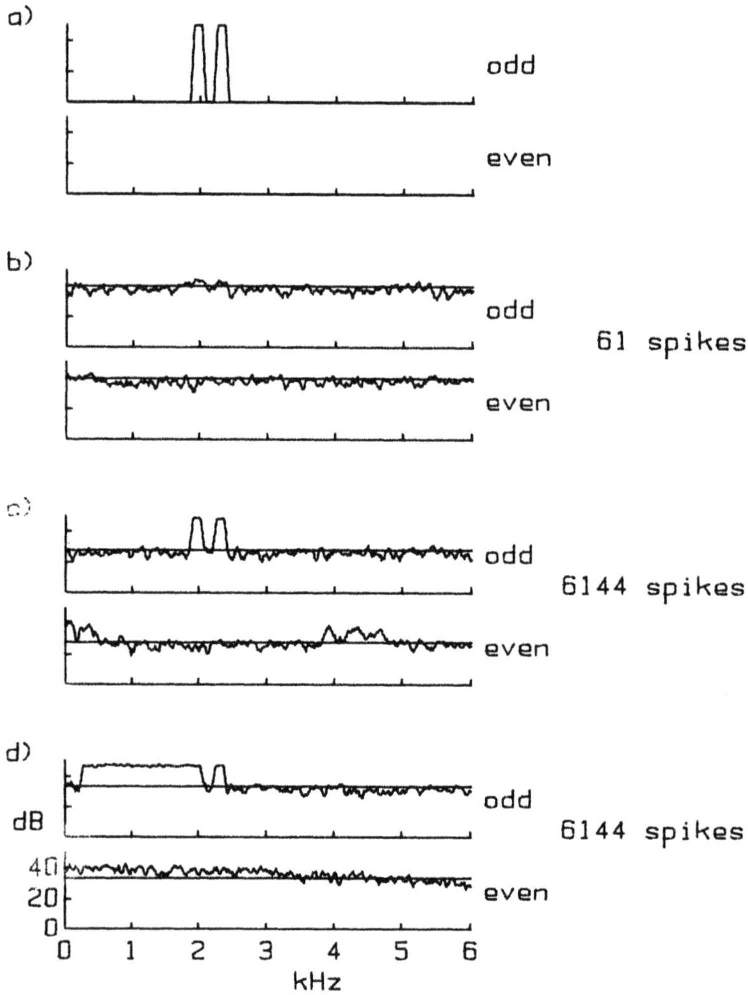

Figure 5. Influence of stochasticity on the spectrum. (a): Spectrum of narrow-band primary signal. (b): Spectrum of the signal generated by one simulated nerve fibre. (c): Spectrum of 100 independent signals as in panel b. (d): The same as (c) but for a primary signal with larger bandwidth.

After having treated even-order distortion we come to the subject of odd-order distortion, like the process of production of CDTs. In general, the amplitude of a distortion product varies more rapidly with the amplitude of the primary components than with the first power. To circumvent this problem we choose a type of distortion that does not have this property, namely, *infinite clipping* of the signal. We consider a narrow-band signal, the one of which the spectrum is shown by Figure 3-a. Panel c of that figure shows the spectrum of the signal when it is infinitely clipped. The spectrum now contains only odd-numbered components, the two primary bands are well visible but the gap between them, as well as the spectral regions just outside them, are partially filled by distortion components. From Fig. 4 it is evident that the distortion is larger for a primary signal with a larger bandwidth. Quite generally, we conclude that odd-order nonlinear distortion can be well studied by the use of stimulus signals with a spectral gap.

4. STOCHASTIC MULTIPLICATION

The response of a nerve fibre is intrinsically stochastic, hence, apart from nonlinear distortion, *stochasticity* of firing is another source of new frequency components in the response. We will assume that *the firings occur with a probability density that is proportional to the rectified effective stimulating signal (LES).* We thus extend our simulation procedure to include a stage with a probabilistic pulse ('spike') generator. The simulated spike signal is a non-homogeneous Poisson process, see Appendix C for more details. Properties of the spectra of simulated spike trains are illustrated by Figure 5. Part a of the figure shows the primary-signal spectrum (which is the same as that of Figure 3-a). Figure 5-b shows the - moderately smoothed - spectrum of the response of one simulated nerve fibre. For this record 61 spikes were used. The simulation is performed by computer with the aid of a random-number generator as described in Appendix C. The primary spectrum with the gap is entirely lost because of the newly generated components due to stochasticity. Note that the condition used roughly corresponds to the response of one fibre to one period of the stimulus.

Figure 5-c illustrates the spectrum obtained when one hundred of independent stochastic responses to the same stimulus are added. Or, equivalently, when 6144 spikes are used in the computation - a situation comparable to that in the experiments to be described later on. In this spectrum the original spectrum of Figure 5-a is recognisable but the gap is partially filled with noise components. In the simulation procedure only positive values of the primary signal give rise to firings: rectification of the primary signal is therefore included. And indeed, even-numbered distortion products are evident in Figure 5-c; they show the features we observed before (Figure 3-b) but they do not contribute to the components filling the gap. Actually, the level, L_{st}, of the newly generated components can be estimated (see Appendix C), it is indicated in Figures 5-b and -c by horizontal lines. It is seen that the prediction is good, only in the region of the second harmonics do we find stronger components; obviously, these can be attributed to rectification.

The theory predicts that the average power density of the new components is proportional to the bandwidth (see appendix C, eq. C.19). This effect is shown by Figure 5-d where the primary signal is the same as the one used in Figure 4, the bandwidth being four times larger than before. For actual responses of nerve fibres in the healthy cochlea this point is not relevant because the bandwidth of the LES is limited by the cochlear filter. A damaged or badly functioning cochlea may well have a cochlear filter with an enlarged bandwidth and the effect demonstrated may be important for people with a corresponding hearing defect.

5. EXPERIMENTAL TECHNIQUE - A FEW DETAILS

Single auditory-nerve fibre recordings were made with glass microelectrodes in five Mongolian gerbils (Meriones unguiculatus, 12-18 months old). The auditory nerve of the right ear was exposed by a ventral approach (Sokolich, 1977; Chamberlain, 1977). Animals were anaesthetized by a combination of Nembutal and Thalamonal®. The characteristic frequency (CF) of each fibre was taken to be the frequency with the lowest firing threshold as estimated with an automated scanning routine.

All stimuli were based upon pseudo-random noise centred around the CF, the spectrum extended from 0.1 CF to 3.0 CF. Since the stimuli were generated digitally this was accomplished simply by adjustment of the sample frequency which was set equal to 8.192 times the characteristic frequency. The stimulus spectra contained a spectral notch with a width equal to 0.05 CF, 0.1 CF, 0.15 CF, or 0.2 CF, symmetrically placed around CF, and a depth of at least 50 dB below the overall spectral level. The spectral edges of the notches have slopes of more than 1000 dB per octave. The stimuli were presented at three levels of intensity: 20, 40, and 60 dB SPL/third octave. The lowest level at which a stimulus was presented had to be at least 10 dB above the pure-tone threshold at CF.

As in the preceding sections neural responses are processed in the form of PST histograms. This procedure simulates the processing of responses of many neurones as is supposedly done by the CNS.

(a)

(b)

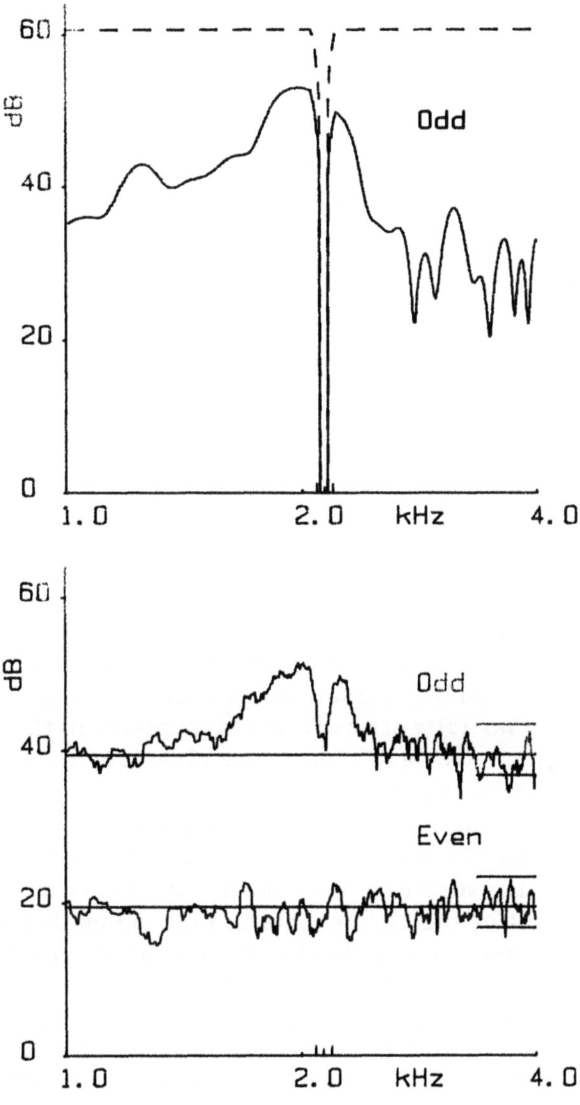

Figure 6-a: Spectrum of primary signal (dashed line) and Linear Excitatory Signal (solid line) for an auditory nerve fibre (CF: 2.142 kHz). (b): Spectrum of PST histogram, plotted with even- and odd-order components separated. Value of even components is lowered 20 dB for clarity. Stimulus level: 20 dB SPL per third-octave.

The responses (spikes) to continuous presentation of the stimuli were accumulated in period histograms with a length of 4096 bins corresponding to the period of the stimulus. For each histogram 6144 spikes were collected. From these histograms spectra were obtained by taking the Fourier transform and computing the square of the absolute value of each component. In the figures spectral values are plotted as moving averages over 7 adjacent (odd or even) components, this corresponds to a spectral resolution of 0.03 CF.

6. RESULTS

The dashed line in Figure 6-a shows the (smoothed) spectrum of one of the sound stimuli that were used as stimulus in the experiments. The spectrum is flat with a gap centred around the characteristic frequency (CF) of the nerve fibre under study. The spectrum of the linear excitatory signal (LES) for that fibre (see the introduction section and Figure 2) is shown by the solid line in the same figure. The LES spectrum reflects the linear filter properties of the outer and inner ear with respect to the fibre. The gap in the spectrum remains empty because the LES is a linear transform of the stimulus signal, that is, no distortion is involved. The influence of stochasticity stems from inaccuracies in the revcor function, in the peak region it is small since more than 12000 spikes were used for the determination of the revcor function.

The spectrum of a nerve-fibre response that was obtained with this stimulus is shown in Figure 6-b. In the peak region this spectrum resembles the LES spectrum very much: shoulders and gap are clearly visible. The level of the components in the regions far outside the pass-band of the cochlear filter is, on the average, about equal to the predicted level L_{st} (the horizontal line). Detection of a gap is a function of the signal-to-noise ratio, i.e., the component level in the peak relative to L_{st}. Figure 7 shows the signal-to-noise ratio, in this case the ratio of the power of an odd component at the frequency 0.85 CF and the power corresponding to L_{st}, plotted as a function of elapsed time. The data are averaged over 38 recordings. For one component the ratio of the power of that component and its expectation value has a chi-square distribution with 2 degrees of freedom. If components are uncorrelated, that ratio for the average power of seven components has a chi-square distribution with 14 degrees of freedom. The horizontal line in the Figure denotes the 95% confidence interval for the latter distribution. As a rough estimate, we can conclude from this Figure that a difference of 5 dB between 'signal' and 'noise' can be detected with 95% confidence only when spikes over at least 15 sec. are accumulated. For a depth of 10 dB this becomes 60 sec.

The gap in the stimulus spectrum is also visible in the spectrum of the response but it is partially filled up. In this case the level of the odd-numbered spectral components in the gap is seen to exceed the predicted level of the stochastic components slightly. To test the significance of this finding the average power of 7 odd- or even-numbered components with adjacent frequencies in the middle of the gap was determined for 85 records (obtained from 7 fibres in 5 animals). The ratio of the power of a component and its expectation value, averaged over 7 components and 85 records, has a chi-square distribution with 1190 ($7 \times 85 \times 2$) degrees of freedom. The mean power ratio obtained from the records was 2.11 (3.24 dB) for odd- and 1.18 (0.71 dB) for even-numbered components. In both cases the deviation from 1.0 is significant ($p < 0.01$). This finding suggests that newly generated components in the gap are partly caused by something other than stochasticity of the nerve firings, and it is likely that they are distortion components. However, measuring them in this way is impossible.

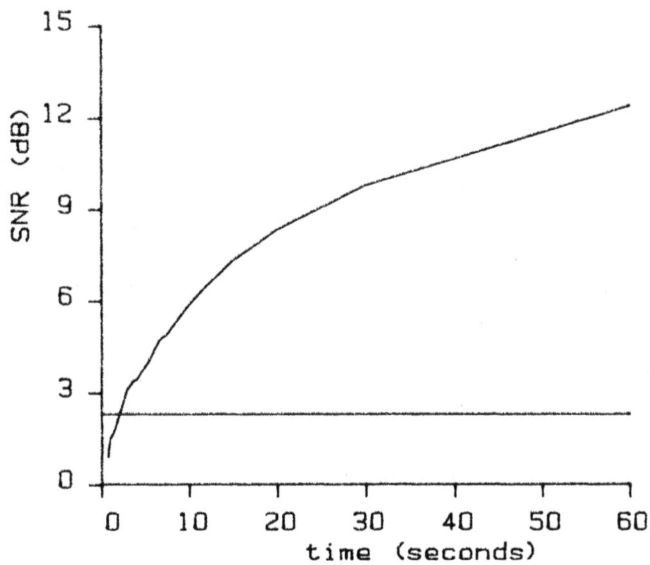

Figure 7. 'Signal' and 'noise' levels as functions of recording time. For explanation, see text and Appendix C.

7. COMMENTS AND CONCLUSIONS

The basic question investigated in this work is: how much information about the spectrum can be transmitted to the brain by auditory-nerve fibres? Distribution of the rate of firing over the fibres is the simplest possible indicator of spectral content of the sound stimulus. For a given fibre the rate of firing is closely associated with the magnitude of the LES for that fibre and we have observed that the LES spectrum can be predicted on the basis of the stimulus spectrum and the spectrum of the revcor function for the fibre. In terms of rate of firing, it is, therefore, the effective width of the revcor spectrum which delimits the resolution in the frequency domain.

Still finer details of the stimulus spectrum are present in the temporal firing pattern of a fibre, and what we have been investigating is under which conditions this information could be utilized by the CNS. The results show, then, that details of the shape of the spectrum are indeed preserved in the temporal response pattern but this is only true when *a very large number of firings* is used in the computation. Responses of a single fibre are not sufficient. Stochasticity of the firing process appears to be the major limiter of temporal encoding, and not nonlinear distortion in the cochlea.

We have made use of the PST histogram of the firings which is an average measure of the firing pattern over a number of responses of one fibre to the same (repeated) stimulus. However, the averaging could equally well have been performed over responses of more than one fibre to a single stimulus - and we consider here the 'ideal' case where latency differences are compensated and the fibres have identical revcor functions. An estimate of how much cooperation would be involved is derived as follows. The auditory system appears to determine differential thresholds with a time constant between 50 msec (for modulation of wide-band and fluctuating signals, Zwicker, 1956, Figure 12) and 200 msec (for detection of tone bursts in noise or quiet, Plomp and Bouman, 1959). Thus, we can imagine a decision is made after spikes have been accumulated over approximately 50 msec. One fibre produces, on the average, about 100 spikes per second, or, about 5 spikes in 50 msec. On the basis of Figure 7 one may expect that a gap in the spectrum with a width of 0.03 CF (seven components) and a depth of 5 dB can be detected with 95% confidence when the equivalent of minimally 15 seconds of recording time can be used. With the same average rate of 100 spikes per second, this leads to 1500 spikes. According to this reasoning at least 300 nerve fibres should be cooperating for detecting a gap. Given the width of the revcor spectrum and the density of innervation, there will certainly be several hundreds of nerve fibres with their CF in the pertinent region.

8. APPENDIX A: Correlation functions

The average temporal relation between two random signals x(t) and y(t) can be studied from the cross-correlation function. Assume both signals are members of two ensembles of statistically stationary and ergodic signals. When E[.] denotes the ensemble average, the cross-correlation function (ccf), $\phi_{xy}(\tau)$, is defined by

$$\phi_{xy}(\tau) = E\,[x(t)\; y(t-\tau)]. \tag{A.1}$$

Due to the stationarity, the result does not depend on time (t). For signals that are ergodic (Davenport and Root, 1958) the result will be the same when the average is taken over time, and it is on this property that experimental procedures are based: all averages are averages over time. A special case of the ccf is the auto-correlation function (acf), here the two signals x(t) and y(t) are the same. The acf of y(t) is:

$$\phi_{yy}(\tau) = E\,[y(t)\; y(t-\tau)]. \tag{A.2}$$

An important case for the ccf obtains when y(t) is a linearly filtered version of x(t). For a filter with impulse response h(t), the response y(t) to the input x(t) is given by the convolution integral

$$y(t) = \int_0^\infty x(t-\tau)\; h(\tau)\; d\tau. \tag{A.3}$$

When x(t) is a white-noise signal, its acf is a delta function $\delta(\tau)$. In that case the ccf of input and output signals has a simple form:

$$\phi_{xy}(\tau) = h(-\tau). \tag{A.4}$$

We prove a more general relation for the case where x(t) has an arbitrary acf:

$$\phi_{xy}(\tau) = E\,[x(t)\; y(t-\tau)] =$$

$$= E\,[\int_0^\infty h(\sigma)\; x(t-\tau-\sigma)\; x(t)\; d\sigma\,]$$

$$= \int_0^\infty h(\sigma)\, E\,[x(t)\, x(t-\tau-\sigma)]\, d\sigma$$

$$= \int_0^\infty h(\sigma)\, \phi_{xx}(\tau+\sigma)\, d\sigma. \qquad (A.5)$$

Obviously, when $\phi_{xx}(\tau) = \delta(\tau)$, the result is eq. (A.4). The interpretation of this property is that the filter's transfer function, h(t), can be recovered from the ccf of two on-going signals x(t) and y(t). In quite another vein, the Fourier Transform of eq. (A.5), which can be written as

$$\Phi_{xy}(\omega) = \Phi_{xx}(\omega)\, H(\omega), \qquad (A.6)$$

can be interpreted to show that spectral details of the x-signal are preserved in the ccf. For the acf of y(t) the relation reads:

$$\Phi_{yy}(\omega) = \Phi_{xx}(\omega)\ |\, H(\omega)\, |\ ^2. \qquad (A.7)$$

10. APPENDIX B: The reverse-correlation procedure

That the filter's transfer function can be recovered by the cross-correlation procedure, remains true, to some extent, when one or both of the signals are distorted by a no-memory nonlinearity. The most general expression of this property is Price's theorem (Price, 1958), valid for Gaussian signals; here we will use the version where only y(t) is distorted by a memoryless function F(.):

$$z(t) = F\ [\ y(t)\ ]. \qquad (B.1)$$

Then, the 'new' ccf $\phi_{xz}(\tau)$ is proportional to the original cross-correlation function $\phi_{xy}(\tau)$:

$$\phi_{xz}(\tau) = A\ \phi_{xy}(\tau), \qquad (B.2)$$

where A is determined by the distortion function F(.) and $p_y(y)$, the probability density function of y(t):

$$A = \int_{-\infty}^{\infty} y\, F(y)\, p_y(y)\, dy \,/\, var_y, \qquad (B.3)$$

and where var_y is the variance of y(t). This theorem has been proved by Bussgang (1952) and it can easily be derived from Price's theorem. Note that in this case only x(t) is gaussian.

Consider now cochlear responses. Assume that for the i-th fibre of the auditory nerve the stimulus x(t) undergoes a linear filtering process and that the filter output signal $y_i(t)$ controls the probability density $p_i(t)$ that a spike occurs in the fibre in a monotonic and instantaneous but certainly not in a linear way. Then the ccf between the x(t) signal and $p_i(t)$ should be proportional to the ccf over the filter. When x(t) is a broad-band noise signal, the cross-correlation procedure thus produces the time-inverted filter's impulse response $h_i(t)$, apart from a multiplicative constant. The ccf between x(t) and $p_i(t)$ is known as the reverse-correlation function, abbreviated: revcor function. Details about the measurement technique and the properties of revcor functions are described in de Boer and de Jongh (1978). Further applications of the reverse-correlation technique are reviewed in Eggermont et al. (1984).

11. APPENDIX C: Homogeneous and inhomogeneous Poisson processes

The generation of action potentials (spikes) by the sensory neurones of the inner ear is stochastic and can be approximated by a Poisson process. Such a process has the property that the instants at which spikes occur are random, and independent of one another. Similarly, the number of events in a certain time interval is independent of the number of events in another non-overlapping time interval, this is known as the property of independent increments. The approximation by a Poisson process is very good due to the relative sparsity of spikes. After each spike there is a short period with a reduced firing probability but this period is such a small fraction of the average time between spikes that it has little effect.

For the simulation of a Poisson process use is made of the exponential distribution of the intervals between events. Random samples u_n are generated of a variable that is homogeneously distributed between 0 and 1. The u_n are transformed to samples e_n of a time interval with an exponential distribution: $e_n = -\ln(u_n)$. Now the time instant θ_m of the m-th spike is defined as the sum of m intervals e_n:

$$\theta_m = \sum_{n=1}^{m} e_n. \qquad (C.1)$$

Counting the number of events in the time interval from 0 to t yields a Poisson process:

$$N_1(t) = \int_0^t \sum_{m=1}^{\infty} [\delta(\tau - \theta_m)]\, d\tau, \qquad (C.2)$$

where δ is the impulse or dirac function. This is a *homogeneous* Poisson process, i.e., a Poisson process with a constant firing probability density of 1 per unit of time.

We consider two variables x_1 and x_2 defined by the arbitrary, ordered, time instants $t_1 \ldots t_4$ ($t_4 > t_3 > t_2 > t_1$):

$$x_1 = N_1(t_2) - N_1(t_1), \qquad (C.3)$$
$$x_2 = N_1(t_4) - N_1(t_3). \qquad (C.4)$$

Then, the covariance between x_1 and x_2 is zero :

$$COV[x_1, x_2] = E[(x_1 - E[x_1])(x_2 - E[x_2])] = 0 \qquad (C.5)$$

(E is the expectation operator). This relation expresses the property of independent increments.

The next concept is the *inhomogeneous* Poisson process in which the firing probability density $\lambda(t)$ is a given non-negative function of time. If $N_1(t)$ is the number of events of a point process with a constant firing probability density of 1 per time unit, then a process with a time-dependent firing probability density function $\lambda(t)$ can be constructed by defining a new process $N_\lambda(t)$ in terms of $N_1(t)$ by

$$N_\lambda(t) = N_1(\tau), \qquad (C.6)$$

where τ is given by

$$\tau = \int_0^t \lambda(s)\, ds. \qquad (C.7)$$

The inhomogeneous Poisson-process $N_\lambda(t)$ encodes the signal $\lambda(t)$. To recover this signal we are going to count the firings in a certain time interval. We discretize time into segments with length Δ, ordered with index j. The number q(j) of spikes in segment j is given by:

$$q(j) = N_\lambda(j\ \Delta) - N_\lambda((j-1)\ \Delta). \tag{C.8}$$

The variable q(j) has a Poisson distribution and possesses the property of independent increments. It is in the form of q(j) that we have information about the encoded signal $\lambda(t)$. The actually obtained values q(j) will spread around the expected value s(j) defined as:

$$s(j) = E[q(j)]. \tag{C.9}$$

Clearly, when the encoded signal $\lambda(t)$ doesn't vary over the interval Δ:

$$s(j) = \lambda(j\ \Delta)\ \Delta. \tag{C.10}$$

From here on we consider s(j) as the 'signal'. The 'noise' r(j) in the record q(j) is defined by:

$$r(j) = q(j) - s(j). \tag{C.11}$$

What we want to derive is an expression for the power of the noise r(j) and relate this to the power of the signal s(j). We will do this via the spectra of s(j) and r(j) and assume, therefore, that the total time interval contains M segments. The spectral components S(k) and R(k) of s(j) and r(j) are:

$$S(k) = \left| \sum_{j=0}^{M-1} [s(j)\ e^{(-2\pi i jk/M)}] \right|^2, \tag{C.12}$$

and

$$R(k) = \left| \sum_{j=0}^{M-1} [r(j)\ e^{(-2\pi i jk/M)}] \right|^2. \tag{C.13}$$

The average power of the noise r(j) will be distributed equally over all M components because the covariance of q(j) is zero. By using the fact that the variance of q(j) equals the expected value s(j), we find the expected value $<R(k)>$ of the power for each k as:

$$\langle R(k) \rangle = \sum_{j=0}^{M-1} s(j). \qquad (C.14)$$

When M is large, the sum of s(j) can be replaced by the sum of q(j), which equals the total number N_{tot} of spikes observed in the record of length M Δ, hence:

$$\langle R(k) \rangle = N_{tot}. \qquad (C.15)$$

$\langle R(k) \rangle$ can be interpreted as the average level of the frequency components that are due to the stochasticity. In the text the dB level corresponding to it is called L_{st}.

In our simulation the signal s(j) was derived from a gaussian noise signal with a narrow-band spectrum, let σ^2 be the time-averaged value of the square of the latter signal. The original spectrum consisted of K odd-numbered components with equal amplitudes (see Figure 5). The signal s(j) was obtained by rectification whereby the power of each odd-numbered component is reduced to one quarter of its original value. Thus the shape of the spectrum of the odd components is not influenced by the rectification. Then, by the use of Parseval's relation the power S(k) of each component of s(j) is found as:

$$S(k) = \tfrac{1}{4} M^2 \sigma^2 / K. \qquad (C.16)$$

For ideal rectification of a gaussian signal the average value s_{ave} is

$$s_{ave} = \sigma / \sqrt{(2\pi)}. \qquad (C.17)$$

Furthermore, the total number N_{tot} of spikes is M times the same average s_{ave}. Hence the signal-to-noise ratio SNR(k) of the odd components, defined as

$$SNR(k) = S(k) / \langle R(k) \rangle, \qquad (C.18)$$

reduces to:

$$SNR(k) = \tfrac{1}{2} \pi N_{tot} / K. \qquad (C.19)$$

The signal-to-noise ratio is proportional to the number of spikes or, equivalently, to the rate of firing. Combining the spikes of more than one nerve fibre is equivalent to increasing the number of spikes and this will increase the signal-to-noise ratio. The signal-to-noise ratio is inversely proportional to the number of frequency components K in the encoded signal s(j). Increasing the bandwidth of the encoded signal s(j) or decreasing the sample time Δ thus will always result in reduction of the signal-to-noise ratio.

12. REFERENCES

Boer, E. de and Jongh, H.R. de (1978) On cochlear encoding: Potentialities and limitations of the reverse-correlation technique. J. Acoust. Soc. Am. 63, 115-135.

Breed, A.J., Kanis, L.J. and Boer, E. de (1992) Cochlear nonlinearity for complex stimuli. In: Auditory physiology and perception (eds. Y. Cazals, L. Demany and K. Horner) Pergamon Press, Oxford, pp. 189-195.

Bussgang, J.J. (1952) Crosscorrelation functions of amplitude-distorted gaussian signals. MIT, Res. Lab. of Electronics, Techn. Rep. 216.

Chamberlain, S.C. 1977. Neuroanatomical aspects of the gerbil inner ear: light microscopic observations. J. Comp. Neur. 171:193-204.

Davenport, W.B.Jr. and Root, W.L. (1958) *An Introduction to the Theory of Random Processes and Noise*, Mc-Graw Hill, New York, pp. 65ff.

Eggermont, J.J., Johannesma, P.I.M. and Aertsen, A.M.H.J. (1984) Reverse-correlation methods in auditory research. Q. Rev. of Biophys. 16, 341-414.

Plomp, R. and Bouman, M.A. (1959) Relation between hearing threshold and duration for pure tones. J. Acoust. Soc. Am. 31, 749-758.

Price, R. (1958) A useful theorem for nonlinear devices having Gaussian inputs. IRE Trans. Inf. Theory 4, 69-72.

Sachs, M.B. and Young, E.D. (1979) Encoding of steady-state vowels in the auditory nerve: Representation in terms of discharge rate. J. Acoust. Soc. Am. 66, 470-479.

Sachs, M.B. and Young, E.D. (1980) Effects of nonlinearities on speech encoding in the auditory nerve. J. Acoust. Soc. Am. 68, 858-875.

Sokolich, W. (1977). Some electrophysiological evidence for a polarity-opposition mechanism of interaction between inner and outer hair cells in the cochlea. Special Report (ISR-S-15) (Institute for Sensory Research, Syracuse University, NY, USA).

Zwicker, E. (1956) Die elementaren Grundlagen zur Bestimmung der Informationskapazität des Gehörs. Acustica 6, 365-381.

Part Two

Space-Time Aspects of Brain Function: Control of Action

Brain Theory / A. Aertsen (Editor)
1993 Elsevier Science Publishers B.V.

The representation of space and time in the motor system.

C.C.A.M. Gielen

Dept. of Medical Physics and Biophysics, University of Nijmegen,
Geert Grooteplein Noord 21, NL 6525 EZ Nijmegen, The Netherlands.

Abstract

In the transformations of sensory signals into motor commands sensory signals, represented in a reference-frame that is related to the receptor (e.g. a retinal coordinate system for visual information) have to be transformed into motor signals represented in a muscle-coordinate system. Between the peripheral sensory and motor stages multiple sensory signals are often fused in order to obtain a more complete picture of the environment. These transformations raise the issue whether and at what stage the signals in the sensory-motor chain are represented in an intrinsic or extrinsic frame of reference.

When a target position has been selected for a goal-directed movement the available evidence indicates that *a target position in space* is transformed into a *dynamic signal* which causes the movement by activating a set of muscles each with a very specific intensity and with a specific duration.

Frequently, the biological limb has a number of degrees of freedom such that the redundancy causes that the position of the end-effector (e.g. a hand or foot) can be obtained by a large set of different joint angles.

In this chapter we will give a review of recent literature which has appeared on these issues.

1 General introduction

Most biological limbs, like the arm and leg, have multiple joints which make it possible, that a particular position for the hand or foot is reached with a large set of different joint angles. This redundancy in the number of joints adds to a redundancy in the number of muscles, which is available: most joints are crossed by both mono- and bi-articular muscles. As a result a flexion or extension torque can be obtained by a large set of muscle activation patterns. For example, a flexion movement of the arm can be made by activating the mono-articular m. brachialis or m. brachioradialis, or by activating the bi-articular m. biceps, or by any linear combination of activation patterns. Despite of this redundancy, it seems that most studies reveal a more or less consistent activation pattern for a given motor task (see e.g. [1], [2]). This raises the issue why a particular set of joint angles and muscle activation patterns is chosen for a given motor task. This problem receives wide interest not only in the neurobiological and psychological communities, that study the neuronal and behavioural basis for the coordination of movements, but also in the robotics community where the construction of flexible robot systems requires multiple degrees of freedom. In the past 10 years there has been an exchange of ideas and results in both fields which has led to some common principles which are considered to be relevant for dealing with redundancy. In this chapter we will review some of the main ideas and algorithms to which the research efforts in both fields have converged to.

Related to the problem of redundancy is the problem how a desired target position in space can be reached by moving a limb from an initial position to that target position. It is still a matter of debate in what reference frame target position is coded and in what frame of reference motor actions are prepared. When a target position is selected, this target position in space has to be transformed into a dynamic sequence of muscle activations such that each muscle involved is activated at the proper intensity and with the proper timing. In the field of oculomotor research a relatively good understanding of these processes has been obtained in the past decade. The experimental results from the oculomotor field will be related to recent experimental findings in the control of limb movements.

2 From stimulus position to motor program

2.1 Plant mechanics

Contrary to limb movements, where the load is higly variable, the mechanical properties of the oculomotor system are almost constant. For most practical purposes, the mechanical properties of the oculomotor system may be modelled by a second order low-pass system (see e.g. [3]). In the Laplace-domain the eye movement signal y(s) is related to the neuronal motor command signals x(s) by

$$y(s) = \frac{1}{(1+T_1 s)}\frac{1}{(1+T_2 s)}x(s) \tag{1}$$

This implies that for fast eye movements, so called saccades, the input for the oculomotor plant should consist of a step-change in activation in order to provide the change in spring-force which is necessary to keep the eye in the new eye position, and a pulse of activation, proportional to eye velocity in order to overcome viscous force components. These two activation components are necessary to match the visco-elastic properties of the eye in its orbit (see [4]). For a second order system there are also force components related to acceleration and deceleration of the eye. These force components are relatively small for the eye. This can be understood from the fact, that the inertia of the eye is relatively small (the inertia of the eye can be increased by a factor of 100, before oscillations occur). Such as for any second-order system, it is characterized by two time constants. The largest time constant T_1 (about 150 ms) is related to the visco-elastic properties of the eye in its socket. The smaller time constant T_2 (about 10 ms) corresponds to the contractile properties of the eye muscles.

For the limb, the inertial force components are much more important than for the eye. Fast goal-directed limb movements are characterized by a large burst of activity in the agonist muscles, followed by a burst of activity in the antagonist muscle. These bursts of activity are related to the acceleration and decelerating forces of the limb (see e.g. [5], [6]). The contribution of the mechanics of the limb to viscous forces is very limited. The major viscous components are due to muscle mechanical properties and to the properties of muscle spindles which become evident in the muscle activations by reflex actions.

2.2 From target position in space to a dynamic muscle activation pattern.

Some time ago Robinson [7] proposed a model to explain how a difference between target position and eye position (called "motor error") might give rise to a pulse-step activation signal as described above. In this model it was assumed that the motor error signal provides the input for the so-called burst cells, which generate a high-frequency spike rate depending on motor error. This burst of activity is related to saccade velocity and the number of actionpotentials in the burst is proportional to saccade amplitude ([4]). The model proposed that the step-change in muscle activation is obtained by integration of the burst. The integrated activity is added to the burst in order to provide the proper activation for the eye muscles. Since the integral of the burst is proportional to eye displacement, Robinson proposed that it might be used as an efference copy signal, which could be subtracted from the initial motor error in order to provide an up-to-date signal which specifies the distance that separates the eye from the target. When the difference between desired target position and eye position, as monitored by the efference-copy signal, has become zero, the burst cells would stop firing such that the eye movement stopped (see Fig. 1A).

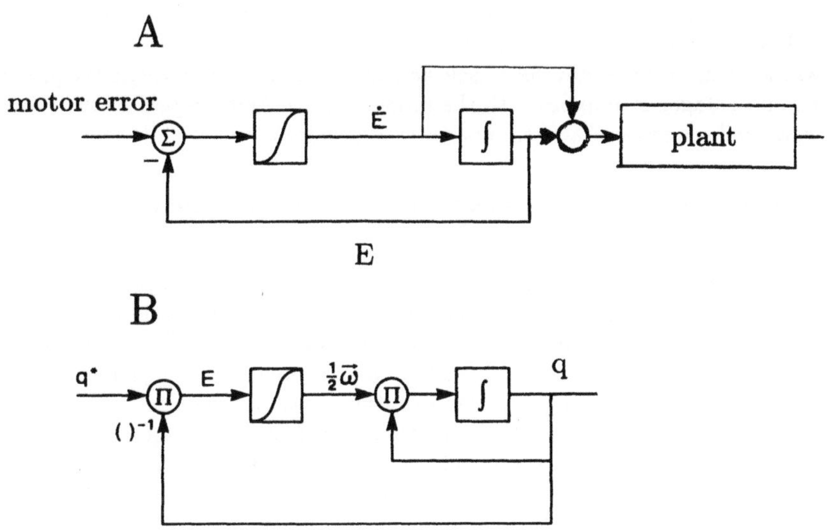

Figure 1. Robinson model (A) and Tweed and Vilis model (B).
See text for further explanation.

In this model, the transformation from a target position in space to a dynamic activation pattern for the eye muscles is obtained by the dynamics of the burst cells, which are thought to provide an output, proportional to eye velocity, depending on instantaneous motor error. This model provided an excellent working hypothesis to investigate several critical aspects of the oculomotor system and has been a source of inspiration to investigate the oculomotor system. The idea of a neural integrator is widely accepted now and there is good neurophysiological and anatomical evidence on which structures are involved in the integration process ([8], [9]). Although it is well accepted that the output of the neural integrator is used in an internal feedback loop, it is not quite sure yet, which neural structures are part of this feedback loop.

The relatively simple mechanics of the oculomotor plant and the fact, that the motoneurons of the eye muscles and their input, the so-called medium-lead burst cells are accessible for electrophysiological recordings, have contributed to a relatively good understanding of the neuronal information processing in the peripheral oculomotor system. For limb movements a good neurophysiological basis for efference copies is lacking so far. There is some evidence from behavioural studies for the existence of an efference-copy control of limb movements. Van der Meulen et al. ([10]) studied fast goal-directed movements. They found that the variability at the initial acceleration of the movement is large relative to the variability in the position at peak velocity and in end-position of the movement and movement amplitude. The variability in the initial acceleration gives rise to a variability in the movement trajectory. However, this variability decreases gradually during the movement which was taken as evidence that variability at the beginning of the movement is reduced due to feedback of efferent command signals.

For limb movements, it is not known yet how and where the conversion of signals coding target position in space to muscle activation patterns takes place. It has been established that cells in motor cortex code movement direction in space (see e.g. [11], [12], [13]). However, it is not known how these cortical command signals in the cortico-spinal pathway are transformed into the muscle activation signals that activate the proper set of agonist and antagonist muscles with the proper timing. With regard to the rubro-spinal pathway, similar questions are still unanswered (see for a review [14]).

2.3 Implications of joint rotations on motor programming.

Recently, some modifications have been proposed for the Robinson model, which were necessary since the Robinson model could only predict saccadic eye movements in one dimension. This is due to the fact, that rotations in 3 dimensions do not commute: the orientation of an object after two non-colinear rotations depends on the order of the rotations (see Fig. 2). Therefore, subtraction of an efference copy signal which represents the rotation of the eye in a particular direction, may not be simply subtracted from the desired rotation vector, which is necessary to fixate the target in space, without loosing control of the torsion component in eye orientation (see [15]). Since rotations in 3 dimensions do not commute, it has been a remarkable observation since long, that the orientation of the eye is the same irrespective of previous eye positions (see [16], [17]). This observation is known in the oculomotor field as Donders' Law. This result implies that during normal eye movements only 2 of the available 3 degrees of freedom are independently controlled. The third degree of freedom (torsion of the eye) is a function of the rotation of the eye in horizontal and vertical direction. The quantitative relationship between fixation direction and the amount of torsion of the eye is described by Listing's law which further states, that a single reference orientation ("primary position") exists, such that all orientations of the eye during fixation may be reached from this reference position by a single rotation, the axis of which is perpendicular to the original gaze direction, to the final gaze direction ([16]). Actually, Listing's Law implies that all rotation axes that describe the eye position relative to some reference vector lie in a plane.

There are good reasons now to think that Listing's law has a neural origin rather than a mechanical basis. One argument comes from the observation that Listing's law is violated during vestibular eye movements induced by torsional rotation of the head ([18], [19]) and also during eye movements, made during sleep ([20]). This fact and the fact that Robinson's model would give rise to violations of Donders' law, has led Tweed and Vilis to propose an alternative to Robinsons model. Their model preserves the internal feedback loop, however, in a different way. In the model by Tweed and Vilis the "motor error signal" has been replaced by a single-axis rotation vector, which brings the eye from the fixation position to the target position. This single-axis rotation vector is calculated from the desired target position and the current eye position vector. If eye position is represented by a rotation vector in quaternion representation $\mathbf{q}$, the instantaneous motor error is given by $\frac{\mathbf{q}^*}{\mathbf{q}(t)}$, where $\mathbf{q}^*$ represents the desired eye orientation in 3-dimensional space and where $\mathbf{q}(t)$ represents the actual 3-dimensional eye orientation. In this model the output signal of the burst cells is a signal coding the angular velocity vector ω:

$$\omega = 2\dot{\mathbf{q}}\mathbf{q}^{-1} \tag{2}$$

The output ω of the burstcells is transformed into $\dot{\mathbf{q}}$ by multiplication with the current orientation of the eye $\mathbf{q}$. The vector $\dot{\mathbf{q}}$ is used to compute the new eye orientation $\mathbf{q}$ by integration.

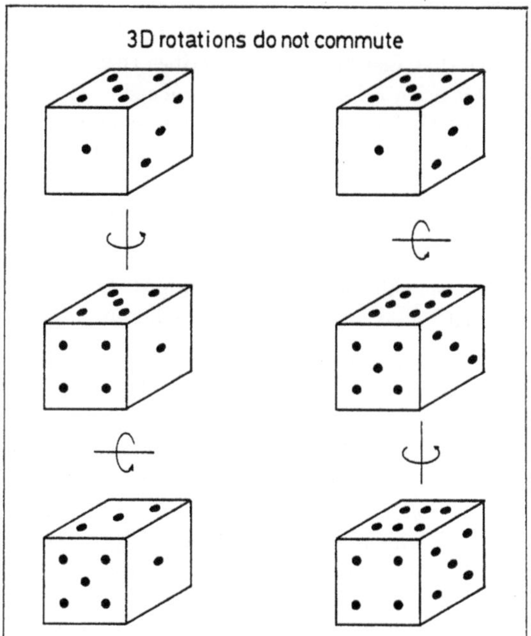

Figure 2. Illustration of the fact that the orientation of an object depends on the order of two non-colinear rotations.

In the model by Tweed and Vilis the input for the oculomotor system is the desired eye orientation in 3-D, specified as $\mathbf{q}^*$ above. Therefore, the input to the oculomotor system is represented in 3-D space. Later, it was hypothesised ([21]) that the input to the oculomotor system in the Tweed and Vilis model may be given by a 2-dimensional, rather than a 3-dimensional signal. If the desired final orientation is represented by the rotation vector $\mathbf{r_B}$ and the initial eye position by the rotation vector $\mathbf{r_A}$, then the axis, when the eye moves from the initial orientation $\mathbf{r_A}$ to the final orientation $\mathbf{r_B}$, is given by the rotation vector

$$\mathbf{r}_{AB} = \mathbf{r}_B - \mathbf{r}_A + \mathbf{r}_A \times \mathbf{r}_B \tag{3}$$

where $\times$ denotes the vector outer product. In this model the motor error is coded as a two-dimensional signal that elicits horizontal and vertical components of saccades corresponding to the desired eye displacement $\mathbf{d}_{AB} = \mathbf{r}_B - \mathbf{r}_A$ from Listing position $\mathbf{r_A}$ to Listing position $\mathbf{r_B}$. Since rotation vectors do not commute, the problem of the implementation of Listing's Law is not solved by this model, but rather the solution to this problem is shifted to a location more peripheral in the oculomotor system. As a consequence of the implementation of Listing's law all rotation vectors $\mathbf{r}$ corresponding to eye orientations lie in a plane, in agreement with Listing's Law.

It was shown by Tweed and Vilis ([22]) that the implementation of Listing's law requires that the angular *velocity* vectors for saccades do not fall in a plane. This may seem to be paradoxical to the fact that Listing's Law prescribes that all eye rotation vectors fall in Listing's plane, even for orientations during the saccade. This paradoxical result can be understood from the fact that a torsional component comes in from the term $\mathbf{r}_A \times \mathbf{r}_B$, which is nonzero and which is orthogonal to both $\mathbf{r}_A$ and $\mathbf{r}_B$ whenever $\mathbf{r}_A$ and $\mathbf{r}_B$ are nonparallel rotation vectors. This torsional component has opposite sign if the initial and final orientations for the eye are interchanged. Therefore, the angular velocity vectors do not fall in a plane.

Experimental observations confirmed the model proposed by [21] and [23]. This implies that Listing's Law is implemented upstream in the oculomotor system.

The principles that underlie the coordination of eye movements seem to be quite similar to those underlying rotations in other joints. The shoulder, for example, has 3 rotational degrees of freedom, just as the eye. When pointing to an object in space, the pointing direction of the hand is important, but the orientation of the hand is not. Therefore, pointing leaves the torsional degree of freedom unspecified. Recent investigations ([24], [25], [26], [28]) have shown that Donders' law is valid for head and arm movements too. This is illustrated in Figure 3, which shows trajectories of the arm during saccadic pointing to dots on a screen. During these experiments, the orientation of the right forearm and upper arm was measured during natural pointing movements with the extended arm. The shoulders of the subject were fixated so as to eliminate movements of the upper arm due to movements of the trunk. Each position of the arm was represented by the hand position as viewed from three orthogonal directions (Fig. 3A) and by the rotation vector that brought the arm to that position starting from the reference position. Then a plane was fitted to the data. Further details on the experimental procedures and analyses can be found in [26]

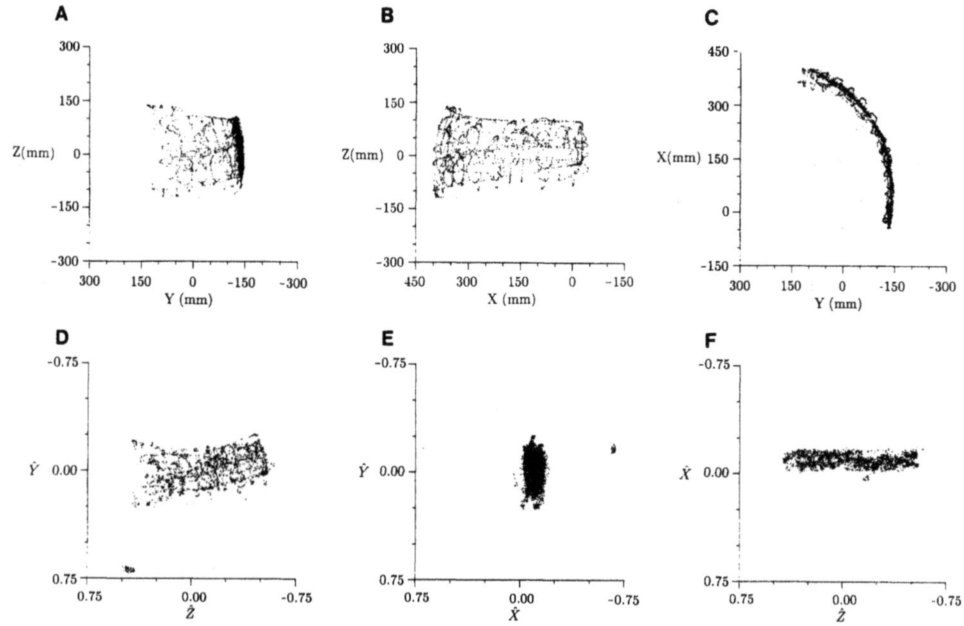

Figure 3. Upper arm position and orientation during pointing movements. Views from behind (A), from the left side (B) and from above (C) the subject. (D-F) Rotation vectors describing the orientation of the upper arm during the same trial. The vectors, shown in a front view (D), a side view (E) and a top view (F) can be reasonably well approximated by a plane. The components of the vectors indicate left/right ($\hat{Z}$), up/down ($\hat{Y}$), and torsional ($\hat{X}$) rotations.

Listing's Law states that there is a reference position for the eye such that when each eye position is described as a single-axis rotation vector relative to the reference position, all rotation vectors corresponding to eye positions in the normal working range, fall in a plane. For any arbitrary reference orientation, the orientation of the rotation plane can be defined by the vector normal to its surface. However, the direction of the rotation vectors, and therefore the orientation of the rotation plane, are dependent on this arbitrarily chosen reference. The primary position defined by Listing's law is the unique reference orientation which yields a rotation plane orthogonal to the pointing direction when the arm is in the primary position. This unique pointing direction we will refer to as "primary direction".

The top 3 panels (A, B and C) represent the Cartesian positions of the hand. The 3 panels represent views from behind the subject (yielding a "front" view of the data), from the left side, and from above the subject. The axes were chosen such that the z-axis is pointing upward, the x-axis forward, and the y-axis to the left relative to the subject. In order to obtain these perspectives, some of the axes have been reversed in this and following figures. Because of the relatively restricted range of the movements in the vertical direction, the view from above (panel C) clearly shows an arc projected from the spherical surface on which the cross moved.

The 3 panels at the bottom (D, E and F) contain rotation vectors expressing the orientation of the arm throughout the experiment relative to the primary position. The $\hat{z}$ component of these vectors represents the magnitude of the rotation of the pointing direction around the vertical $\hat{z}$-axis. Rotations around this axis move the pointing direction primarily left and right. Therefore, this axis was chosen as the horizontal axis in panels D and F. Similarly, rotations around the horizontal $\hat{y}$-axis represent up and down movements, and it was therefore chosen as the vertical axis in panels D and E. The $\hat{x}$ component finally, represents the amount of torsion.

The range of the $\hat{z}$ component for the trial shown in figure 3 was approximately 110 degrees, equal to the full width of the horizontal workspace. The range of the $\hat{x}$ (torsional) component on the other hand, was only about 15 degrees, and the thickness of the plane, expressed as the second moment of the distribution of torsional data points (i.e. the standard deviation) was 4.3 degrees. Considering the rotation axes to lie in a plane seems a reasonable first order approximation, given the significantly larger range of the $\hat{y}$ and $\hat{z}$ components.

A closer analysis including data from arm movements in a lager workspace revealed that the rotation vectors did not fall in a flat plane, but rather in a curved 2-dimensional surface. It was as if the normal to the surface, that contained the rotation vectors in a small workspace, "moved" with workspace. These results were quantitatively different from the results on eye movements, which showed that for all eye orientations the corresponding rotation vectors fall in a plane. However, the basic result is the same for the eye and the arm: there is a reduction in the number of degrees of freedom that is used for fixation (eye) or pointing (arm) movements in a very consistent way. For the eye these rotation vectors fall in a flat plane; for the arm they fall in a curved surface.

Hepp et al. [25] and Miller et al. [27] also studied movements of the head during natural gaze movements. The results appeared to be the same as those for the upper arm and forearm: there is no flat Listing plane that describes the orientations of the head for different directions. Rather, the rotation vector surface has a curvature which is slightly less than that for the arm.

So far it is clear, that for arm movements a Listing plane is not an adequate way to describe the arm rotation vectors. Hore et al. [28] showed that a twisted plane gives a better description. If one has a vertical sheet of paper as if to read it, with the left hand holding the left edge and with the right hand holding the right edge, fold the right hand over the left, then raise the right hand vertically. Then the curved surface is close to the position vector surface that describes the positions that can be achieved by a Fick gimbal. A gimbal is a device in which a system of axes is nested within each other. A gimbal in which the "horizontal" axis is mounted inside the "vertical" axis is called a Fick gimbal. An example of a Fick gimbal is an earth-fixed telescope: the telescope can rotate horizontally about the fixed vertical axis and can rotate vertically around a mobile horizontal axis [20]. A consequence of this system is, that when scanning the environment the horizontal crosshair in the eye piece always lines up with the horizon. Hore et al. speculate, that the functional significance of the resemblance with a Fick-gimbal system is that vertical movements always occur along the shortest path thereby minimizing energy expenditure. In contrast horizontal movements, made when the gimbal points up or down, involve rotations around the "vertical" axis and do not occur via the shortest path.

The curvature in the surface with rotation vectors indicates that the normal to the curved surface moves horizontally in the same direction as the arm in workspace. This raised the question, whether the difference between curvature in horizontal and vertical direction was related to the gravity vector (see e.g. the explanation postulated by Hore et al. [28]). In order to investigate whether the extrinsic coordinate system was related to gravity, Hore et al. ([28]) investigated pointing for subjects that were standing or lying. They found, that the plane with rotation vectors showed a stronger alignment with gravity than with the body: it tilted from the vertical by an average of only 10 deg for body tilts of more than 60 deg. Again this result indicates that arm orientation is determined by extrinsic coordinates, rather than by intrinsic coordinates.

3 What frame of reference is used ?

So far we have used terms as "target position in space" and "eye position" without explicitly specifying in what frame of reference these signals were represented. For sensory stimuli the representation is often quite obvious. For example, the position of a visual target on the retina is represented in a retinal frame of reference: the position of the target on the retina depends on eye position. Similarly, the neural activity in the auditory nerve, which codes an auditory target, depends on head position since the relative phase and relative intensity of the sound at the two ears depends on head position relative to the auditory target.

A problem arises at more central stages of information processing. The different frames of reference, that are used to code various sensory modalities of the same target, have to be merged in a common frame of reference which is used to initiate and coordinate goal-directed movements. A nice example of sensor fusion has been found in the superior colliculus, which is a nucleus in the tectum, which is thought to provide the input to the oculomotor system. The superior colliculus receives both visual and auditory information, each of which is represented in a somatotopically organized map. Quite interestingly, these maps are in register ([29], [30]). When one of the ears is plugged, thereby disturbing the relative intensity of the sound at the two ears, sound localisation is initially perturbed. However, a recalibration process appears to compensate for the perturbation and after a few days, auditory sound localization has recovered to normal again. Electrophysiological studies have revealed, that after ear-plugging, the auditory map in the superior colliculus which was disturbed by plugging, was brought into register again with the visual map. Moreover, the centre of activity in the auditory map depends on eye position: it shifts within the map during saccadic eye movements. This clearly indicates, that the auditory map in the deep layers of the superior colliculus is not in craniotopic coordinates.

The question then is what frame of reference is used to represent target position for motor acts. Initially, like in the Robinson model it was assumed that in the superior colliculus target position was coded with respect to eye position relative to head position. Later studies (e.g. [31]) have indicated that target position is coded not as a motor error for the eye but as gaze error. With gaze error, the same representation of target position can be used for eye and head movements. This indicates that target position may be coded in such a way, that it can be used simultaneously by several motor systems. Evidence for the fact that several motor systems may share the same input signal, which codes target position in space, was provided earlier by Gielen et al. [32]. They found that when two visual targets were presented subsequently, both the eye and arm move towards the same position in space, which for the double-target stimulus could be a position between the two visual targets.

Several models, e.g. the models by Hepp and van Opstal ([25], [21]) postulate a difference vector between target position in space and eye or limb position in space as the input for the motor system. However, it is still an open issue whether the input to motor systems is represented as a difference vector between target position in space and position of the head or trunk, or whether the signals are represented in a body frame of reference. The latter hypothesis comes from experiments by Andersen et al. [33] who recorded from neurons in the parietal cortex and who found that these cells had a receptive field in a retinal coordinate system. However, the amplitude of a visual response in the receptive field was modulated by eye position relative to the head. Therefore, the response of these cells was not determined by a difference vector; rather, the response amplitude was modulated by eye position.

At the motor side it was found that the activity of neurons in motor cortex depends on the direction of a movement. Results from neurophysiological experiments ([11], [12]) have demonstrated that the firing rate of cells in motor cortex is not related to changes in a particular joint angle, but rather to movement direction in space. Each cell has a preferred movement direction in space for which the activity of the cell is largest. The activity FR of a cell can be described simply by the innerproduct between preferred direction $\mathbf{d}_{pref}$ and movement direction $\mathbf{d}$: $FR \propto \mathbf{d}_{pref} \cdot \mathbf{d}$. Since the preferred directions of cells are homogeneously distributed in three-dimensional space, the population vector $\mathbf{D}$ which represents the contribution delivered by all cells in motor cortex, and which is defined by

$$\mathbf{D} = \sum_i FR(i)\mathbf{d}_{pref,i} \tag{4}$$

provides a good estimate of the real movement direction.

Since the same movement in different parts of workspace require different rotations in joint angles and therefore also the use of different muscles, the issue was raised whether movement direction was defined in joint coordinates or in workspace. Recording of single-unit activity of cells in motor cortex ([13]) has shown that the preferred direction of motocortical cells (which is the movement direction in 3-D space associated with the largest firing rate of that cell) is not constant. The preferred direction changes as a function of the position within workspace. When arm movements are made at the left, front, or right side of the body, the preferred direction is not constant but changes in a direction that corresponds to the change in workspace. Caminiti et al. interpreted these results as evidence for the hypothesis that movement direction in motor cortex is coded in external work space coordinates, rather than in an intrinsic joint angle coordinate system. This seems a plausible hypothesis given the observation ([34]) that neurons in motor cortex do not always project to the motoneuron pool of a single muscle, but usually have different projection sites in the spinal cord. This anatomical observation is in agreement with the hypothesis that motor cortex specifies movement direction in space and that the transformation to motor commands in a muscle coordinate system is made at a stage after motor cortex. However, it should be mentioned, that the result by Caminiti et al. is also compatible with the hypothesis, that motocortical cells operate in a muscle-coordinate system (see [35]).

4 Constraints on the control of a redundant set of muscles.

Even for simple flexion movements of the arm several muscles are involved. M. brachioradialis and m. brachialis are mono-articular flexor muscles which contribute mechanically to torque at the elbow joint. Another elbow-flexor muscle, m. biceps brachii, contributes to torques in supination/pronation direction of the hand too, and in addition contributes to flexion torques in the shoulder as well, since m. biceps brachii is a bi-articular muscle acting also across the shoulder joint. Since more than one single muscle acts across the elbow joint, a simple flexion torque at the elbow can be obtained by a large variety of different muscle activation patterns. For example, a larger activation of m. brachialis can be compensated by a smaller activation of m. brachioradialis such that the total elbow torque remains the same. Despite this redundancy, experiments by various authors have revealed a consistent and reproducible activation pattern within and between subjects for both isometric (e.g. [2], [1], [36]) and isotonic ([37]) contractions. Isometric contractions are contractions where the limb is fixated at a fixed position and where force increases gradually: the length of muscles does not change. During isotonic contractions, the external force remains constant while the limb moves usually at a constant velocity. In order to compare muscle activations during isometric and isotonic contractions, the activation is compared at the same force and when the limb moves through the position, that is used in the isometric contractions.

In a series of experiments ([38], [39],[40], [37]) it was shown that the activation pattern of muscles is different for isometric and isotonic contractions. Apparently, the muscle activation pattern is consistent for a given motor task but may be different for different motor tasks. Another consistent movement-dependent type of muscle activation pattern was found by Smith et al. [41] and by Nardone et al. [42], who demonstrated that the recruitment order of motor-units within a single muscle may be different depending on the dynamics of the contraction (slow movements versus fast cyclic movements) or on shortening versus lengthening contractions. Since only very little is known about muscle activation in dynamic contractions, we will only discuss the coordination of relatively slow voluntary contractions and movements.

4.1 Optimal control versus global control using biomechanical and neurophysiological constraints.

Several studies which have dealt with the apparent redundancy for simple motor tasks and which have tried to understand the observed muscle activation patterns, used theoretical techniques from optimal control in systems theory. Nelson [43] investigated aiming movements and analysed the effect of constraints like "minimum energy", "minimum impulse", etc. He found that several optimization criteria predicted approximately the same velocity profiles for the aiming movements.

Later studies investigated optimal criteria for a specific type of movement such as "minimum jerk" for the movement [44], "minimum fatigue" during contractions [45] [46], "minimum total torque" produced by the muscles [47] or "minimum change in torque" [48]. Some of these criteria (e.g. the minimum jerk hypothesis) are only descriptive and are not meant to provide insight in the various stages of information processing in the motor system. From an experimental point of view, Gielen et al. [49] demonstrated that some of the optimization criteria led to results that are in contradiction with experimental findings. For example, it was shown that for supination movements of the forearm, m.biceps brachii is activated. This muscle also has a mechanical contribution in flexion direction. Therefore, in order to have a pure supination torque, m. triceps brachii (an elbow extensor) had to be activated in order to cancel the flexion contribution by m. biceps brachii. This activation pattern clearly is in contradiction with the hypothesis of "minimal total muscle force" since the same supination torque could well have been obtained by exclusively activating m. supinator without the problem of cancellation of forces in flexion/extension direction. The hypothesis of "minimum fatigue" is also in contradiction with experimental observations, since it was found that the relative activation of m. biceps, m. brachialis and m. brachioradialis is not the same (see also [2]). If it is assumed that fatigue is proportional to muscle activation, then a minimum fatigue model would predict an equal activation for all synergistic muscles.

4.2 Theory on the use of mono- and bi-articular muscles.

Recently, new insights in how to find an appropriate muscle activation pattern have emerged from inverse dynamic analyses of human leg movements. As in most robotic devices, joints in biological organisms allow rotations. This implies that most tasks in which an object has to be translated, require a transformation relating changes in joint angles to translation of the extremity. This transformation appears to be subject to some peculiar constraints which were not explicitly recognised before. For example, it was found ([50]) that muscle activation patterns of mono- and bi-articular muscles in man need a very precise timing in order to be able to jump in a purely vertical direction and that small changes in relative timing had a large effect on the height of vertical jumps. This is due to the fact that rotational kinetic energy due to joint rotations has to be transformed into translational kinetic energy. Realising the consequences of these constraints it was possible to explain why the extremities of man and animal are not only equipped with muscles which cross over one single joint, but also with muscles which cross over two or more joints (see [51], [50], [52]).

In most tasks where an arm (or leg) is in contact with the environment, the task requires that the most distal segment exerts a force on an object (or the ground) and thereby displaces that object (or one's own body). In this analysis we will not include the effect of gravity and inertia of the different segments although it would not change the main results. Then it is not difficult to see with simple mechanics that each amplitude and direction of force at the distal segment of the arm or leg can be realised by one particular combination of torques in the joints, since torque $\vec{T}$ is simply the result of the vector product of force $\vec{F}$ and the lever arm $\vec{r}$: $\vec{T} = \vec{r} \times \vec{F}$. However, for displacements of the manipulator endpoint the situation may be more complex since in general, for a multi-joint manipulator there may be a large number of changes in joint angles that produce the same displacement of the end point of the manipulator. In order to eliminate this redundancy for the moment, let us consider a two-joint manipulator acting in a 2-dimensional (e.g.

horizontal) plane (see Fig. 4).

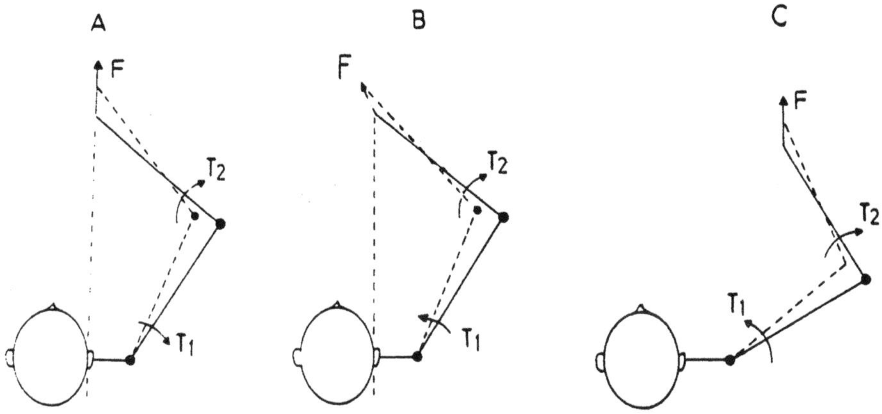

Figure 4. Schematic illustration of a two-joint manipulator showing the effect of the direction of force exerted by the manipulator on joint torques (A versus B) and the effect of two different positions of the same two-joint manipulator in a horizontal plane (A versus C). The arrows in A, B, and C indicate the torque in the joints related to the external force $\vec{F}$. Note that the sign of the torque in joint 1 is different in A versus B and C. Reprinted with permission from [53].

In Fig. 4A the lever arm relative to the external force **F** has the same direction for the elbow and shoulder. Therefore, the torque in elbow and shoulder has the same sign: it is a torque in extension direction. The only way to produce the external force **F** in Fig. 4A is achieved by a combination of an *extension* torque in the shoulder (joint 1) and elbow (joint 2). However, a movement in the same direction as the force in Fig. 4A requires an elbow *extension* and a shoulder *flexion*. This means that at the shoulder there may seem to be a conflict between the requirements for the control of force and displacement since the torque and change in joint angle required for the movement are in opposite directions. In Fig. 4B where the direction of force and displacement are different, this apparent conflict does not exist since the same force vector is obtained by an extending torque in the elbow and a flexion torque in the shoulder, i.e. both the torque and the change in joint angle have the same sign both at the elbow and shoulder joints. The same is true for the situation in Fig. 4C which shows the same limb in a different position. The change in the sign of the shoulder torque is caused by the fact that the line of the direction of force passes the shoulder joint at a different side in Figs. 4A, B, and C. Summarizing, there may be a conflict between the direction of torque and the change of joint angle depending on the direction of force and displacement of the manipulator end-point and depending on the position of the manipulator.

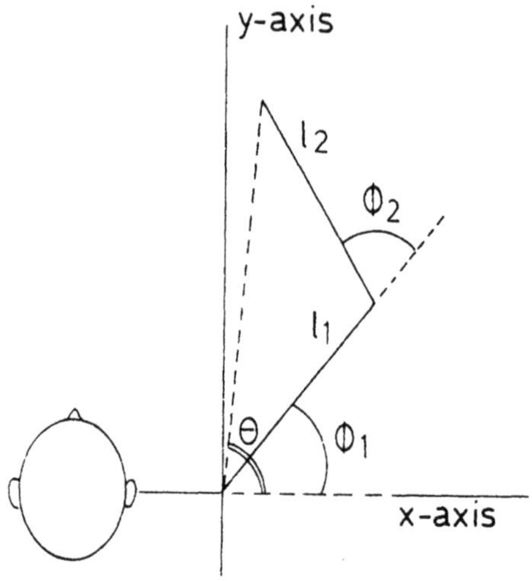

Figure 5. Schematic illustration of a two-joint manipulator illustrating the symbols used in the various formula's. Joint angles are indicated by ϕ_1 and ϕ_2, respectively. Length of the links is l_1 and l_2, respectively. d is the distance of the manipulator endpoint with respect to the origin.

In order to analyse this problem a little further consider a two-joint manipulator with joint angles ϕ_1 and ϕ_2 (see Fig. 2) which allows movements in a 2-dimensional (e.g. horizontal) plane. Assume for convenience that the origin of the cartesian world-space coordinate system is at the shoulder. For small displacements the relation between changes in joint angles $\Delta\phi_1$ and $\Delta\phi_2$ is related to the displacement ($\Delta x, \Delta y$) in world space by the equation

$$\begin{pmatrix} \Delta x \\ \Delta y \end{pmatrix} = J(\phi_1, \phi_2) \begin{pmatrix} \Delta\phi_1 \\ \Delta\phi_2 \end{pmatrix} \tag{5}$$

where $J(\phi_1, \phi_2)$ is the Jacobian for joint angles ϕ_1 and ϕ_2 given by

$$J(\phi_1, \phi_2) = \begin{pmatrix} -l_1 sin(\phi_1) - l_2 sin(\phi_1 + \phi_2) & -l_2 sin(\phi_1 + \phi_2) \\ l_1 cos(\phi_1) + l_2 cos(\phi_1 + \phi_2) & l_2 cos(\phi_1 + \phi_2) \end{pmatrix} \tag{6}$$

The relation between a change in external force $\Delta\mathbf{F}$ and the corresponding changes in joint torques $\Delta\mathbf{T}$ is given by

$$\Delta\vec{T} = J^T \Delta\vec{F} \tag{7}$$

One should note that in general, the direction of force and the direction of motion in Equation (5) will not be the same. This is due to the fact, that in a non-redundant system, joint displacements map into end-point displacements via a Jacobian matrix (equation 5). However, joint torques map into endpoint forces via the inverse of this Jacobian matrix (see Equation 7). This inverse relationship is responsible for the general lack of equivalence between directions of forces and motion.

As a consequence it follows that when the displacement $\Delta \mathbf{x}$ and the change in force $\Delta \mathbf{F}$ are in the same direction, the direction of the resulting change in joint angle $\Delta \phi$ and that of the change in torque $\Delta \mathbf{T}$ may be different in some joints. Even the sign of the corresponding components in $\Delta \phi$ and $\Delta \mathbf{T}$ may be different. If this is the case there may seem to be a conflict in the sense that the rotation in a joint is opposite to the direction of the torque acting in that joint. As a result the actuator in the joint is *dissipating* work, instead of *delivering* work. Since the total amount of mechanical work is the same when expressed in joint or endpoint coordinates, this result implies that if there is dissipation in one joint of a two-joint manipulator, another joint should deliver the total work delivered by the manipulator plus the dissipated work in the other joint. If the dissipated energy is lost such as in muscles, such a movement might become very inefficient with mono-articular muscles only.

However, most mammalian extremities contain two different types of actuators: Muscles that cross over one single joint (usually referred to as mono-articular muscles) and muscles that cross over two or more joints (called bi- or poly-articular muscles). For the human arm and leg, which can be considered as two-bar systems such as illustrated in Fig. 5, these actuators can be summarised to four mono-articular muscle groups and two bi-articular muscles, as outlined in fig. 6. In Fig. 6 mono-articular muscles that act across the same joint (such as m. brachialis and m. brachioradialis) are lumped together into one single mono-articular muscle. The bi-articular muscles (nr. 5 and 6) refer to the long head of m. triceps and biceps, respectively.

The movement illustrated in Fig. 4A which also requires a force as indicated in this figure, would not be possible without internal energy losses if it had to be performed by mono-articular muscles only. The elbow extensor (muscle 3 in Fig. 6) would have to deliver considerably more work than externally required, in order to compensate for the negative work which would be the inevitable result of the lengthening of the activated shoulder extensors. These mono-articular shoulder extensors would have to be activated in order to warrant the desired direction of the external force. What actually appears to occur in this phase (see [37]) is that the net extension torque in the shoulder is achieved by activation of the bi-articular triceps muscle which acts over the elbow as well as over the shoulder joint (muscle 5 in Fig. 6). The extending torque and the elbow extension are realised by the activation of the mono-articular elbow extensor muscles (muscle 3) and the bi-articular part of m. triceps (muscle 5). The bi-articular long head of m. triceps also provides the shoulder extension torque. The shoulder flexion is realised by a mono-articular shoulder flexor (muscle 2). Since the elbow is extending and the shoulder is flexing, lengthening of the bi-articular triceps is relatively small (if present at all). The advantage of this bi-articular muscle action is that now the required torque in the shoulder can be established without (or with only small) energy dissipating lengthening contractions (see [53] for a more detailed analysis). Similar actions of bi-articular muscles have been observed in vertical jumping ([50]) and cycling ([52]).

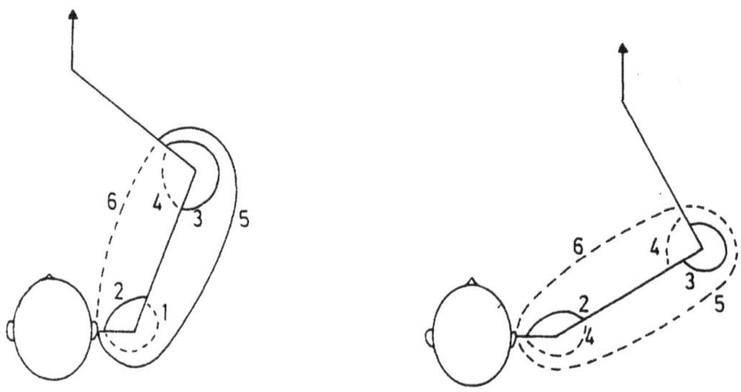

Figure 6. Schematic representation of mono- and bi-articular muscles in the human arm. Muscles that would be activated according to the ideas explained in the text are represented with solid lines. Muscles that would be relaxed are represented by broken lines.

Since lengthening contractions give rise to negative work production (which implies energy dissipation) and thereby are inefficient, any general strategy for muscle coordination should avoid lengthening contractions as much as possible when the organism has to produce positive work. As previously shown, lengthening contractions cannot be avoided by using mono- articular muscles exclusively and in this context bi-articular muscles may be useful. The use of biarticular muscles greatly improves the efficiency of multi-joint movements ([53]).

In terms of the resulting net torques and net joint work, one might state that bi-articular muscles appear to distribute the mechanical output (torque and work) of mono-articular muscles over the joints that are crossed in such a way that the specific task demands (expressed in desired direction of external force) are satisfied. From the analysis and from the presently available data, it appears that mono-articular muscles are activated mainly if they are allowed to shorten, thereby delivering positive work.

The constraint which follows from this study does not eliminate all redundancies in a redundant manipulator. For example, the manipulator in Fig. 3 with the 6 muscles is redundant and the redundancy is not completely removed by the constraint.

4.3 Experimental data on task-dependent activation.

The above considerations on the different functional role of mono- and bi-articular muscles have stimulated several experiments in order to study muscle activation patterns in various motor tasks. The first detailed studies were by Tax et al. ([38], [39], [40]), who recorded the activity from single motor units in human arm muscles. In these experiments subjects were seated comfortably in a chair with the upper arm 80 deg. abducted and with an angle of 90 deg. flexion in the elbow. Subjects were instructed to make isometric flexion contractions or isotonic flexion movements in the elbow joint. These authors found that the recruitment threshold of motor units was not fixed, but dependent instead on the motor task. For motor units in m.biceps brachii, the recruitment threshold was lower in isotonic contractions than it was in isometric contractions (see Fig. 7). This result was significant and could not be explained by the well-known force-velocity relationship since movement velocities in the isotonic contractions were very low. Any effect of the force-velocity relationship would be much smaller than the effects obtained. For motor units in m.brachioradialis opposite results were obtained. For motor units in this muscle the recruitment thresholds were increased. This result indicates that the relative activation of muscles is not constant, but is dependent on the motor task.

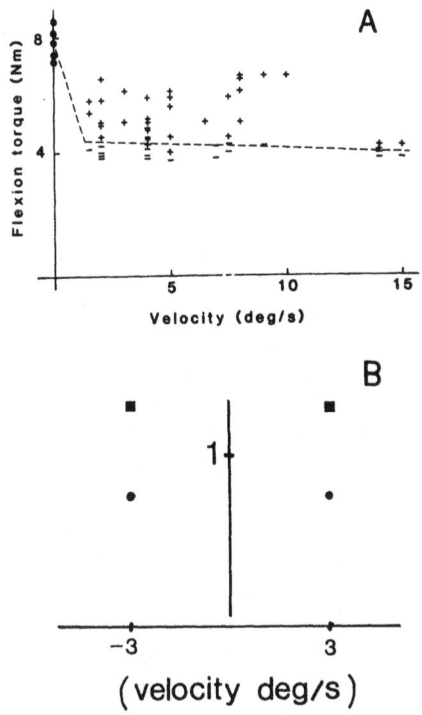

Figure 7. Part A shows the absence ('-') or presence ('+') of motor-unit activity for different torques at the elbow as a function of shortening velocity. Part B shows averaged recruitment thresholds for motor units in m. biceps (circles) and m. brachioradialis (squates) for isometric contractions (v=0) and for voluntary flexion (v=3 deg/s) and extension (v =-3 deg/s) movements. (Part A reprinted with permission from Tax et a., 1989).

Later studies ([37]) have shown, that the increased activation of m. biceps and the decreased activation of m. brachioradialis in isotonic contractions, also depend on the direction of the isometric force or movement (see also [53]). These direction-dependent differences in isometric and isotonic contractions were qualitatively in agreement with the predictions based on the different role of mono- and bi-articular muscles, that was described above (see also [54]).

5 From joint rotation to limb displacement

When we make goal-directed movements with the hand in space in a horizontal plane at relatively low velocity, e.g. to grasp for an object, the hand moves along a more or less straight trajectory ([55], [56]). For movements in vertical direction this trajectory is slightly curved ([57]). Such a trajectory in space has to be made by rotations in joints. Evidently, this imposes some constraints on the coordination of changes in the joint angles, that contribute to the desired movement (see [52]). The available experimental data in the literature suggest that movements are planned in 3-D external space. These data come both from behavioural [58] and from neurophysiological experiments [13],[11],[12]. Soechting and Ross ([58]) have shown that variability of arm position is minimal for certain directions, which are related to world space coordinates, not to specific joint angles. This was interpreted as evidence for the idea that movements are planned in an extrinsic coordinate system.

Morasso ([55]), Flash ([59]) and Abend et al. ([60]) have suggested approximately straight hand trajectories between initial and target position. Later, Hogan and Flash ([61]) gave a theoretical basis to explain the experimentally observed bell-shaped velocity profiles for the end effector during aiming movements. It was found by Abend et al. [60] that when an obstacle is placed between start and target position, the velocity profile had two peaks with a clear dip in the middle suggesting that the movement was built-up by two separate movements, partially overlapping in time.

When a trajectory in space has been selected, the problem becomes which joint angles to select in order to generate the desired trajectory. This problem is known as the "inverse kinematics" problem. Most theoretical approaches that have been proposed in the literature are based on the concept of the "equilibrium hypothesis". In order to understand this concept, it is important to realise that muscles can be considered as (nonlinear) springs (see [62]). Therefore, each joint has spring-like properties depending on the activation of the mono- and bi-articular muscles that act across that joint. Although different versions of the equilibrium hypothesis can be found in the literature (e.g. [63], [64], [62], [44], [56], [65]), the basic idea is that muscles behave like spring-like elements and that the position of the end effector corresponds to that point in 3-D space where external forces (e.g. due to gravity) balance the total force exerted by all muscles together. In this view, the desired position in space corresponds to the equilibrium position where external and muscle forces are in equilibrium. According to this model a movement is made by shifting this equilibrium point. If muscles behave like spring-like elements, the spring constants of all muscles involved, which describe the relation between joint torques and joint angles, can be easily transformed into a stiffness matrix in workspace which describes the relation between external force and displacement for various directions in world space ([66]). When several muscles act across the same joint, each muscle with it's own stiffness, then joint stiffness is the result of the total contribution of stiffness from all muscles. The precise relationship between muscle stiffness K_m, joint stiffness R and stiffness in cartesian world coordinates K can be found in Mussa-Ivaldi et al. [67] and is given by

$$\begin{aligned} R &= J^T K \\ R &= J_m^T K_m \end{aligned} \tag{8}$$

where J is the Jacobian of the transformation from joint coordinates to world space coordinates ($d\mathbf{x} = Jd\phi$), see also Equations 5 and 6) and where J_m is the Jacobian of the transformation from joint space to muscle space ($d\lambda = J_m d\phi$), where λ is a vector referring to muscle lengths. The matrix element R_{ij} represents the stiffness in joint i due to a change in joint angle j. Since the matrix R is in general not diagonal, there have to be projections from muscle receptors, that measure changes in joint angle j, to muscles that act across joint i. This can be achieved by muscle receptors in bi-articular muscles or by heteronymous projections from receptors in a muscle to motoneurons of another muscle.

Mono-articular muscles contribute to stiffness in a single joint. Therefore, their contribution to stiffness in world space depends on the joint angles of the limb. Bi-articular muscles act across two joints. Therefore, their contribution to stiffness in worldspace is more complicated. For example, a change in the position of one joint will cause a change in length of the bi-articular muscle and thereby cause a change in force output of that muscle. This change in muscle force has also an effect on the torque in the other joint that is crossed by the muscle. In fact, the available evidence suggests that humans have a very limited control on the shape and orientation of stiffness. This is compatible with the hypothesis proposed by Hogan [66] that bi-articular muscles may be necessary in order to maintain an isotrope stiffness in 3-dimensional space for different positions of a multi-joint limb. The role of the bi-articular muscles would be to relate changes in torque in one joint due to changes in the angle of the other joint in a flexible way. This idea would be correct if stiffness were the result of muscle mechanical properties alone rather than of reflex components. Since it is generally accepted that the major component of stiffness is due to reflexes (e.g. [68], [69]) bi-articular muscles do not seem to be necessary from this point of view, since heteronymous projections from mono-articular muscles to motoneurones of a mono-articular muscle crossing another joint may be sufficient. Such a heteronymous projection precisely does the job of changing torque in a joint due to stretch of a muscle crossing another joint. The existence of such (mono- or polysynaptic) heteronymous projections has been demonstrated ([70], [71], [72]). However, reflex-mediated stiffness itself might not be appropriate to insure limb stability given the inherent delays unless there is some damping. There is some experimental evidence in the literature for viscous damping (see [73]). In the next paragraphs we will see another reason why bi-articular muscles are really helpful in order to be able to make movements in 3-D space in an efficient way with joints that allow rotations.

When the number of joints is equal to the dimension of the work space, each position in workspace corresponds to a unique set of joint angles. However, when the number of joints exceeds the dimension of workspace, the problem of inverse kinematics (finding a set of joint angles that corresponds to a desired position in work space) has become an ill-defined problem, since the inverse of the Jacobian $J(\phi) = \frac{\partial \mathbf{x}}{\partial \phi}$ does not exist because of the fact, that the kernel of J is not zero. The standard approach to solve this problem is to use a generalized (or pseudo) inverse J^+ of the Jacobian. Given a desired end point displacement $\mathbf{dx}$, the corresponding change in joint angles $\mathbf{d}\phi$ is underdetermined. In this case, a solution is found by chosing a matrix J^+ such that $\mathbf{d}\phi = J^+\mathbf{dx}$ by

$$J^+ = CJ^T(\phi)[J(\phi)CJ^T(\phi)]^{-1} \tag{9}$$

The matrix C is a weight matrix, which parametrizes the pseudo-inverse. By chosing the identity matrix for C, one obtains $J^+ = J^T(JJ^T)^{-1}$, which minimizes the norm of the configuration change $|\mathbf{d}\phi|^2$ compatible with the desired end-point displacement $\mathbf{dx}$. The generalized inverse correponding to this choice is called the Moore-Penrose inverse.

A well known problem (see [74]) is, that the set of joint angles before and after a movement, in which the end point effector is moved along a closed path in workspace, is not the same. As a consequence, the joint configuration reveals a hysteresis, depending on past trajectories. A solution can be obtained if one starts from the expression $\mathbf{T} = J^T(\phi)\mathbf{F}$ which relates force $\mathbf{F}$ in workspace to joint torques $\mathbf{T}$. The relationship between a change in torque $\Delta\mathbf{T}$ in joint space and a change in force $\Delta\mathbf{F}$ at the end-point of the manipulator in world space is given by the relation

$$\begin{aligned} d\mathbf{T} &= J^T(\phi)d\mathbf{F} + dJ^T(\phi)\mathbf{F} \\ &= J^T(\phi)d\mathbf{F} + \Gamma d\phi \end{aligned} \tag{10}$$

where Γ is a matrix expressing the "curvature" of the kinematics:

$$\Gamma_{ij} = \sum_k \frac{\partial^2 x_k}{\partial\phi_i\partial\phi_j} F_k \tag{11}$$

where the summation is over all cartesian coordinates of workspace. Mussa-Ivaldi and Hogan [76] have shown, that when a compliance is attributed to each joint, a unique solution for a set of joint angles for a given position of the end effector always exists. This unique solution corresponds to a condition of minimal potential energy in the springs, that cause the joint compliances, and it corresponds to a particular generalized inverse.

A few words should be spent on the ideas proposed by Pellionisz [77]. The ideas of Pellionisz are based on the notion, that all movements have to be learned and calibrated using sensory information and that a transformation is required to convert the sensory signals into motor commands. Therefore, the coordination of movements can be viewed as a transformation from sensory into motor activity. The problem then is how to select a muscle activation pattern out of the many available activation patterns ? Pellionisz used a mathematical approach by using a procedure, which is equivalent to finding that set of muscle activations which minimizes the square of differences in muscle activation patterns that is necessary in order to reach the desired posture. In terms of joint angles, it predicts that set of joint angles, which requires the smallest changes from the initial set of joint angles. Later studies [36] have indicated some weaknesses of this approach, since it was shown that this theory predicts sometimes negative activation for muscle and, more seriously, that the relative activation of the muscles that was predicted by the theory was not in agreement with the experimental observations. Gielen and van Zuylen [36] proposed some modifications that gave theoretical predictions which were very close to the experimental data. However, they were not able to explain why other muscle activation patterns, that were also allowed by the (modified) theory, were not observed.

6 Discussion

In this chapter we have discussed how a target position in space can give rise to goal-directed movements of the eyes and the arm. Arm movements were discussed more extensively, especially the fact that biological limbs are redundant for simple movements. In the following paragraphs we will discuss a few related issues in more detail.

Implications for neural control

There are two common teleological explanations for the origin of Listing's law for the eye. One is sensory in nature, having to do with achieving orientations of the eye which minimize changes in orientation of objects on the retina for all possible eye positions. [16]. The other is motor in nature, and suggests that the restriction of eye movements to two degrees of freedom is advantageous in that it simplifies the necessary control system. Recently it has been suggested that during pointing to a visual target, gaze, head and arm all move within parallel, 2-dimensional rotation spaces, and that the primary direction for each of these systems moves in the yaw direction approximately with the center of the workspace ([25], [24]). This would effectively reduce the 9 degree of freedom system to the 2 dimensions necessary to specify the direction to the visual target. Our results for the arm movement system are consistent with these earlier results, but demonstrate additional complexities in the system as well.

We have shown that the rotation space for the arm is not actually a flat plane, but rather a curved surface in 2 dimensions. Other studies on head movements (unpublished results) indicate that the rotation surface for the head is qualitatively similar to that of the arm. It is likely that these 3 systems, gaze, head, and arm, all move on similarly curved rotation surfaces. If this is true, then the planar surface generally reported for the rotation vectors of the eye in head is unlikely to have a sensory basis. Generally humans make large gaze shifts with a combination of eye and head movements. Any visual advantages conferred by a flat rotation surface would be lost for gaze movements requiring head movements along a curved surface. Instead the restricted geometry of eye movements is more likely due to the simplified motor control strategies which result. Similar advantages probably explain why arm orientation during pointing is also approximately constrained to a (curved) 2 dimensional surface.

The relation between movement direction and neuronal activity has been the topic of considerable study ([11], [12]). Many neurons in the proximal arm region of motor cortex have direction- dependent activity during the interval prior to and during arm extension. The preferred direction of individual neurons was also found to depend on the location of the workspace. When examined at the population level, preferred directions tended to move with the center of the workspace albeit with a slope somewhat less than 1.0 (Caminiti et al. 1990). However, for particular neurons, there was considerable variation in the slope of this relation. Occasionally the movement of primary direction was opposite that of the workspace location. There is a striking similarity between this description of the dependence of preferred direction on workspace location, and our description of the dependence of torsion on mean pointing direction. However, neither that study nor any others has considered possible torsional components related to cell firing rate. It would be worth examining whether a significant amount of the unaccounted variability in firing rate might be related to torsional movements of the monkey's limb.

Our results have demonstrated that even during a simple pointing task, the control of arm orientation is significantly more complicated than the simple model of Listing's law applied to eye movements. Even a description of a primary direction which moves with the workspace is not adequate. Rather, we must consider a curved 2-dimensional rotation surface to describe the limb orientation during movement. The shape of the surface suggests there is some tendency among all subjects to keep the hand level, although not to the extent of a Fick gimbal system. In general, the required torsion occurred primarily as a result of endo / exo-rotation of the shoulder. However, for long, vertical workspaces, shoulder rotation was relatively restricted, and most subjects made additional supination / pronation movements.

Acknowledgements

This research was supported by the Foundation for Biophysics (NWO) and by the ESPRIT Basic Research Actions MUCOM (nr. 3149) and MUCOM2 (nr. 6615). I gratefully acknowledge the stimulating discussions with my colleagues dr. J.A.M. van Gisbergen, dr. A.J. van Opstal, dr. L.E. Miller and dr. M. Theeuwen.

References

[1] T.S. Buchanan, D.P.J. Almdale, J.L. Lewis, W.Z. Rymer, J. Neurophysiol. 56, 1225-1241 (1986)

[2] E.J. van Zuylen, C.C.A.M. Gielen, J.J. Denier van der Gon, J. Neurophysiol. 60, 1523-1548 (1988).

[3] D.A. Robinson, In: Handbook of Physiology, Section 1: The Nervous System, Vol. II Motor Control part 2. Am. Physiol. Soc., Bethesda, Maryland (1981).

[4] J.A.M. van Gisbergen, D.A. Robinson, S. Gielen, J. Neurophysiol. 45, 417-442 (1981).

[5] C.C.A.M. Gielen, K. van Oosten, F. Pull ter Gunne, J. Motor Behav. 17, 421-442 (1985).

[6] W. Wadman, J.J. denier van der Gon, R.H. Geuze, C. Mol J. Hum. Mov. Stud. 5, 3-17 (1979).

[7] D.A. Robinson, In: Basic Mechanisms of Ocular Motility and their Clinical Implications (Bach-y-Rita P. Lennerstrand G, eds.) pp. 337-374, Pergamon Press (1975).

[8] A.K. Moschovakis, C.A. Scudder, S.M. Highstein. J. Neurophysiol. 65, 203-217 (1991).

[9] A.K. Moschovakis, C.A. Scudder, S.M. Highstein. J. Neurophysiol. 65, 218-229 (1991).

[10] J.H.P. Van der Meulen, R.H.J.M. Gooskens, J.J. Denier van der Gon, C.C.A.M. Gielen, K. Wilhelm, J. Motor Behav. 22, 67-84 (1990).

[11] A.P. Georgopoulos, J.F. Kalaska, R. Caminiti, J.T. Massey J Neurosci 2: 1527-1537 (1982).

[12] A.P. Georgopoulos, R.E. Kettner, A.B. Schwartz J Neurosci 8: 2928-2937 (1988).

[13] R. Caminiti, P.B. Johnson, A. Urbano J Neurosci 10: 2039-2058 (1990).

[14] P.R. Kennedy TINS 13, 474-479 (1990).

[15] D. Tweed and T. Vilis. J. Neurophysiol. 58, 832-849 (1987).

[16] H. von Helmholtz, Handbuch der Physiologischen Optik. 3rd edn, vol.3, Leopold Voss Verlag, Hamburg (1910).

[17] R.H.S. Carpenter Movements of the eyes. Pion Press, London (1977).

[18] R.J. Leigh , E.F. Maas , G.E. Grossman , D.A. Robinson Exp Brain Res 75: 221-226 (1989).

[19] J.D. Crawford, T. Vilis J Neurophysiol 3: 407-423 (1991).

[20] K. Nakayama In: Lennerstrand G, Bach-y-Rita P (eds.) Basic mechanisms in ocular motility and their clinical implications. Pergamon Press, New York pp 193-207 (1975).

[21] A.J. van Opstal, K. Hepp, J.M. Hess, D. Straumann, V. Henn, Science 252, 1313-1315 (1991).

[22] D. Tweed, W. Cadera, T. Vilis Vision Res. 30, 97-110 (1990).

[23] K. Hepp. Commun. Math. Phys. 132, 285 (1990)

[24] D. Straumann, T. Haslwanter, M.C. Hepp-Reymond,K. Hepp Exp Brain Res. 86, 209-215 (1991).

[25] K. Hepp, T. Haslwanter, D. Straumann, M-C Hepp-Reymond, V. Henn In: "Control of arm movement in space: Neurophysiological and computational approaches" (ed R. Caminiti), Exp Brain Res Suppl, Springer, Heidelberg pp. 307-320 (1992).

[26] L.E. Miller, M. Theeuwen, C.C.A.M. Gielen Exp. Brain Res. 90 415-426 (1992).

[27] L.E. Miller, M. Theeuwen, C.C.A.M. Gielen J. Motor Behav., submitted for publication (1993).

[28] J. Hore, S. Watts, T. Vilis, J. Neurophysiol., in press (1992).

[29] M.F. Jay, D.L. Sparks. J. Neurophysiol. 57, 22-31 (1987)

[30] M.F. Jay, D.L. Sparks. J. Neurophysiol. 57, 35-55 (1987)

[31] D.P. Munoz , D. Guitton, D. Pelisson J. Neurophysiol. 66, 1642-1666 (1991).

[32] C.C.A.M.Gielen, P.J.M. van den Heuvel, J.A.M. van Gisbergen Coordination of fast eye and arm movements in a tracking task. Experimental Brain Research, 56, 154-161 (1984).

[33] R.A. Andersen, G.K. Essick, R.M. Siegel. Science 230, 456-458 (1985).

[34] H. Humphrey, Federal Proceed. 45, 2687-2699 (1986).

[35] F.A. Mussa-Ivaldi Neurosci. Lett. 91, 106-111 (1988).

[36] C.C.A.M. Gielen, E.J. van Zuylen, Neuroscience, 17, 675-684 (1986)

[37] S. Gielen, G.J. van Ingen Schenau, T. Tax, M. Theeuwen In: Multiple Muscle Systems: Biomechanics and Movement Organization Eds. J.M. Winters & S.L.-Y. Woo. Springer Verlag, New York. pp. 302-311 (1990).

[38] A. Tax, J.J. Denier van der Gon, C.C.A.M. Gielen, and C.M.M. van den Tempel, Exp. Brain Res., vol. 76, pp. 55-63, (1989).

[39] A.A.M. Tax, J.J. Denier van der Gon, C.C.A.M. Gielen, M. Kleyne Exp. Brain Res. 79 138-142 (1990).

[40] A.A.M. Tax, J.J. Denier van der Gon, C.J. Erkelens Exp. Brain Res., 81, 567-572 (1990).

[41] J.L. Smith, B. Betts, V.R. Edgerton, R.F. Zernicke J. Neurophysiol. 43, 612-620 (1980).

[42] A. Nardone, C. Romano, M. Schieppatti J. Physiol. 409, 451-471 (1989).

[43] W.L. Nelson Biol. Cybern. 46, 135-147, (1983).

[44] N. Hogan J. Neurosci. 4, 2745-2754 (1984).

[45] Dul, J, Johnson JE, Shiavi, R and Townsend, MA, J. Biomech. 17, 663-673 (1984).

[46] J. Dul,J.E. Johnson, R. Shiavi, M.A. Townsend J. Biomech. 17, 675-684 (1984).

[47] B.P. Yeo J. Biomech. 9, 413-416 (1976).

[48] Y. Uno, M. Kawato, R. Suzuki Biol. Cybern. 61, 89-101 (1989).

[49] C.C.A.M. Gielen, E.J. van Zuylen, J.J. Denier van der Gon In:Int. Series on Biomechanics, Vol. 7A. Eds.: G. de Groot, A.P. Hollander, P.A. Huying, G.J. van Ingen Schenau. Free University Press, Amsterdam, pp. 155-166 (1988).

[50] M.F. Bobbert, G.J. van Ingen Schenau, J. Biomech., 21, 249-262, (1988).

[51] G.J. van Ingen Schenau, M.F. Bobbert, R.H. Rozendal J. Anatomy, 155, 1-5 (1987).

[52] G.J. van Ingen Schenau Hum. Mov. Science 8, 301-337 (1989).

[53] C.C.A.M. Gielen, G.J. van Ingen Schenau IEEE Trans. Syst., Man, and Cybern., 22, in press (1992).

[54] L.E. Miller, C.C.A.M. Gielen, M. Theeuwen, C. Doorenbosch In: Control of arm movement in space. Neurophysiological and Computationally Approaches. R. Caminiti, P.B. Johnson, Y. Burnod (Eds.) Springer Verlag Berlin Heidelberg. pp. 1-16 (1992)

[55] P. Morasso Exp. Brain Res. 42, 223-227 (1981).

[56] N. Hogan Can.J. Physiol.Pharm. 66, 508-517 (1988).

[57] C.G. Atkeson, J.M. Hollerbach J. Neurosci. 5, 2318-2330 (1985).

[58] J.F. Soechting, B. Ross Neuroscience 13, 595-604 (1984).

[59] T. Flash Biol. Cybern. 57, 257-274 (1987).

[60] W. Abend, E. Bizzi, P. Morasso Brain 105, 331-348 (1982).

[61] N. Hogan, T. Flash Trends Neurosci. 10, 170-174 (1987).

[62] A.G. Feldman J. Motor Behav. 18, 17-54 (1986).

[63] E. Bizzi, N. Accornero, W. Chapple, N. Hogan J. Neurosci. 4, 2738-2744 (1984).

[64] A.G. Feldman Biophysics 11, 565-578, (1966).

[65] E. Bizzi, N. Hogan, F.A. Mussa-Ivaldi, S. Giszter Behav. Brain Sci., in press (1992)

[66] N. Hogan, Biol. Cybern., 52, 315-331 (1985).

[67] F.A. Mussa-Ivaldi, Morasso P, Zaccaria R Biol. Cybern. 60, 1-16 (1988)

[68] J.A. Hoffer, S. Andreassen J. Neurophysiol. 45, 267-285 (1981).

[69] M.H. Vincken, C.C.A.M. Gielen, J.J. Denier van der Gon Neuroscience 9, 529-534 (1983).

[70] P. Cavallari, R. Katz Exp. Brain Res. 78, 465-478 (1989).

[71] N. Fritz, M. Illert, S. de la Motte, P. Reeh, P. Saggau J. Physiol. 419, 321-351 (1989).

[72] J.B.J. Smeets, C.J. Erkelens J. Physiol. 440, 455-465 (1991).

[73] C.C.A.M. Gielen, J.C. Houk J. Neurophysiol. 52, 553-569 (1984).

[74] C.A. Klein, C. Huang IEEE Trans. Man Systems and Cybernetics SMC-13, 245-250 (1983).

[75] F.A. Mussa-Ivaldi, N. Hogan, and E. Bizzi, J. of Neurosci. 5, 2732-2743, (1985).

[76] F.A. Mussa-Ivaldi, N. Hogan Proc. IEEE Int. Conf. on Robotics and Automation (1989).

[77] A.J. Pellionisz J. Theor. Biol. 2, 185-211 (1983).

Brain Theory / A. Aertsen (Editor)

Intricacies of movement control. An essay

Valentino Braitenberg[a b]

[a]Max-Planck-Institute for Biological Cybernetics, Spemannstr. 38, 7400 Tübingen, Germany, and

[b]IRST, 38100 Trento - Loc. Pantè di Povo, Italy

The synthetic approach of robotics and the analytic approach of motor physiology have not yet reached much common ground. To be sure, some ideas first formulated in engineering have entered the vocabulary of the physiologist (e.g. position feedback, bang-bang control, stability, frames of reference) thereby making the writings of each camp more palatable to the other. However, the engineers still regard the biological system as just one possible solution among many, almost certainly not the most efficient one. Conversely, to the physiologist, the robot looks much too clumsy to be ever taken seriously as an approximation of the real thing.

I believe that we could come closer to each other if each would actively invade the other's territory, the physiologist inventing machines unhampered by restraints of technical feasibility, and the engineer explaining the brain in ignorance of the (indeed scanty) physiological evidence. I will do a little bit of both, putting myself in the role of both kinds of trespassers, combining so to say two kinds of ignorance in the hope of achieving some fruitful synthesis.

1. PHENOMENOLOGY OF ANIMAL MOVEMENT

Movement is a change of the position of parts of the animal relative to each other. We shall distinguish this from locomotion (often also termed movement in ordinary language) which is a change in the position of the animal relative to external coordinates. In most cases locomotion implies movement, but it may occasionally be almost independent from changes of the animal's configuration, as in the case of a medusa drifting in the ocean.

Neglecting forces generated within the animal tissue, we may distinguish *movement against an opposing (external) force* from simple *deformation*. Both categories admit of further classification.

Deformation may or may not change the position of the center of mass within the body. In the first case it is *asymmetrical deformation*, as in the case of an amoeba protruding a single pseudopodium from the bulk of its body. Extending or flexing one arm is of this kind too. In the other case, when the deformation is *symmetrical* with respect to the center of mass, the latter keeps its position within the body. We may think of a human (in space) thrusting his arms upward and extending his legs just in the right proportion. Note that movement symmetrical with respect to the median plane, the plane of bilateral symmetry, is not a sufficient condition: lifting both arms symmetrically will raise the position of the center of mass.

Extending a pseudopodium (or an arm) may not only change the position of the center of mass within the body, as it would if the force producing the extension were radial with respect to the center of mass, but may also produce a *rotation* of the bulk of the body if it is not. There is also the case of a rotation of the body without movement of the center of mass, as in the antisymmetric extension of two pseudopodia along separate lines lying on either side of the center of mass.

The principal form of movement in animals with either exo- or endoskeleton is rotation of one part with respect to the other around a hinge or joint. It is remarkable how jointed appendages (limbs) are more similar to pseudopodia than would appear at first, since *extension* and its opposite *flexion* are the dominant mode of action in limbs consisting of two main (i.e. more bulky) segments such as upper arm and forearm, thigh and lower leg. When in extension or in flexion the two segments rotate in opposite directions, as they mostly do, the principal effect is that of extending or shortening the limb, i.e. of shifting the center of mass of the limb away or towards the body like in a pseudopodium protruding or contracting. The angular momenta of the two segments tend to cancel each other, thereby minimizing the torque transmitted to the rest of the body.

Swinging an appendage on its hinge (e.g. an arm around the shoulder) will produce an opposite rotation of the bulk of the body around a correspondingly smaller angle. The change in the orientation of the animal would seem to depend on a change in its conformation, since the body returns to its original orientation when the arm is swung back to its previous position. Can an animal perform any movement such that it ends up in a new orientation with the original relative position of the moving part and the bulk of the body? A machine with an internal fly-wheel can of course do this if the fly-wheel turns an integral number of times around its axle.

A body appendix may act as a *fly-wheel* if the hinge connecting it to the body has two degrees of freedom. This is the way a cat thrown into the air lands on its feet after it regained its upright position by swinging its tail around appropriately, roughly describing the surface of a cone. It is an interesting question whether the same effect can be achieved by movement of a limb in one plane. We have already seen that a rigid limb will not do, but a pseudopodium could swing its center of mass in a circle by an appropriate combination of contraction/expansion and swinging. A jointed limb such as our arms and legs can do the same. It is noteworthy that this depends on the body being heavier than the two segments of the limb. A *"worm"* made of three equal segments connected by joints with motors cannot change its position in space by any combination of movements in the two joints: starting from a stretched conformation it will always end up stretched in the same orientation as before. However, if one of the segments (except the middle one) is longer, or heavier (= has a greater moment of inertia) than the others, there are patterns of movement that will result in a change in orientation, as can be readily shown by computation of elementary mechanics. Of course, a worm consisting of four equal segments can also do the trick, since it can double over two of its segments to make one heavier one as it were.

Note that a flexible body as a whole can perform a similar motor strategy. Think of a skater threatening to fall over forward or backward. Since he cannot exert any (horizontal) force against his support, he is in a situation similar to a man in space. He will shorten his body by bending his knees, then stretch out again and at the same time bend his body, the whole performance designed to regain the vertical position (of course in this case we cannot exclude the role of forces exerted vertically against the support).

A kind of movement where the opposing forces are mostly internal is *oscillation*. Periodic to and fro movement occurs in many biological contexts (a dog scratching itself, a woodpecker pecking etc.). Of particular importance is harmonic, pendular oscillation. In it the kinetic energy of the movement is stored in the restoring forces instead of being dissipated, as it is in other kinds of movement.

Deformations of the body *against external forces* are of a different kind. We may distinguish the following cases: An important category is movement against or in the direction of gravity. This is important especially in keeping the upright *posture*. Gravity is much less important, however, in limb movement where, at the speed of ordinary motor performance (such as bringing a glass of wine to one's mouth) the inertial forces generated by the movement always exceed gravity. An impressive demonstration is swinging a pendulum around. The angular velocity at which the pendulum hangs at 45°, i.e. the centrifugal force equals gravity, is much below the speed of so-called "fast voluntary movements". With the elbow bent at 90°, and $r = 0.2\,\mathrm{m}$ (the length of the upper arm), the centrifugal force on the forearm $m\,r\,\omega^2$ exceeds gravity $m\,g$ for all speeds beyond 1.1 rotations per second, a fairly moderate speed, comparable to that of an orchestra conductor beating adagio.

Pushing and *pulling* are movements in which the active animal becomes involved in a mechanical system comprising other bodies. Pushing away or pulling towards are clearly defined by changes in the relative position between the two partners. But these terms apply also to situations in which the relative position does not change. Pushing a cart is exerting a force against a force that tends to shorten the distance, without actually changing it, pulling against one that tends to lengthen it. Pushing or pulling downhill, however, demonstrates that the position in front or behind the object being pulled also plays a role.

Pushing and pulling become highly involved when both partners, being animate, participate actively in the interaction. In such a situation there is room for complex planning of motor actions with the reactions of the partner homogeneously involved in the computation of one's own movement. Pushing (or pulling) against frictional forces is at the basis of locomotion.

Jumping and *throwing* are special cases of pushing where the contact between the agent and the object is interrupted at some point. The obvious difference between the two is the relative bulk of the object, much bigger than the agent (= the earth) in the case of jumping, smaller in the case of throwing. They have in common the dynamics of the limb acting on the object. In order to achieve sufficient velocity the force has to be applied for some time while the contact is kept between the agent and the accelerating object. For this either the motor must accelerate while it applies the force, or there must be a variable coupling. The rolling joint of the human knee may have something to do with this.

Touching and *holding* are the gentlest forms of movement, if indeed they can be called that. In touching the position is controlled by the elastic forces arising in the deformation of the tissue at the point of contact. Holding is a more complicated performance and involves the readiness to react to unforeseen disturbances. Holding a chopstick between the index finger of both hands while moving both arms remains one of the most astonishing motor performances.

There are other situations in which the concept of movement as it is mostly held becomes elusive. Think of a *piano player.* Most of the time his arms do not move, only his fingers execute tiny movements in precisely predetermined sequences and combinations. Energy is hardly of concern here, nor are the masses of the fingers such that inertia would play a significant role. As in handwriting or in typing, in the execution of music the individual movements are completely at the service of a channel of communication, the entire performance being of the nature of information rather than of the mechanical interaction between an animal and its environment.

Locomotion, the other main category of movement beside deformation, occurs in a great variety of types, ranging from the crawling of a snake to the gaits of a horse, not to mention various techniques of swimming and flying. These have all been well analysed and offer fewer conceptual challenges than the types of deformations mentioned before.

2. MOTOR CONTROL

The preceding excursion into the taxonomy of movement was aimed at conveying a feeling for the complexity of the task facing the nervous system at its output end. Studies on motor control have often concentrated on isolated simple situations such as movement of a human arm in one or two joints, or have given up on an exhaustive analysis in more complex cases, such as that of a human maintaining his upright posture while being shaken on a movable platform. I believe that situations considered in isolation may lead the interpretation astray. For instance, the tendency of the hand to follow a straight trajectory when reaching a target away from the body (Wadman et al. 1980, Morasso 1981) and less so, when the movement involves swinging the whole arm sideways is only puzzling if the fundamental mode expansion - contraction is not taken into account. As we have seen, this involves rotation of two segments in opposite directions, annulling angular momentum and having the straight trajectory as a trivial geometrical consequence.

Another case in point is the movement of the man keeping himself upright on skates, or being shaken on a platform. Once we have realized that for any fast movement the inertial forces are much more important than gravity, we will not be surprised to observe correcting forces directed against the support which seem actually to destabilize the situation (Dichgans and Diener 1987). They are part of a motor strategy which the man would use if he wanted to change his orientation in free space, devoid of any support. He uses some part of his body as a fly-wheel to generated the angular momentum which brings the body into the desired position. The complex motor reactions of the man on the platform have not been analyzed with this idea in mind, but a simpler situation may illustrate the point. A beginner on skates leaning backward beyond the point of stability regains control by rotating his extended arms around a transversal axis through his shoulder, in the direction up in front and down in the back. In the up-in-front phase, considering only linear acceleration, he helps the action of gravity and would seem to contribute to his fall. But gravity is unimportant in the correction although it is in the threat of falling. The correction is a question of playing angular momenta against moments of inertia.

A third example. Soechting et al. (1986) wondered about the differences in the results of drawing a circle (or rather a succession of superimposed circles) on a board transversally in front of the subject as opposed to drawing on a board oriented in the sagittal plane. In the first case the outcome is a good circle, while on the sagittal plane a very distorted figure is obtained. Here we must remember the tendency of limbs to adopt energy saving pendular motion. With the arm stretched toward the blackboard, the center of mass of the arm is roughly in the elbow. A combination of two pendular motions at right angles to each other and at the right phase angle will make the center of mass describe a circle, and the hand too. In the other situation the arm has to bend during the motion, with the center of mass of the arm moving to different positions on the bisectrix of the elbow angle depending on the angle. If the center of mass is again allowed to follow the energy (and control) saving circular trajectory, this time in a sagittal plane, the trajectory of the hand will be something very similar to the shape Soechting et al. have observed (Braitenberg 1988).

My plea is for the simplest possible principles of mechanics to be invoked in the explanation of movement since the motor system, in order to be efficient and economical, cannot but conform to such principles.

3. INVENTING THE CEREBELLUM

Most of the motor performances mentioned are quite complex and would involve a great deal of computing power for their control. On the other hand, if only the goals of the movements are considered, such as reaching a certain object with an arm or turning the body around a certain angle, the description is much simpler and the interplay of the movement with the sensory input turns into a more homogeneous affair.

If I were to invent a control system for movement, I would therefore distinguish two main tasks and assign each to a separate mechanism. The first task is movement as the deliberate influence on the state of the world whenever the sensory input does not conform to the desired picture. This we may call *action*. It is best described as the transition from one internal representation of the world, including the agent's position in it, to another more desirable one, with the motor output arising as a consequence of the transition.

The other task is keeping the body as a whole stable. *Stability* is threatened not only by perturbations coming from the outside, but even more so by the mechanical disturbances which every intended movement, when it is executed, necessarily communicates to the rest of the body. These disturbances are of various kinds. One is simply actio = reactio: every active movement of one part of the body provokes an opposite movement in the support. Another is connected with inertia: movements tend to overshoot. Another again is due to elastic elements which are involved in the machinery and which release their energy after the force ceases to act.

The mechanism which takes care of the first task, *action*, if we consider it in isolation, is of the simplest kind. We provide a cognitive machine, whose complex internal structure shall not detain us at this point, and see to it that all relevant states of the world are represented in states of activity within the mechanism. These internal states include representations of the animal's position in the world, and of the relative positions of its body appendages. Now we only have to arrange connections between the internal elements representing parts of the body and the parts themselves. The connections are of two kinds, efferent and afferent. The efferent connections translate the states of the internal elements into signals to the motors which activate them exactly to the degree that corresponds to their state of contraction in the internal representation. The afferent connections, on the contrary, signal deviations from the internal representations and can then be used to assess external forces that are in the way of the motor realization. In this view, the position of the body is an automatic consequence of "thinking" that position. Movement is an equally automatic consequence of rethinking one's position, of changing one internal representation into another one (for whatever reason this may happen).

This scheme corresponds to an idea which was proposed as "equilibrium point control" in motor physiology by Feldman (1986) and by Bizzi et al. (1982). It is excellent for slow movements and quite plausible in terms of neural mechanisms. With fast movements, however, all the inertial, elastic and other complications come into play. Here we must add a more down-to-earth mechanical computer to the cognitive machine which proposes the action.

The *computer of stability* could in principle be informed about every transition planned by the cognitive machine. Incorporating knowledge of the laws of mechanics and of the quantities (mass, shape, elasticity, viscosity) involved, it could in principle compute all the deviations from the planned action and produce the corrections in real time. In praxis this does not seem to be feasible, judging from the reluctance of mechanical engineers to tackle simple multi-jointed and multi-motor-systems even with the aid of powerful computers. Hence for the problem of stability I would resort to a more empirical approach based on learning. The conditions for my machine are the following.

Since many different situations have to be dealt with for which there is no common solution, I would compartmentalize my machine. Each compartment I would make responsible for the corrections associated with one particular movement.

The trigger for each of the corrections is the planning of a particular movement. Since we decided to represent movement in the cognitive machine not as such but as a sequence of static representations, the trigger signal coming from there and reaching the stability machine will be a particular sequence of signals in time. The stability machine thus has sequences as input and must associate to each a particular output sequence. My machine would work in a sequence-in, sequence-out mode.

The best way to display a sequence in a machine (if it is not supposed to incorporate any moving carriers such as disks or tapes) is to display them in linear arrays in which different locations stand for different phases in the sequence. The compartments of my computer would have an elongated shape, with the long axis representing time. Many such elongated compartments arranged in parallel would give my stability machine a very characteristic architecture.

Signals from the cognitive machine would reach different compartments in different combinations and at different locations (signifying different phases). Thus each compartment would display a particular sequence of signals. I should like each compartment to react to the corresponding sequence. This can be achieved very simply by connecting the input points within the compartments through delay lines which produce exactly the delays between the elementary signals as they occur in the sequence. This way, when the elementary signals reach the compartment at different times in different places, through the delay lines they add up to a synchronous powerful signal. In other compartments, the spatial distribution of the elementary signals and therefore the delays being different, the same signal would hardly produce any effect.

Having thus mapped sequences on compartments, all we need to do is to let the compartments produce the desired output sequences. With all the delay lines in the compartments already there, we have a device which can easily produce sequences by letting signals sweep through the delay lines in order to trigger appropriate output elements. The best way to do this is to use the signals set up by the input in the delay lines to activate the output elements.

The output elements would have to be connected to the delay lines at appropriate locations, dictated by the phase at which the correction must come in during the movement. I would let the choice of the phase happen in the course of a learning process. This would transform the clumsy, angular, overshooting movements of the beginner into the elegant performance of the accomplished violinist or gymnast. The learning process would be under the guidance of afferent input monitoring all the slinging, all the stresses and strains which are generated locally by less than perfect movement.

Finally, I would try to minimize expenditure of mechanical energy. What comes to mind is to make the corrections as much as possible by diminishing motor action rather than by counteracting the error actively. With all the passive inertial forces arising during movement, it would be wise to diminish the action of the motors that act in the direction of an inertial (e.g. centrifugal) force, rather than to counteract it by further expenditure of motor energy.

The reader will have observed that the machine I have been inventing has many features in common with the cerebellum. The cerebellar cortex is a large array of elongated compartments, the *folia* or *beams of parallel fibers.* They acquire their individuality by virtue of the system of inhibitory fibers (*basket* and *stellate cell* axons) which are arranged at right angles to the beams and therefore functionally isolate one beam from another. A large proportion of the input to the cerebellum is derived from the cerebral cortex (the "*cognitive machine*") via the pontine nuclei. The bulk of the internal connections in the cerebellar cortex is provided by the *parallel fibres* which, being long and slow may well act as delay lines. The output cells of the cerebellar cortex, the *Purkinje cells* are connected to the parallel fibers via dendritic spines, a trait which we take as an indication of *plasticity*, i.e. learning. The action of the Purkinje cells on the cerebellar nuclei and further in the motor output is *inhibitory*, in agreement with the energy saving principle we had postulated.

REFERENCES

1. V. Braitenberg: Some Types of Movements. In: Artificial Life, Studies in the Sciences of Complexity, Vol. VI, C. Langton (Ed.), Addison-Wesley, p. 555-565 (1988).
2. E. Bizzi, N. Accornero, N. Chapple and N. Hogan: Arm trajectory formation in monkeys. Experimental Brain Research 46, 139-143 (1982).
3. J. Dichgans and H.C. Diener: The use of short- and long-latency reflex testing in leg muscles of neurological patients. In: Clinical aspects of sensory motor integration. A. Struppler, A. Weindl (Eds.), Springer-Verlag Berlin (1987).
4. A.G. Feldman: Once more on the equilibrium-point hypothesis (lambda model) for

motor control. Journal of Motor Behavior 18, 17-54 (1986).
5. J.F. Soechting, F. Lacquaniti and C.A. Terzuolo: Coordination of arm movements in three-dimensional space. Sensorimotor mapping during drawing movement. Neuroscience 17, 295-311 (1986).
6. P. Morasso: Spatial control of arm movements. Experimental Brain Research 42, 223-227 (1981).
7. W.J. Wadman, J.J. Denier van der Gon and R.J.A. Derksen: Muscle activation patterns for fast goal-directed arm movements. Journal of Human Movement Studies 6, 19-37 (1980).

ACKNOWLEDGEMENTS

I am very grateful to Hubert Preißl and Stefan Rotter, especially for discussions about the worm in space, and no less to Prof. Matthias Schramm who identified the problem as one that had agitated the French academy in the last century. Hubert Preißl was also my companion in some of my adventures into a cerebellar theory. Margarete Ghasroldashti was, as always, the coordinator of our efforts.

Brain Theory / A. Aertsen (Editor)

Specific responses of the cerebellar cortex to moving stimuli

Detlef Heck

Max-Planck-Institut für biologische Kybernetik, W-7400 Tübingen, FRG

1. Introduction

The anatomical arrangement of parallel fibers and Purkinje cells in the molecular layer of the cerebellar cortex is such that on every Purkinje cell there is convergence of input from an elongated strip of granular layer, extending a few millimeters to either side of the cell. There should be delays between the activation of a granular cell and its effect on the Purkinje cell which depend on distance, the longest being of the order of 10 ms judging from the known propagation velocity in parallel fibers of about 0.5 m/s. A trivial consequence of this is that a train of input pulses reaching the cerebellar cortex at successive locations (in the direction of the parallel fibers) should produce a maximal effect if the 'movement' of the input occurs at the velocity of propagation in parallel fibers, since then every new input would add to the wave already travelling in the parallel fiber system. The summation of successive inputs would be less effective if the 'movement' occurs at velocities higher or lower than that of propagation in parallel fibers. Of course such a travelling wave can affect a Purkinje cell only if the input 'moves' toward it, not in the opposite direction.

It should be noted that not only the Purkinje cells would profit from such dynamic effects, but also the inhibitory interneurons which also receive their main input from the parallel fibers system.

There have been doubts, however, on the effectiveness of the parallel fiber to Purkinje cell synapses which to other authors[6,1] seemed much less significant than the synapses on the ascending axons of granular cells. These findings are astonishing in view of the much larger number of synapses in the parallel fibers than on the ascending axons before their bifurcation. A possible explanation may well reside in the dynamic effects mentioned which do not come into play for static input. I tested this idea in the following experiments.

2. Material and Methods

Acute slices of 400 μm thickness were prepared from 3 to 4 week old rat brains (Sprague-Dawle) by means of a vibratome (Campden, UK). Cuts were oriented in parallel direction to the long axis of the folia. An array of 11 stimulating electrodes (epoxy-insulated tungsten wire A-M Systems Inc., USA), regularly spread at distances of about 130 μm, was inserted into the granular cell layer of rat cerebellar slices which lay fully submerged in ACSF[4] at 33 °C. On appropriately selected slices several long sections through the entire cerebellar cortex between the surface and the white matter would appear, cut at various angles to the plane of the slice.

The recording electrodes (micropipettes filled with 3 M NaCl, with tip diameters of less than 1 μm and a DC resistances of 4 to 9 MΩ) were inserted into the molecular layer under microscopic control. The distances between the nearest stimulating electrode and the recording site varied between 200 and 800 μm. Short current pulses of 50 μs duration and a current strenth between 200 - 500 μA were sequentially passed through the 11 stimulating electrodes, controlled by DAP-cards (Microstar Lab., USA) installed in a PC-486. Extracellularly recorded signals were digitized at a sampling rate of 25 kHz and analyzed off line using a CED-1401 (CED, Cambridge, UK). Fig. 1 shows the arrangement of the stimulating and recording electrodes in the cerebellar cortex.

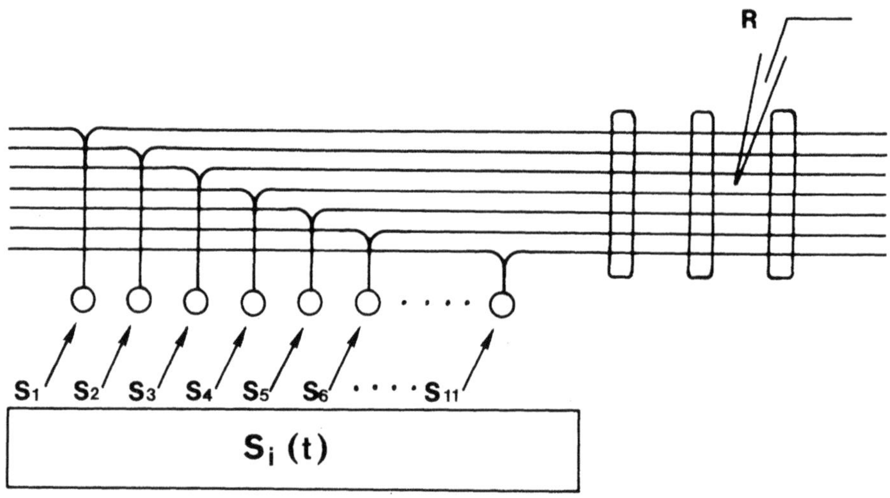

Figure 1: Schematical illustration of the arrangement of stimulating ($S_1 - S_{11}$) and recording (R) electrodes in the cerebellar cortex. Circles symbolize the granule cells with their axons bifurcating in the molecular layer to form the so called parallel fibers. The rectangles oriented perpendicular to the parallel fibers indicate Purkinje cell dendritic arborizations. The box below represents the digital control of the sequential current application.

3. Results

Extracellular responses had the shape of short lasting negative pulses riding on the decaying phase of the stimulus artifact when the stimulus 'moved' toward the recording site (Fig. 2A). The amplitudes of these peaks were taken as the intensity of the response.

The responses for 'movement' toward the recording site were maximal for a velocity of aboud 0.4 m/s in Fig. 3A, and 0.3 m/s in Fig. 3B. Responses decay rapidly for larger and

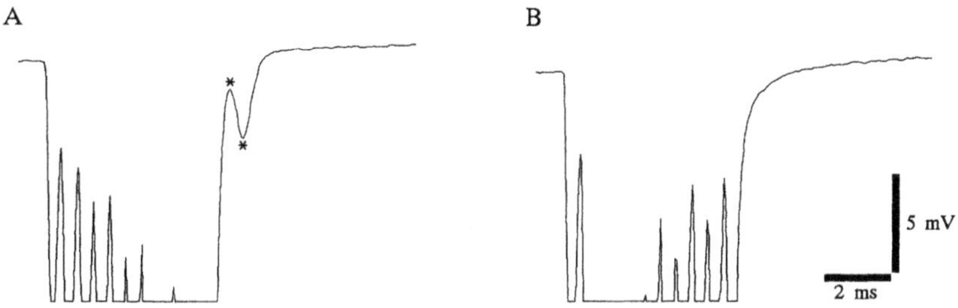

Figure 2: Stimulus artifacts and responses produced by stimuli 'moving' with the same velocity (0.26m/s) either toward the recording site (A) or in the opposite direction (B). The stimulus artifact is reflected in a series of negative pulses, clipped by the recording system. A long lasting negative deflection which has not quite returned to the baseline at the time when the response becomes apparent is also part of the stimulus artifact. A) Movement toward the recording site elicits a considerable response of about 2.5 mV (measured as the difference in hight between the points indicated by the asterisks). B) Sequences moving away from the recording site have no detectable effect beyond the stimulus artifact.

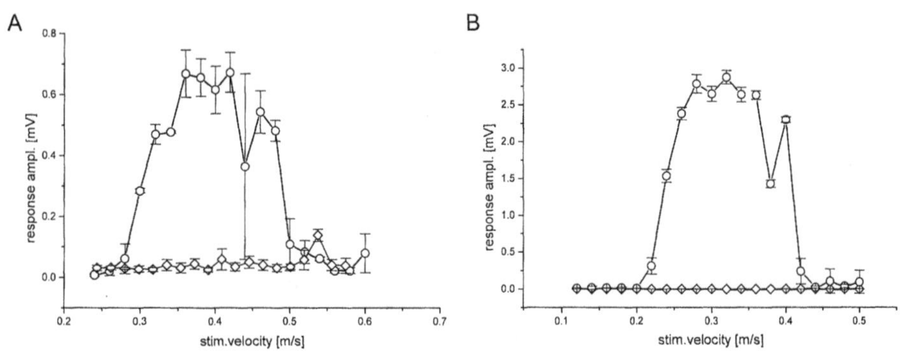

Figure 3: A,B) Dependence of response amplitude (the potential difference between the two points marked by asterisks in Fig. 2A) on 'movement' velociy and direction. Results are obtained in two different slices taken from different animals. The intensity of the extracellular response is plotted as a function of the velocity at which the pulses 'move' through the array of electrodes (○: movement toward recording site, ◇: movement away from recording site). Each data point represents the average of three successive stimulus presentations. Error bars indicate the standard deviation. Note the extremely small variability of responses in B.

smaller velocities, becoming no longer detectable for velocities above 0.45 m/s or below 0.2 m/s in the example shown in Fig. 3B.

Near the velocity yielding the maximal responses, in most cases there was a local minimum (in Fig. 3A at about 0.45 m/s and in Fig. 3B at about 0.37 m/s), the significance of which shall not be discussed here.

In every case 'movement' in the direction away from the recording site did not produce any detectable response (Fig. 2B). We are presently investigating the nature of the recorded potentials, which are most likely reflecting activity in the population of parallel fibers[5]. The local minimum probably requires an explanation in terms of inhibitory interneurons which may come into play only when the activity in the parallel fibers is maximal. If the recorded signals are not due to any postsynaptic events in the molecular layer, possible candidates for this inhibition are the Golgi cells which act on the granular cells from which the parallel fibers originate.

4. Discussion

The cerebellar cortex responds specifically to 'moving' stimuli applied at the granular layer. These responses are both a function of 'movement' velocity and of direction.

These experiments show that the dynamic aspects of the cerebellar input may play a crucial role in the further processing of signals there. The special role of ordered sequences of pulses in the cerebellar input gives the cerebellar cortex the ability to act as a movement detector. An interpretation of the cerebellar structure in these terms was proposed by Braitenberg[2,3] (see also Braitenberg, this Volume).

References

1. J.M.Bower and D.C.Woolston, Congruence of spatial organization of tactile projections to granule cell and Purkinje cell layers of cerebellar hemispheres of the albino rat: vertical organization of cerebellar cortex, J.Neurophysiol., 49 (1983) 745
2. V.Braitenberg, The cerebellum revisited, J. Theoret. Neurobiol., 2 (1983) 237
3. V.Braitenberg, The cerebellum and the physics of movement: some speculations. In: Cerebellum and Neuronal Plasticity, M.Glickstein, Ch.Yeo and J.Stein, eds., Plenum Publishing Corp. (1987) 193
4. F.Crepel et al., Morphological and electrophysiological characteristics of rat cerebellar slices maintained in vitro, J.Physiol., 316 (1981) 127
5. J.C.Eccles et al., Parallel fibre stimulation and the responses induced thereby in the Purkinje cell of the cerebellum, Exp. Brain Res., 1 (1966) 17
6. R.Llinás, Radial connectivity in the cerebellar cortex: a novel view regarding the functional organization of the molecular layer. In: The Cerebellum, New Vistas, S.L.Palay and V.Chan-Palay eds., Springer, New York (1982) 189

Brain Theory / A. Aertsen (Editor)

On connecting syntax and the brain

Friedemann Pulvermüller

Department of Applied Linguistics, UCLA, Los Angeles, CA 90024, USA

Abstract

It is obvious that certain brain mechanisms allow humans to use and understand sentences. However, sentences are rather complicated beasts and some of their underlying regularities are hard to translate into the structure and function of brain-like devices. For example, in the center of every sentence another sentence can be embedded; this rule can be applied repeatedly, resulting in "center-embedded" strings like "[The man [whom the girl [that the dog belongs to] met] is a butcher]". Here the sentence started first is finished last, whereas the last sentence started is finished first. Within the Theory of Formal Languages, it has been shown that abstract automata of a certain kind - so-called pushdown automata - can handle center-embedded strings, whereas simple finite automata can not. The question of how such abstract devices relate to the brain has hardly been asked and never been answered. However, it is possible to specify the pushdown store of the human brain if we assume Hebbian cell assemblies that slowly lose activity after they have been activated. Electrophysiological data are discussed that provide evidence for the proposal. Some speculations follow for why syntactic computation is possible in the human brain but not in other animals' brains.

1. INTRODUCTION

What makes the human brain able to process language? What are the neuronal mechanisms underlying the production of sentences? A brain theory that attempts to explain the behavior of humans must answer these questions. Therefore, linguistic data become relevant to brain theories. If a theory offers no perspective on the explanation of language phenomena, it can not provide an adequate model of brain function. In this paper, I will consider a rather significant linguistic problem, the problem of center-embedding. A brain theory based on Hebb's concept of cell assemblies will be used to solve the problem.

2. THE PROBLEM

The problem of center-embedding can be divide into two subproblems. Subproblem one is the following. It is possible to interrupt every sentence and to start a subordinate clause; it is only when the subordinate clause is finished that the superordinate sentence can be continued. Interestingly, the subordinate clause can vary in length considerably, as the following examples illustrate (for clarity, clause boundaries are marked by brackets) "[The man [whom the girl met] is a butcher]", "[The man [whom the tall girl met after she had put on her nice dress] is a butcher]", "[The man [whom the tall girl met after she had put on her nice dress, the high-heeled shoes, a fine summer hat and a new diamond ring] is a butcher]". So in a string AA' consisting of two complementary parts (A = "the man", A' = "is a butcher"), another string BC... is embedded so that ABC...A' results. It is not easy to conceive a machine which produces or accepts strings in which subordinate clauses of arbitrary length can be inserted. This seems to be very difficult or even impossible for both multilayer perceptrons [1] as well as so-called Marcovian devices [2]. The latter store the probabilities with which an element (word) i follows a particular string of k elements. If the embedded string BC... contains k words or more, a Marcovian device can not determine that the string A influences the selection of the elements of A'. A possible solution of the problem would be to chose a rather large k, say 100 words, so that it would be very unlikely for subordinate clauses of more than k words to occur. However, this strategy has two disadvantages. First, a subordinate clause of 101 words can be made up easily, and second, such a device would need an unattractively large storage capacity [2].

There is, however, a more elegant solution of the problem. We could assume something like a "memory switch" which represents A'. This switch is switched on by the occurrence of A and remains in its active state until A' has been completed. Even after an extremely long embedded chain B, the fact that A' has a high probability of occurrence would be stored by means of the "activated" switch. Of course, such a switch can be realized in any kind of object that has two stable states. It can be realized in a finite automaton consisting of neuron-like elements [3]. Perhaps a brain-internal "switch" can be postulated in order to answer the first question of center-embedding, which can be formulated in the following way:

(i) Given a finite set of units A, B, C, ..., and A', how could a device consisting of nerve cells be organized that produces strings of arbitrary length
- in which the units B, C, ... may occur more than once, and
- in which A' occurs at the end of the string if and only if A was part of the string?

Part two of the center-embedding problem reads as follows. The occurrence of the initial parts of the clauses in a sentence with center-embedding determine the sequence of the complementary parts. Consider the following examples. "[The man [whom the girl met] is a butcher]", "[The man [whom the girl [that the dog belongs to] met] is a butcher]", "[The man [whom the girl [that the dog [which everybody is afraid of] belongs to] met] is a butcher]", "[The man [whom the girl [that the dog [which everybody [whom I know] is afraid of] belongs to] met] is a butcher]". Also here, the sentence started first has to be finished last, the second clause started is to be finished before the first one etc., and the clause started last is the first to be completed. If the initial sentence parts are coded as A, B, C, ..., and the complementary final parts as A', B', C', ..., we arrive at the following simple representations of the above sentences: ABB'A', ABCC'B'A', ABCDD'C'B'A', ABCDEE'D'C'B'A'. Whereas a string ACBB'C'A' is also possible, ABCB'C'A' would be ill-formed. So in each case, a string of elements has to be followed by the inverse sequence of their complements. Call this kind of sentences *mirror image strings*. Representing each of these strings separately in a store would require a lot of memory space. It is attractive to assume a mechanism which combines given pairs of elements only in such a way that mirror image strings result. In a finite automaton, A, B, ... could each turn on a "memory switch". However, these switches would only store the information that A', B', ... have to be produced later, but not in which order they must occur.

Also for part two of the story an elegant solution exists. The symbols A', B', ... representing the complements can be written on small pieces of paper which are entered into a matchbox, one after the other. So if A occurs, the symbol A' representing the missing part of the sentence is written down, and with B and C, B' and C' are entered into the store. Finally, the order of paper pieces in the matchbox contains the information about the sequence in which the initial sentence parts had occurred. If the string is to be completed, the pieces can be taken out of the store one after the other, so that the uppermost - in this case C' - would be used first, followed by B' and finally by A'. So a matchbox-like store is helpful for generating mirror image strings. The matchbox-like store is usually referred to as a *pushdown store*. In the Theory of Formal Languages, it has been postulated that automata with pushdown stores of unlimited capacity are needed to represent grammars [4, 5]. The storage capacity of these devices is unlimited, since linguists have argued that the human language faculty allows to center-embed an arbitrary number of clauses into each other [4]. This would require a matchbox of unlimited depth. However, as can be seen when considering the above examples, sentences with three or more center-embedded subordinate clauses are very hard to produce or understand. Psycholinguistic evidence supports this view [6]. So it seems sufficient to postulate a cognitive matchbox or pushdown store in which a very limited number of entries can be made. A limited pushdown store could be realized in some way in the brain. In this case, we could refer to it while answering the second center-embedding subquestion:

(ii) Given a finite set of initial units A, B, C,..., K and their complements A', B', C',..., K', how could a device consisting of neurons be organized that produces strings
- in which a complement occurs if and only if the corresponding initial unit occurs,
- in which initial units precede complements, and
- in which the sequence of complements is the mirror image of the sequence of initial units?

Note that the problem of center-embedding represents only one beast out of the zoo of phenomena linguists have been dealing with during the past 40 years. However, it is a rather significant one. Since Chomsky had shown that strings with an unlimited number of center-embeddings can not be generated by finite automata but only by pushdown automata, the latter were used for implementing the "core" of grammars. Early generative grammars included a core component called Phrase Structure Grammar (PSG) which is equivalent to a pushdown automaton [4, 5]. More recent grammars include components which show close similarities to a PSG, e.g., the X-bar theory component of the Government and Binding Theory [7]. These systems of rules or principles are hard to translate into the structure and function of brain-like devices. They are even not brain-compatible, since they are based on the assumption of a store of unlimited capacity which our brain obviously lacks. It seems worthwhile to develop a brain-compatible model which allows for the generation of mirror image strings and, therefore, for a solution of both center-embedding (sub-)problems.

At first glance, humans do not have either switches or matchboxes in their heads. Nevertheless, their brains are capable of producing center-embedded sentences. Which neuroanatomical structures and neurophysiological processes are the basis of that capacity?

3. THE FIRST QUESTION AND ITS BRAIN-THEORETICAL ANSWER

One of the many theoretical approaches to brain function postulates that elements like objects, words, and concepts are represented in the brain by means of strongly interconnected neuron ensembles, so-called *cell assemblies* [8]. We define a cell assembly as a set of strongly connected cortical neurons; if the assembly is cut into two parts there will be reciprocal connections between them [9]. Assemblies can have two important functional characteristics: first, if an assembly is activated in part, its internal excitatory connections may lead to an activation of all assembly neurons, i.e., to an *ignition* of the assembly; second, if a certain amount of activity is present in the assembly, it may circulate within the assembly so that the activity is retained, i.e. the assembly *holds*. These functional characteristics of cell assemblies have been postulated on the basis of brain-theoretical considerations [9, 10]. Meanwhile, some empirical evidence for these hypotheses is available [11]. It is assumed here that particular assemblies have either the capacity to ignite or the capacity to hold. So *ignitable* and *holding* assemblies are distinguished.

It is important to note that the presence of several ignitable cell assemblies in a network requires a control device that prevents several assemblies from igniting at the same time. Otherwise, the network containing the assemblies would overshoot in its activity, and an epileptic condition would result. Therefore, a regulation device, a so-called "threshold control mechanism", was postulated which allows only one assembly to ignite at a given time [9, 10]. Only the assembly which receives the largest amount of activity ignites; the regulation device suppresses competing assemblies. The device could be based on inhibitory connections between cell assemblies, possibly on inhibitory neurons of the cortex. The loop of connections from the cortex to the basal ganglia and back to the cortex is also a plausible substrate for such a regulation device, since the striatal network shows strong lateral inhibition. If every cortical cell assembly is connected to few striatal neurons, the lateral inhibition in the striatum could guarantee that only one assembly ignites at a given time and the other ones get inhibited [12].

Assemblies that hold can be activated by excitatory input to the assembly and can be "switched off" by inhibitory input. So they represent bistable elements, like a flip-flop or a light switch. It has been assumed that cell assemblies play a crucial role in language processing [13, 14]. If this is the case, holding assemblies would be exactly the elements needed in order to solve the first problem of center-embedding: a holding assembly could store the information about the missing part of an incomplete sentence until the embedded clause has been finished.

We may assume that the neurobiological counterparts of words are Hebbian cell assemblies. Several word assemblies could form higher-order assemblies that correspond to larger parts of sentences. Assume, for simplicity, that the brain correlate of the sentence "The man is a butcher" consists of two cell assemblies, one for its initial part ("the man" = A) and one for its final part ("is a butcher" = A'). The assemblies must be connected in such a way that the ignition of the first assembly, call it α, is followed by the ignition of the second assembly, α'. If a subordinate clause B is inserted in between A and A', a bistable element is needed in order to store the information about A' while B is uttered. This is possible if α is connected to a holding assembly α^* which, in turn, feeds into α'. If the string AA' is produced, the ignition of α activates α^* which finally ignites α'. After A, a string B can interfere if the corresponding assembly, ß, gets active before α'.

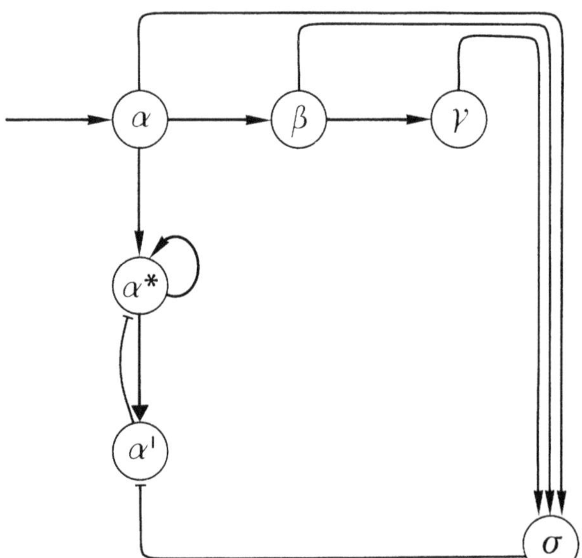

Figure 1. A device that generates the sequence $\alpha ß \gamma \alpha'$ of assembly ignitions

Figure 1 illustrates a cell assembly machine that generates a string in which the complement A' of the initial A occurs only after a string BC, so that the corresponding assemblies ignite in the order $\alpha ß \gamma \alpha'$. The labelled circles stand for cell assemblies. Arrows represent excitatory connections and lines ending in a T shape represent inhibitory connections. In contrast to all other assemblies in the network, α^* holds after stimulation (notice the self-stimulating loop). All other assemblies ignite after stimulation and get inactive thereafter. Suppose that at an initial time step, α receives activity. In the following time step, α ignites and stimulates both ß and α^*. Consequently, in the third time step ß ignites and α^* is also active. This leads to a stimulation of γ and α' at time step four. However, these assemblies do not ignite at the same time, since α had also activated σ (at time step 3) which, therefore, provides α' with inhibitory input at time step four. Thus, only γ can ignite. After α, ß, and γ had been active, also the inhibitory σ will go back to its inactive state, so that α' which still receives stimulation from the holding assembly α^* ignites. This happens at time step six. Finally, α^* is switched off by inhibitory input from α'. Thus, the assemblies $\alpha ß \gamma \alpha'$ ignite in the desired order.

A machine like this can lead to a sequence of ignitions in which an "embedded" sequence of arbitrary length is preceded by the ignition of α and followed by the ignition of α'. If the assembly γ would activate further assemblies that stimulate σ, these would ignite before α' as well. Suppose that before the onset of speech production a "selection process" takes place first and determines which assemblies can ignite. In this case, the connections between the assemblies could determine in which order the ignitions take place. (Note that assemblies may ignite more than once if selected.) The connections from α to α^* to α' and the bistable element α^* guarantee that α' will ignite if α has already ignited - regardless of how many assemblies have ignited in between. If α is not selected (and α' receives no input) neither element ignites. Thus, a cell assembly machine can generate strings like ABC...A' or BC..., where BC... can be of arbitrary length.

This answers the first center-embedding question only in a, to some degree, metaphorical way. Some elements of the network of Figure 1 still need to be related to brain structures and mechanisms. The circles labelled α, ß, γ, and α' stand for cortical cell assemblies that can ignite. α^* is a cortical assembly that holds if activated. In contrast, σ stands for an inhibition mechanism that allows only one assembly to ignite at each time step. It is more realistic to assume inhibition between all ignitable assemblies which may be based on cortico-cortical or striatal connections. In this case, all ignitable assemblies would inhibit each other (through inhibitory interneurons). The diagram can be changed easily in this way. In this case it would be necessary to establish stronger connections between α, ß, and γ. Such a more realistic device (which would be equivalent to the one in Figure 1) has the disadvantage of requiring a large number of connections. So Figure 1 is preferred here for clarity.

4. NEUROPHYSIOLOGICAL EVIDENCE FOR MEMORY MECHANISMS

The crucial element in Figure 1 is the holding assembly α^* which stays active once stimulated. Is it realistic to assume such bistable elements? Is there neurophysiological evidence that supports this assumption?

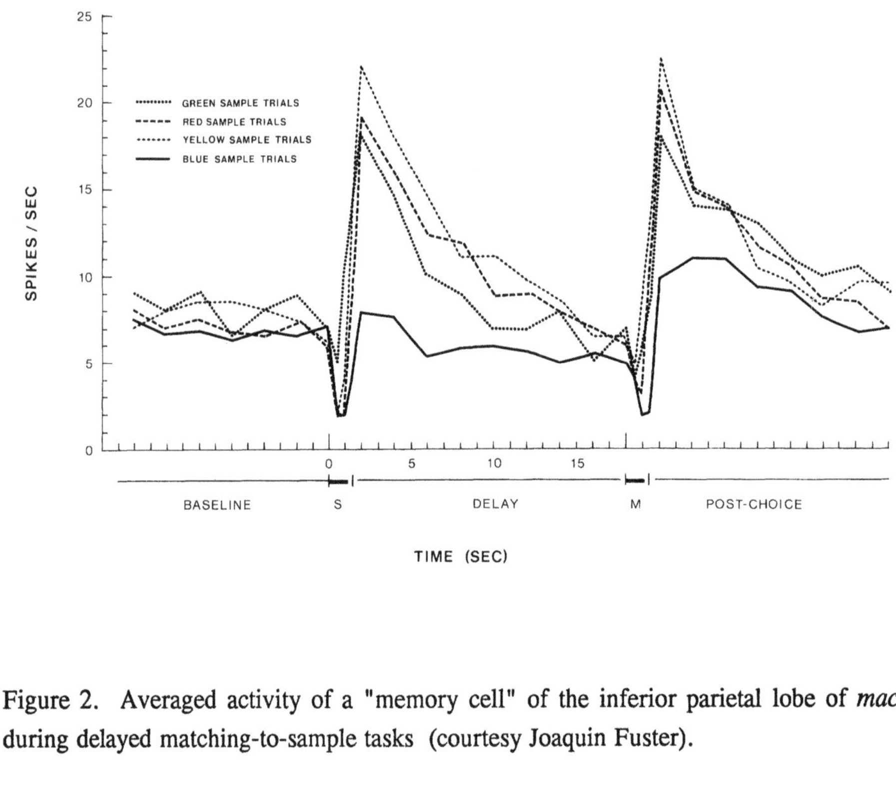

Figure 2. Averaged activity of a "memory cell" of the inferior parietal lobe of *macaca* during delayed matching-to-sample tasks (courtesy Joaquin Fuster).

In higher mammals, in particular in monkeys, so-called "memory cells" have been detected that can stay active for some time [15]. In "delayed matching-to-sample" tasks, the monkey is presented with a sample stimulus (S) and has to select it later from a matching stimulus (M) that includes the sample. The matching stimulus is presented after a delay of several seconds. During that delay, the animal has to keep the sample in its memory. Neurophysiological recordings from individual neurons show that with presentation of S particular cells increase their firing rate and stay in an enhanced activity mode until M occurs and the animal makes its selection. It could be argued that the increased firing rate is due to some unspecific arousal mechanism or that it is related to the fact that the animal is preparing for an action. This, however, is unlikely, since many memory cells respond to rather specific stimuli. For example, there are cells that prefer stimuli of a certain color or shape. Their activity does not highly correlate with an arousal state or with a state of readiness to perform a certain action, but with features of the stimulus the animal has to store in its mind and brain. A possible explanation is that a memory neuron is part of a holding cell assembly in which activity circulates. The findings suggest that the monkey's brain includes cell assemblies that are, to some degree, similar to a light switch. These assemblies may represent objects and features of objects.

However, these neurons are not really bistable. In Figure 2, the activity (counted in spikes per second) of one particular "memory cell" in a delayed matching-to-sample task is plotted against time. This cell in the inferior temporal lobe responds best to green stimuli, less to red and yellow stimuli, and stays on its resting level after blue stimuli. This differential activity shows up in the delay period between the sample and the matching stimulus. Most importantly, the memory cell loses activity after activation. So it does not stay on the active level, like a light switch, but it goes back slowly to its resting level. This decrease of activity is perhaps approximated best by an exponential function.

At first glance, the decrease of activity with time is an unwanted property of memory cells. After a certain time, the activity level of the cell will have reached the resting level. If the cell and the cell assembly it is part of have gone back to their resting levels the encoded information is lost. Thus, the decrease of activity suggests some limitations of the brain internal storage mechanism.

The reported neurophysiological data make it likely that cell assemblies that slowly lose activity can store information for a finite period of time. If such assemblies are not only present in the monkey's brain but also in that of humans, they can be part of the machinery underlying syntax. In this case, they may play the role of α^* in Figure 2. However, the neurophysiological evidence also indicates that it is not realistic to assume a memory mechanism which has two stable states and allows storage for an unlimited period of time. If decreasing activity of cell assemblies underlies center-embedding, the embedded string can not exceed a certain length. If the assemblies of our brain are similar to those of the monkey's brain, information could only be stored for some 20 seconds. Given a mean syllable duration of 200 msec. this predicts that a center-embedded construction with an embedded subordinate clause of more than 100 syllables can not be either produced or understood. Clearly, this is an empirical question which has to be addressed in on-line psycholinguistic experiments rather than by asking native speakers to judge the grammaticalness of strings with center-embedding.

In summary, a modified version of the first question of center-embedding can be answered as follows. A cell assembly that holds after activation and loses activity with time may guarantee that after a unit A occurring at the beginning of a finite string a unit A' will occur at its end.

5. THE SECOND QUESTION AND ITS ANSWER IN TERMS OF ASSEMBLIES

The production of mirror image strings seems to require a matchbox-like mechanism. Given that cell assemblies that constantly lose activity are crucial for the brain-internal syntax machine, it is easy to see how a pushdown mechanism based on such assemblies could work. Assume that there are several assemblies that lose activity in an almost exponential way. Assume further that all of these reach the same activity level after stimulation and their activity decreases in the same way. After the assemblies have been stimulated in a particular sequential order, all of them will later hold different activity levels. The assembly stimulated first has been active for the longest time period. Therefore, its activity has decreased more than that of all other holding assemblies. The assembly stimulated second is on the second lowest activity level etc. The assembly which was activated last is the one with the highest level of activity. Thus, the sequence of assembly stimulations is coded in the hierarchy of activity of holding assemblies. This is a possible, though speculative, brain-internal pushdown store.

Figure 3 illustrates this "pushdown" mechanism. The activity levels of three holding but decreasing assemblies α*, ß* and γ* are shown. These assemblies are activated at t_1, t_2 and t_3. Some time later, at t_4, the last assembly activated is the most strongly active, the second is on the second highest level and the one activated first is on the lowest activity level. An additional mechanism guarantees that at each time step only the most active assembly can lead to an ignition.

Let us, again, consider a network consisting of cell assemblies which can produce a center-embedded string. The diagram in Figure 4 includes the ignitable assemblies α, α', ß, ß', γ and γ' and, furthermore, the holding but decreasing assemblies α*, ß* and γ*. The wiring determines that, after α has been stimulated, ß ignites, followed by γ. In addition, α*, ß* and γ* get stimulated one after the other so that finally a situation is reached similar to the hierarchy in Figure 3 (at t_4). Since γ^* is the most active holding assembly, γ' receives the largest amount of excitatory input. After inhibition by σ has ceased, γ' will ignite first. It is followed by ß' and finally by α'. This requires a "threshold control mechanism" which allows only the most active assembly to ignite at a given time. Note that the ignition sequence of α, ß and γ determines the activity hierarchy of the holding assemblies and, thus, the ignition sequence of the prime-assemblies. If ß ignites before α, also α' must ignite before ß'. Thus, a system of cell assemblies with constantly decreasing activity can lead to mirror image ignition sequences like αß$\gamma\gamma'$ß'α' or ß$\alpha\gamma\gamma'\alpha'$ß'. As discussed above, production of center-embedded sentences of the form ABCC'B'A', like "[The man [whom the girl [that the dog belongs to] met] is a butcher]", can be considered to be analogous to such mirror image activation sequences.

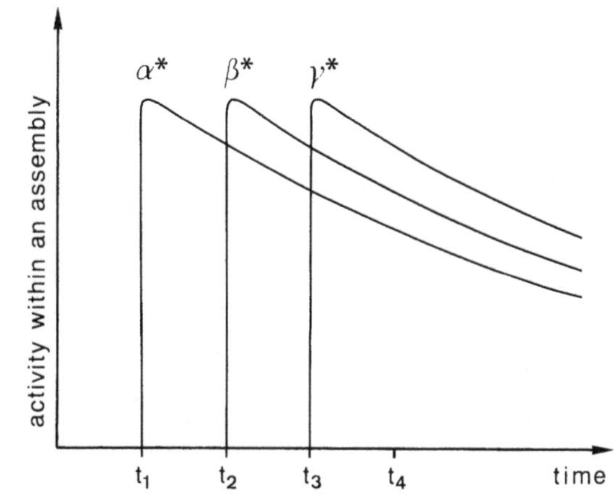

Figure 3. Activity levels of three cell assemblies that have been stimulated one after the other and constantly lose activity. The information about the activation sequence is stored in the hierarchy of the activity levels.

In order to achieve more neurological plausibility, the inhibitory "assembly" σ could be related to inhibitory cortical or striatal mechanisms (see discussion above). The lateral inhibitory connections between the prime-assemblies and the asterisk-assemblies could also be related to cortico-cortical inhibition. (The latter are optional in this scheme; they are supposed to lead to a larger contrast of activity levels of the asterisk assemblies.) What is not shown in Figure 4 is the mechanism that facilitates ignition of the most active ignitable assembly, respectively, and, therefore, accesses the information in the pushdown store. For example, it can be assumed that strong excitatory connections within prime-assemblies lead to ignition after the inhibitory influence of σ has ceased.

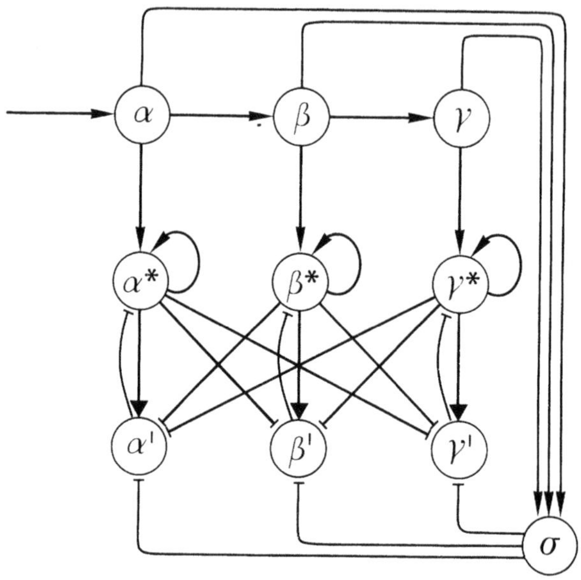

Figure 4. A device that generates the sequence αβγγ'β'α' of assembly ignitions

In summary, the second center-embedding question can be answered as follows. A system of holding assemblies that constantly lose activity represents a pushdown store of finite capacity. If the neuronal counterparts of initial units and their complements are connected through such assemblies, a particular sequence of initial units leads to the inverse sequence of complements. Mirror image strings like ABCC'B'A' or "[The man [whom the girl [that the dog belongs to] met] is a butcher]" can result.

6. SOME PROBLEMS AND PERSPECTIVES OF THE PROPOSAL

It is necessary to address some of the restrictions and problems of the proposal. First, it only accounts for the production of strings, not for the fact that speakers can understand sentences and can even tell a well-formed sentence from an ill-formed sequence of words. We may just assume that the same neuronal machinery underlies production and as well as perception or comprehension. However, dealing with the perception issue in greater detail may make a couple of modifications necessary.

From a linguistic point of view, the proposed approach might seem somewhat unsatisfactory, since the "infiniteness of language" is neglected. However, humans can only understand sentences of limited length and with a small number of center-embeddings. Our brain is definitely not equipped with a store of unlimited capacity. It makes little sense to postulate grammar mechanisms in our mind which can never be housed in our brain.

From a linguistic point of view, there are further and more serious objections to be raised. The proposed models can only produce mirror image strings. They are not capable of capturing further details of sentences with center-embedding. Take, for example, the sentence "[The man [whom the girl [that the dog belongs to] met] is a butcher]". The initial units of the subordinate clauses contain either a wh word (whom) or a complementizer (that), whereas the superordinate sentence lacks such an element. A more sophisticated analysis would also have to account for these differences. In addition, there is a variety of sentences that suggest underlying regularities that can not be captured by a pushdown store (or an equivalent core grammar). This has led linguists to the assumption that, besides the core grammar, a grammar component is necessary that moves the elements of a string which has been generated by the core. How such a shunt component could be realized in the brain has not been addressed here. So the proposal has to be extended in order to capture further syntactic phenomena.

From a neurobiological point of view, the proposed theory makes a couple of very strong assumptions. The cell assemblies that form the pushdown store must lose activity in the same way, i.e., according to the same exponential function. This is a prediction which can, in principle, be tested by electrophysiological studies of neurons of the perisylvian cortex in humans. However, at present the neurophysiological evidence available demonstrates that memory cells in the monkey vary with regard to their activity loss with time (Fuster, p.c.). A pushdown mechanism made up of several assemblies that lose activity in the same way, therefore, would be specific to the human brain. In humans, the postulated exponential decrease of holding assemblies might be related to a very fundamental property of neurons, namely the fact that they show habituation during strong and enduring activity. In contrast, the variable decrease in assemblies of the monkey could be due to different strengths of internal connections in relatively weakly connected assemblies. Habituation could play a role in human cell assemblies if they have very strong internal connections, so that their neurons fire very frequently. This suggests that neurons making up assemblies in humans fire with much higher frequencies than those in non-human monkeys.

The proposal leads to further predictions that are testable in electrophysiological experiments. Given the activity of holding assemblies shows up in slow cortical potentials, the amplitude of these potentials must be related to the number of assemblies activated. Thus, the amplitude of the signal should covary with the degree of center-embedding of stimulus sentences.

It is obvious that the proposal leads to predictions that can be tested in behavioral and physiological experiments. While being still hypothetical, it offers a basis for neuroscientific investigations of complex language phenomena. Furthermore, it opens a neurobiological perspective on the solution of the problem of serial order in behavior.

7. REFERENCES

[1] A. N. Jain and A. H. Waibel, in *Advances in neural information processing systems 2*, D. S. Touretzky (ed.), Morgan Kaufmann, San Mateo, CA, 1990.

[2] G. A. Miller and N. Chomsky, in *Handbook of mathematical psychology. Vol. 2*, R. D. Luce, R. R. Bush and E. Galanter (eds.), Wiley, New York, London, 1963, p. 419.

[3] W. S. McCulloch and W. H. Pitts, *Bulletin Math. Biophys. 5* (1943) 115; S. C. Kleene, in Automata studies, C. E. Shannon and J. McCarthy (eds.), Princeton University Press, Princeton, NJ, 1956, p. 3.

[4] N. Chomsky, in *Handbook of mathematical psychology. Vol. 2*, R. D. Luce, R. R. Bush and E. Galanter (eds.), Wiley, New York, London, 1963, p. 323.

[5] W.J.M. Levelt, *Formal grammars in linguistics and psycholinguistics. Vol. 1: an introduction to the theory of formal languages and automata*, Mouton, The Hague, 1974; A. V. Abo, J. E. Hopcroft and J. D. Ullman, *Data structure and algorithms*, Addison-Wesley, Reading, MA, 1987.

[6] E. Bach, C. Brown and W. Marslen-Wilson, *Language and Cognitive Processes 1* (1986) 249.

[7] N. Chomsky, *Lectures on government and binding*, Foris, Dordrecht, 1981; N. Chomsky, *Barriers*, MIT Press, Cambridge, MA, 1986.

[8] D. O. Hebb, *The organization of behavior. A neuropsychological theory*, John Wiley, New York, 1949.

[9] V. Braitenberg, in *Theoretical approaches to complex systems. (Lecture notes in biomathematics, vol. 21)*, R. Heim and G. Palm (eds.), Springer, Berlin, 1978, p. 171.

[10] G. Palm, *Neural assemblies*, Springer, Berlin, 1982.

[11] G. Palm, in *Real brains, artificial minds*, J. L. Casti and A. Karlqvist (eds.), North-Holland, New York, 1987, p. 165.; G. Palm, *Concepts Neurosci. 1* (1990) 133.

[12] R. Miller and J. R. Wickens, *Concepts Neurosci. 2* (1991) 65.

[13] V. Braitenberg, in *L'accostamento interdisciplinare allo studio del linguaggio*, G. Braga, V. Braitenberg, C. Cipolli, E. Coseriu, S. Crespi-Reghizzi et al. (eds.), Franco Angeli Editore, Milano, 1980, p. 96.; V. Braitenberg and F. Pulvermüller, *Naturwiss. 79* (1992) 103; V. Braitenberg and A. Schüz, in *Language origin: a multidisciplinary approach*, edited by J. Wind, B. Chiarelli, B.H. Bichakjian, A. Nocentini and A. Jonker (eds.), Kluwer, Dordrecht, 1992, p. 89.

[14] F. Pulvermüller, *Concepts Neurosci. 3* (1992) 157; F. Pulvermüller and H. Preißl, *Network* 2 (1991) 455.

[15] J. M. Fuster, *The prefrontal cortex: anatomy, physiology, and neuropsychology of the frontal lobe*, Raven Press, New York, 1989; J. M. Fuster, *J. Neurophysiol. 64* (1990) 681.

Part Three
Space-Time Dynamics of Neuronal Activity in the Working Brain

Brain Theory / A. Aertsen (Editor)

Integration, Synchronicity and Periodicity

M. Abeles[a], Y. Prut[a], H. Bergman[a], E. Vaadia[a] and A. Aertsen[b]

[a]Department of Physiology and the Center of Neural Computation, The Hebrew University of Jerusalem, School of Medicine, P.O. Box 1172, Jerusalem 91010, Israel

[b]Institut für Neuroinformatik, Ruhr-Universität, P.O. Box 10 21 48, 4630 Bochum, Germany

Introduction

The basic properties of nerve cells and connections between nerve cells are such that very precise timing of firing or delays does not seem to be a principal objective. In spite of this, men and animals can perform with amazing time accuracy. A person listening to a sound delivered to both ears can detect when the sound in one ear is delayed by as little as 10μs relative to the other [Moore, 1982]. When throwing a small rock at a 20cm target from distances of more then 7m, the launch window (accuracy of release time) is less then 1 ms [Calvin, 1983]. Some of the mechanisms for achieving such high time accuracy are well understood, while others are still enigmatic.

This contribution deals with ways by which a neuron, acting as a leaky integrator in a noisy environment can achieve a high degree of time accuracy. It reviews the general concepts of timing, synchrony, and oscillations in the mammalian brain with a special emphasis on cortical networks. Special concern is given to theoretical considerations and experimental evidence associated with synchronized activity in cell assemblies, consisting of sets of neurons connected with multiple diverging and converging links.

1. TIMING BY EXTERNAL EVENTS

Assessing of brain responses is usually done by measuring the neural response around an external event. This may be a stimulus, in which case one speaks of *evoked activity*, or it may be a motor action, in which case one speaks of *event related activity*. Typically one measures a number of realizations of the same process and then averages the resulting activity. When the brain activity is evaluated by measuring a potential (from the scalp or from within the tissue) an average response is computed, when activity of a single unit is measured the average firing rate is computed by constructing a histogram (PETH - Peri Event Time Histogram[Gerstein and Kiang, 1960]).

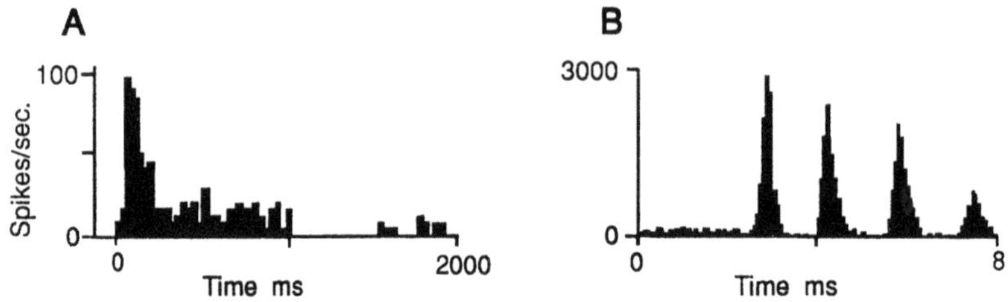

Figure 1. Comparison of response timing in the Visual and Auditory systems.
A. The response of an Y-type ganglion cell in the retina of the cat. A grating at the optimal position, was flashed on, at time 0. Adapted from Hochstein and Shapley 1976, with permission.
B. Response of an Auditory nerve fiber in the cat to a 0.1 ms click. Note that since the first spike time is concentrated within less than 1 ms, its peak instantaneous firing rate is higher than 1000/s. Adapted from Kiang et al. 1962 with permission. Note the drastically different time scales.

Different sensory systems may show very different types of responses. In the visual system, for example, the processes at the transduction level are quite sluggish and therefore the neural responses develop slowly and after a considerable delay. In the auditory system the peripheral delay is very brief and the spike timing is very accurate. Compare the response of a visual neuron shown in Fig. 1A to the response of an auditory nerve fiber shown in Fig. 1B.

Timing of activity plays a major role in localizing sound sources in the auditory space. Sound from a source placed directly in the front of the face at a distance of 1m arrives simultaneously at both ears. A movement of 1 deg. to the right will result in the sound to the right ear leading the sound to the left ear by 12μs. However, comparison between the two ears is not based directly on the activity in the acoustic nerve axons. It happens only at a later processing stage (to which the activity arrives after being relayed through at least one synapse). Accurate comparison needs accurate transmission times and these are assured by an appropriate anatomy.

Fig. 2 illustrates a drawing of the synaptic connection between an axon delivering information from the cochlear nucleus onto the cell body of a neuron in the trapezoid nucleus. The axon thickens as it reaches its target, branches and embraces the cell body. A thin section through this region shows the axon forming a "cup" around the cell body. This structure is named after its discoverer the "calyx of Held". A small piece of such a section may look like Fig. 2C. The synaptic contacts formed by the incoming axon (marked S1) have round synaptic vesicles and a-symmetric membrane thickening. Such a-symmetric synapses are generally excitatory. (In contrast to the symmetric thickening and the pleomorphic vesicle of the synapses marked S2 which are usually inhibitory.) The

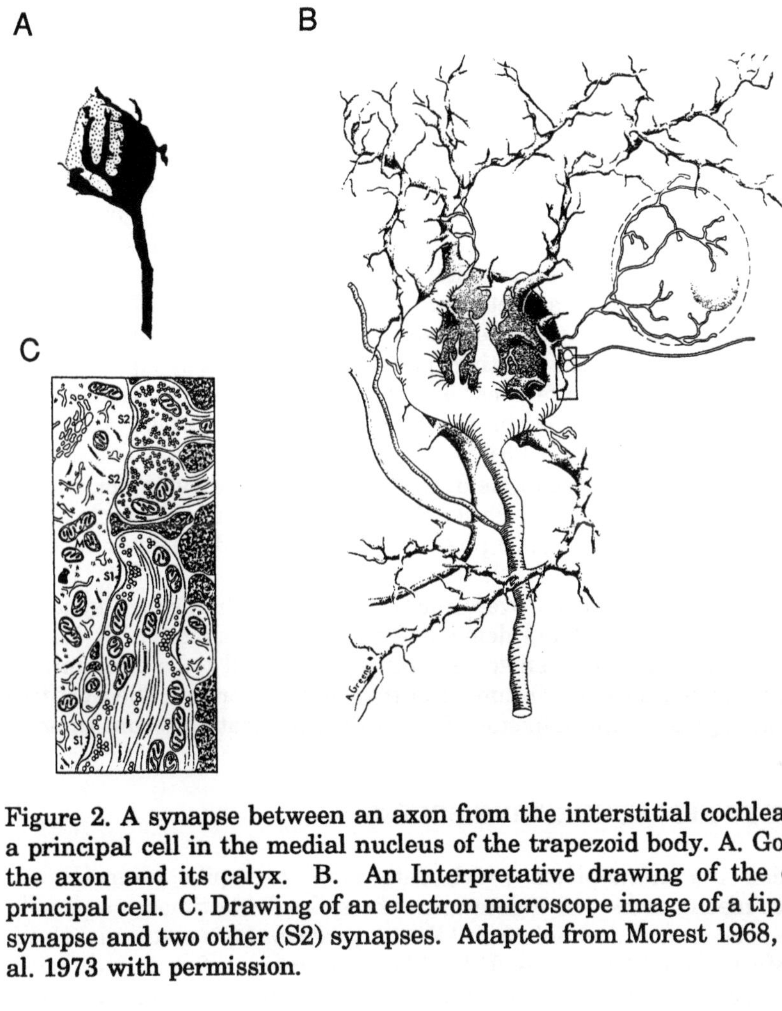

Figure 2. A synapse between an axon from the interstitial cochlear nucleus and a principal cell in the medial nucleus of the trapezoid body. A. Golgi staining of the axon and its calyx. B. An Interpretative drawing of the calyx and the principal cell. C. Drawing of an electron microscope image of a tip of the calyctic synapse and two other (S2) synapses. Adapted from Morest 1968, and Morest et al. 1973 with permission.

calyx contains hundreds of synaptic sites which ensure very reliable transmission. Note that before the information coming from the two ears has any chance to interact in the superior olivary complex it must be transferred through one or two such synapses. The arrival timing of the impulses to the superior olive is critical for producing the acute sensitivity to interaural delays. The demand for precise transmission timing in the auditory system is fixed throughout life and persisted throughout the development of the vertebrate animals. Nature provided the auditory nervous system with the a device that assures reliable and time-wise accurate transmission of impulses from the ear to the superior olive.

In the general case, if activity has to pass through N synaptic stations each producing a delay of D with a variance of σ^2 the total delay will be:

$$N{\cdot}D \pm\sigma\sqrt{N}$$

Typical synaptic delays (D) contain the conduction time along the axon, the synaptic delay (0.5-1. ms), and the delay from the start of the EPSP until the spike itself is generated (1.-3. ms). Among these events, conduction times along axons and synaptic release have very small jitter in the range of few tens of µs[Lass and Abeles, 1975; Cope and Mendell, 1982]. Thus most of the variability is expected to come from the delay between the beginning of the EPSP and the generation of a spike. This delay can be understood by considering the following situation: The EPSP rides on the "resting" potential of the post-synaptic cell. It shows gradual buildup lasting a few ms. Due to the random fluctuations of the "resting" membrane potential it may hit threshold at various time delays after its initiation. If the rise time of the EPSP is very short, and the amplitude is very big , as is the case for the calices of Held, then the synaptic delay jitter will be very small. However for smaller and slower EPSP's, as prevails in the cortex, this jitter is appreciable. [See Abeles, 1991 pp241-249, for a more detailed analysis of jitter in transmission timing.]

While nature found simple ways to overcome the unreliable character of synaptic transmission when there is a fixed demand for secure transmission, it had to resort to more intricate tricks when secure transmission had to be dynamically turned on and off in the noisy environment of the cerebral cortex. The next two chapters deal with possible mechanisms of generating accurate internal timing.

2. TIMING BETWEEN SUCCESSIVE SPIKES

In this section the mechanisms which determine interspike intervals are discussed. Two mechanisms which determine the spike firing times are usually described: The *pace-maker* mechanism and the *integrate and fire* mechanism [see, for instance, MacGregor, R.J., Lewis, E.R., 1977; and many references therein]. In the pace-maker, internal ionic processes generate periodic oscillations of membrane potentials, while in the integrate and fire neuron all synaptic currents add up and are integrated by the leaky membrane capacitor. When the membrane potential which is generated by these processes hits the threshold, the neuron fires and the membrane potential is reset to some initial value.

2.1. Pace Makers

The usual mechanism of action-potential generation is meta-stable. It contains one or more positive feed-back mechanisms and several negative feed-back mechanisms. Its behavior depends on the various gains and time constants of these mechanisms. As most of the processes are non linear, it is very easy to switch from the stable, resting behavior, to an unstable, periodic behavior and back.

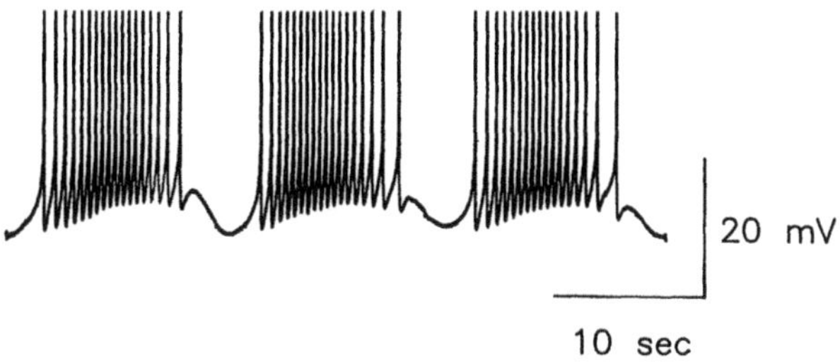

Figure 3. Pace-maker activity recorded from an unidentified neuron in the abdominal ganglion of Aplysia californica. The top of the action-potentials are clipped. Courtesy of Ruth Oren.

During the action potential the sodium conductivity is very high and the membrane potential rises very fast towards the sodium equilibrium potential. However, after a very short delay (much less then 1 ms in most mammalian neurons) the sodium conductivity starts to fall back and the potassium conductivity rises, which pulls the membrane potential back towards its resting value. This repolarizing stage of the action potential is also fast (approximately 1 ms). Its speed is assured by the high levels of potassium conductivity which then recovers slowly towards its resting level. The time course of this potassium conductivity causes the membrane potential to hyperpolarize (become more negative as compared to the resting level) immediately after the action potential and then slowly depolarize towards a steady state level. If the firing threshold of the neuron is below this steady state level, the neuron will fire again and, hence, would never reach a stable resting membrane potential. In addition, activation of calcium conductance by depolarization and delayed potassium conductances can cause oscillations of membrane potentials in many neurons. These are the classical *pacemaker* mechanisms. Fig. 3 shows periodic pacemaker activity in which several such mechanisms produce periodic bursts of action potentials.

2.2. The Motor Neuron of the Spinal Cord.

The motor neuron of the spinal cord is the classical example of an *integrate and fire* neuron. Its properties were described by the pioneering work of Eccles [1957] and served ever since as the model for the mammalian neuron. In the living animal this neuron is usually firing periodically, albeit with variable firing rates.

Periodic firing can be demonstrated in most neurons. Typically they behave like pace makers when depolarized by a constant current. The motor neuron, with its huge dendritic tree collects the input currents generated by about 100,000 synaptic boutons which together generate an approximately constant current. Fig. 4 shows the membrane potential of a motor neuron in the spinal cord of the cat when firing at ten spikes per second. The figure shows a superposition of 5 traces, each composed of two successive spikes with the hyperpolarizing wave that follows them. The actual spikes are clipped, being approximately 100 mV above

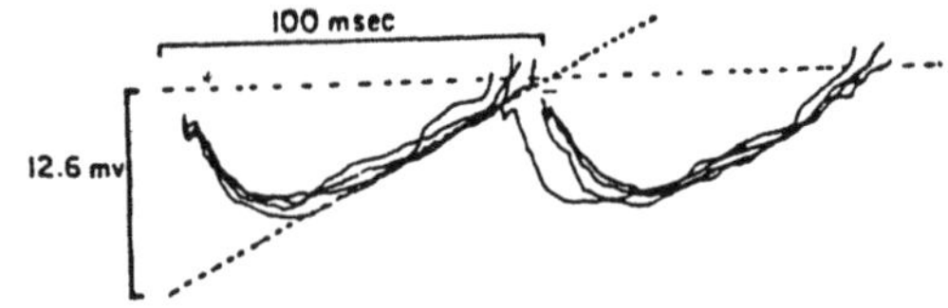

Figure 4. Repetitive firing of a motor neuron in the spinal cord of the cat. Further explanations in text. From Calvin and Stevens 1968, with permission.

threshold. We see in Fig. 4 that after each spike the motor neuron is deeply hyperpolarized, then it recovers gradually. On top of the internal processes leading to recovery, the incoming EPSP's add a (noisy) depolarization; both processes build up until the membrane potential hits threshold again. The whole procedure then starts again, whereby an almost periodic firing is generated. The remaining jitter in the inter-spike intervals is, according to Calvin and Stevens [1968], mainly due to the synaptic noise.

2.3. The Cortical Neuron.

Whereas the motor neuron shows periodic firing in its every day performance, this type of behavior is seen only seldom in cortical neurons. Most of them become periodic only at high firing rates. A major cause for this difference is probably the absence of a deep after hyper-polarization in the cortical neurons. This can be seen in the intracellular records shown in Fig. 5, taken from the work of Douglas et al. [1991] on the visual cortex of the cat.

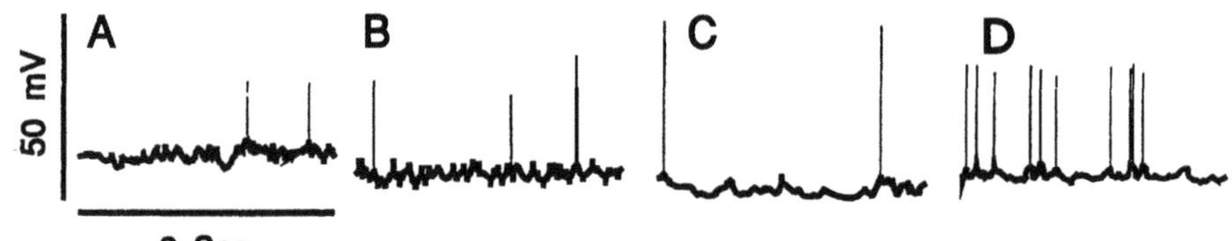

Figure 5. Membrane potentials and spikes in 4 neurons from the visual cortex of the anesthetized cat in the absence of visual stimuli. A. Unknown neuron in layers 2-3, B. Spiny stellate cell in layer 4, C. Pyramidal cell in layer 5, D. Pyramidal cell in layer 6. Adapted from Douglas et al. 1991, with permission.

Not all cortical neurons behave in that way, some do show deep and prolonged hyperpolarization after each spike. Such neurons would reach periodic firing pattern at lower rates. The changes in excitability which follow each spike may be studied by means of the *auto-correlation* function. This function shows how the firing rate of a single neuron varies after it has just fired (this is equivalent to the *renewal density* function of renewal point processes [Cox, 1962]). Fig. 6 illustrates several examples of such auto-correlation functions in cortical neurons.

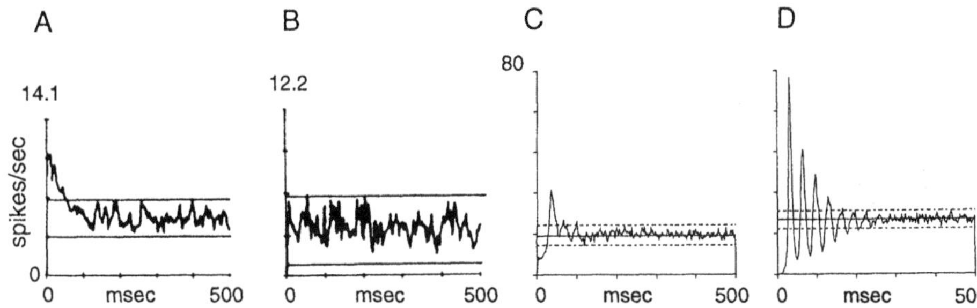

Figure 6. Auto-correlation functions of 3 neurons in the cortex of unanesthetized animals. A. and B. Two neurons from the auditory cortex of the cat, C. and D. from one neuron in somatosensory area II of the monkey. C and D by courtesy of E. Ahissar.

Fig. 6A shows the most common shape of cortical auto-correlation functions - after a very brief (1-5 ms) period of refractoriness the cell is hyper-excitable and then returns to its resting firing level. Fig. 6B shows the next common shape in which after the initial refractoriness the cell returns to its basal Poisson-like firing level. Fig. 6C shows the least frequent type of auto-correlation in which prolonged refractoriness is exhibited. This refractoriness is probably brought about by prolonged hyperpolarization following each spike, similar to what is known for the motor neurons of the spinal cord. While recording the activity of the neuron depicted in Fig. 6C it elevated its firing rate spontaneously (i.e. without any deliberate external stimulus) and started to fire periodically. A more detailed discussion of these neuronal properties can be found in Abeles [1991 ,pp 118-139].

While most neurons in the cortex of the awake animal do not fire periodically, they do show periodicity when placed in a dish. This is demonstrated in Fig. 7 taken from the work of [McCormick et al., 1985]. They categorized 3 main types of cortical cells: The regular spiking (Fig. 7A), the fast spiking (Fig. 7B), and the bursting (Fig. 7C). Regular spiking neurons are usually small pyramidal cells, while fast spiking neurons are probably smooth stellate cells, finally, the bursting neurons, which comprises only 5% of the population, are probably the huge pyramidal cells of layer V. Note that the three types of cells initially respond to a steady depolarizing current in a variety of ways, after which they settle down to fire periodically as if they were pace makers.

Why then does one not observe this periodic firing in cortical neurons in vivo?

The reason is not clear. There are proposals of special intracellular mechanisms which prevent periodic firing [e.g. Softky and Koch, 1993], however it is probably not necessary to invent any new mechanisms to explain this property. If the membrane potential is driven by strong noisy synaptic processes, then the cell would fire only when the random membrane fluctuations happen to hit threshold. Elevated firing rates would mean increased probability of hitting threshold either by reduction of the distance between the average membrane potential and the threshold or by increasing the variance of the membrane fluctuations. This situation holds as long as the membrane potential looks like a noisy wave around a steady mean level. That is, as long as the after hyperpolarization is short relative to the mean interspike intervals (which doesn't seem implausible in view of both the small hyperpolarization in cortical neurons and their typically low firing rates), the spike events will tend to occur stochastically, rather than in a periodic fashion.

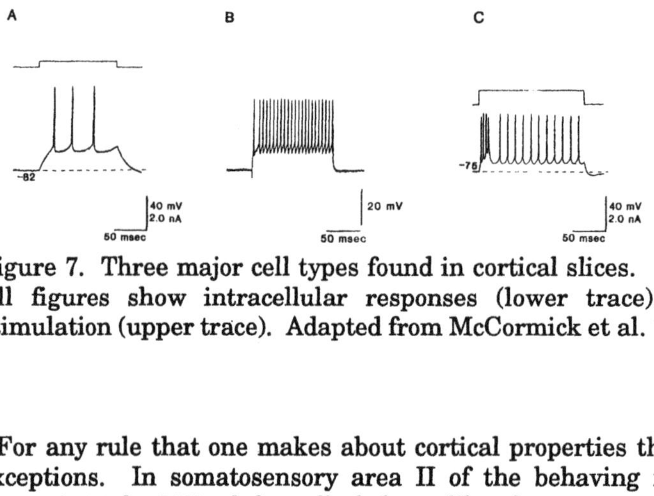

Figure 7. Three major cell types found in cortical slices.
All figures show intracellular responses (lower trace) to constant current stimulation (upper trace). Adapted from McCormick et al. 1985, with permission.

For any rule that one makes about cortical properties there are many notable exceptions. In somatosensory area II of the behaving monkey, for instance, approximately 30% of the cells behave like the motor neurons [Ahissar et al., 1990]. However the same group of researchers, using the same recording techniques, in the same animals (behaving monkeys) did not find this kind of oscillations in auditory cortex, pre-motor cortex, pre-frontal cortex, or posterior-parietal cortex. There are several more reports on periodic oscillations of cortical neurons that are probably due to network behavior (and not internal properties of single neurons) which will be discussed below.

In summary, almost any neuron would show periodic (pace maker like) firing when appropriately hyper/de-polarized. This is true also for most cortical neurons. However in their natural cortical setting this property is usually not expressed, so that by themselves they would reach periodicity only at high firing rates.

2.4. Oscillations in Networks

Periodic oscillations of cortical activity are known since Berger [1929] recorded the *alpha* EEG waves (8-12 Hz) from the scalp. Attention was brought to the relevance of *gamma* (30-70 Hz) waves to processing of information in the brain several times in the past [e.g. Sheer, 1989]. These higher frequency oscillations were brought to focus recently when two groups [Eckhorn et al., 1988; Gray and Singer, 1989] independently found that the local field potential (LFP) recorded by a metal micro-electrode in the visual cortex of anesthetized cats shows strong oscillations in the γ range when the appropriate stimulus is shown. Further support of the physiological relevance of these oscillations was provided by the observation that such oscillations in two regions may become phase locked when stimuli pass through the receptive fields of both regions in a coherent manner, much like the movement of a real object in visual space would bring about [see also Engel et al., 1992 for a review of these experiments]. Similar oscillations were recently observed in the motor cortex of behaving monkeys [Murthy and Fetz, 1992].

These oscillations are certainly not a result of the intrinsic properties of individual neurons but, instead, depend on the neural network of the brain. Indeed, it has been demonstrated that neural network may easily achieve stable oscillations. For instance, Sompolinsky et al. [1988] have shown that even in a random network of very simple neurons one can observe persistent oscillations. In their network the synaptic effects last only one ms, at each step the neuron adds all its inputs and then fires (or not) according to a sigmoid threshold curve $tanh(gv)$ where v is the difference between the membrane potential and the threshold, and g is an arbitrary constant. They were able to show analytically that the network's behavior depends on the variability of the synaptic strengths. When the variance of the synaptic strength (σ^2) was small such that $g\sigma<1$, the network exposed bistable behavior - either all the neurons are almost quiet, or all of them fire at a high rate. When there is a great dispersion of the synaptic values, that is $g\sigma>>1$, the network exhibits chaotic behavior. For intermediate values, that is $1<g\sigma<2$, the network oscillates starting with simple harmonic oscillations when $g\sigma$ is slightly above 1, and with more and more complex oscillations as $g\sigma$ grows. Moreover, in another study along similar lines [Erb and Aertsen, (1992)], it was observed that, again depending on the connectivity statistics, the initially coherent network activity may break up into separate subgroups of neurons ("assemblies"), with the firing of the various neurons exhibiting different and more or less independent correlation dynamics, characterized by the different time constants involved.

Thus, even a random network of very simple neurons with no intrinsic oscillatory tendency whatsoever can exhibit stable periodic oscillations. When the behavior of a network made of neurons that do exhibit a tendency to fire periodically (as was described in section 2.2) is analyzed, even richer behavior can be shown. Hanzel and Sompolinsky [1992] studied the behavior of a network of coupled oscillatory neurons, they were able to show analytically four regimes. When the coupling among the neurons is weak each of them oscillates at its internal frequency. As the coupling strength increases, the whole network starts to oscillate at the same frequency in a phase locked manner. With further increase of coupling strength the behavior becomes chaotic, however the different neurons are not uncorrelated, as exhibited by the fact that the average activity of the net shows chaotic oscillations which do not vanish as the number of neurons becomes very large. When the coupling strength becomes even bigger, each neuron still shows chaotic behavior, but the neurons become uncorrelated.

Hence, the local cortical network has all the ingredients required for exhibiting such complex cooperative behaviors, however usually it doesn't. When a piece of cortical tissue (of approximately 1 by 1 cm) is isolated from all other brain structures, the activity within the piece stops altogether [Burns, 1958; Burns and Webb, 1979]. The same holds true for a cortical slice in a physiological salts solution. Thus the cortical activity depends on incoming excitation derived from other (cortical as well as subcortical) brain regions. The source of 8-12 Hz α activity is generally attributed to the thalamus, although the amplitude of the thalamic α activity is greatly reduced after cutting the cortico-thalamic connections [Morison and Dempsy, 1943]. The source (thalamic or cortical) of the 30-70 Hz γ activity is still the subject of debate.

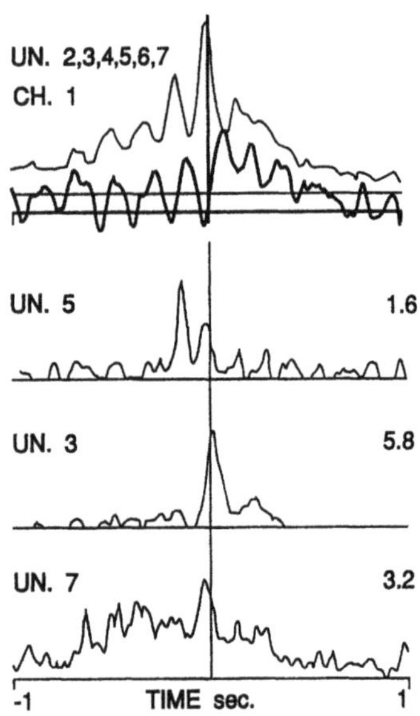

Figure 8. Cross-correlations and Local Field Potentials (LFP) in the cat. Lower trace, the cross-correlation between unit 7 and unit 1. Next trace the cross-correlation between unit 3 and unit 1. Next trace, the cross-correlation between unit 5 and unit 1. Top trace, cross-correlation between the superposed activity of units 2,3,4,5,6,7 and unit 1, and the spike triggered average of the LFP around the firing times of unit 1 (heavy trace). Adapted from Arieli 1992, with permission.

Whatever the source of the oscillatory EEG waves, they certainly affect the firing times of the cortical neurons. During periods of drowsiness or shallow sleep when the EEG exhibits spindles (wave-packets) of α activity, the cortical neurons tend to synchronize their activity with the negative peak of the α wave. This can be seen in Fig. 8, taken from the work of Arieli [1992], in which both LFP and single unit activity were measured through several microelectrodes inserted into the supra-sylvian cortex of an unanesthetized cat. The three lower traces show crosscorrelations between unit 1 and 3 other units. The thick trace at the top shows the averaged LFP which was recorded from electrode no. 1 and averaged around the firing times of unit no. 1, recorded through the same electrode. The amplitude of the averaged LFP was of the same order of magnitude as the ongoing (non averaged) α waves in the LFP, suggesting that a considerable fraction of the spikes of unit no. 1 occur around the negative peak of that wave. The top trace in Fig. 8 shows the cross-correlation between the spike trains of unit no. 1 and a composite spike train generated by the superposition of the spike trains of six other single units recorded within 0.6 mm from unit no. 1. Clearly, unit no. 1 is firing in a coherent way with the LFP, and so do the other units.

A rough estimate of the time window in which cortical cells fire during α activity and, hence, a measure of the quality of phase locking of individual spikes to the ongoing field potential, is a quarter of a cycle, i.e. approximately 25 ms during 10 Hz oscillation. If a similar coherence exists also for the γ oscillations, then a timing accuracy of 5 ms can be expected. However, in some cases one observes much more tightly coupled firing times. These instances will be the subject of section 3.

3. TIMING of ISOLATED SPIKES

The picture drawn in the last section is of a cortical neuron with fluctuating membrane potential around a steady level which is well below the threshold. From time to time the fluctuations reach threshold and the neuron fires. This situation is idealized in Fig. 9, [Abeles, 1982a] which shows a waveform generated by post synaptic potentials arriving at random (Poissonian) times at a rate of

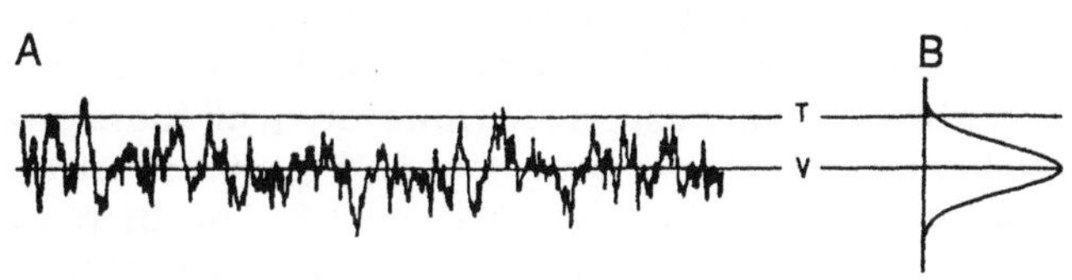

Figure 9. The concept of the noisy membrane potential and a fixed threshold.
A. Noisy potential, which was generated by adding positive and negative "post synaptic potentials" arriving at random (Poissonian) times at a rate of 100,000/s.
B. The probability of the potential to be above threshold (T) is equal to the area under the curve above T.

100,000 per s. The probability density function (PDF) of the membrane potential amplitude is drawn at a right angle on the right of Fig. 9. If the threshold is represented by the line marked *T*, then the area of the PDF curve above *T* is the probability to observe a membrane potential above threshold. For low values, the firing rate is approximately proportional to this probability. According to this approach, the exact time of appearance of an action potential is random, and the spike train generated by a neuron may be roughly treated as an non-homogenous Poisson process, i.e. with a variable firing rate which can be estimated from peri-event-time-histograms, as illustrated in Fig. 1.

3.1. Delayed Synchronization.

There are many reports in the literature describing neurons that fire with very precise timing. Among the first to report such accurate timing were Toyama et al. [1981], who recorded the activity of two neurons within a vertical column in the visual cortex of the anesthetized cat. They reported many cases in which both cells fired with exact (less then 1 ms) synchrony, both spontaneously and while a visual stimulus was presented to the cat. Allum et al. [1982] reported such tight synchronization in the motor cortex of behaving monkeys. Very recently, Nelson et al. [1992] reported tight synchronization between neurons in area 17 and area 18 of the cat. DeRibaupierre and Abeles observed occasionally tight synchronization between two neurons in the primary auditory cortex of the cat [unpublished observations]. Tight synchronization does not seem to be consistent with the random firing model presented above. In some cases (like two neurons within a column of V1) it may be explained by assuming an extremely strong common drive from the thalamus. But even there it is not likely to be a satisfactory explanation since, in order to explain cases where up to 50% of the spikes of the two neurons are synchronized [as in Toyama et al. 1981, Fig. 1D], one would need to assume that the effects of a single thalamic axon on the cortical neuron are strong enough to evoke a post-synaptic firing with a probability of 0.7.

While such tight synchronization between neurons could, at least in principle, be explained along these lines, another type of tight synchronization is much harder to explain: tight timing after a long delay as in Fig. 10.

Fig. 10 shows several cross-correlograms among pairs of cortical neurons in the frontal cortex of behaving monkeys. In all the illustrated cases the firing rates of the neurons are low, so that when a narrow bin (1-2 ms) is used for constructing the correlogram, only few counts per bin are expected. However, in all these cases there is a single bin in which a much higher count is found. Fig. 10A illustrates a case in which the expected count per bin was <0.5 while near 372 ms there was a bin with 7 counts (the probability of obtaining such an event by chance is one in a million). Fig 10B shows the same data in an expanded scale. It reveals that all 7 counts happened within 2 ms. Frostig et al. [1985] reported such events in the forebrain of unanesthetized cats, while Landolt et al. [1985] reported such events in the visual cortex of the cat.

How does one account for such phenomena? In the cortical milieu it is extremely difficult to do so. Long delays must involve transmission through many synapses in tandem. In the cortex there are no synapses of the Calyx of Held type (Fig. 2),

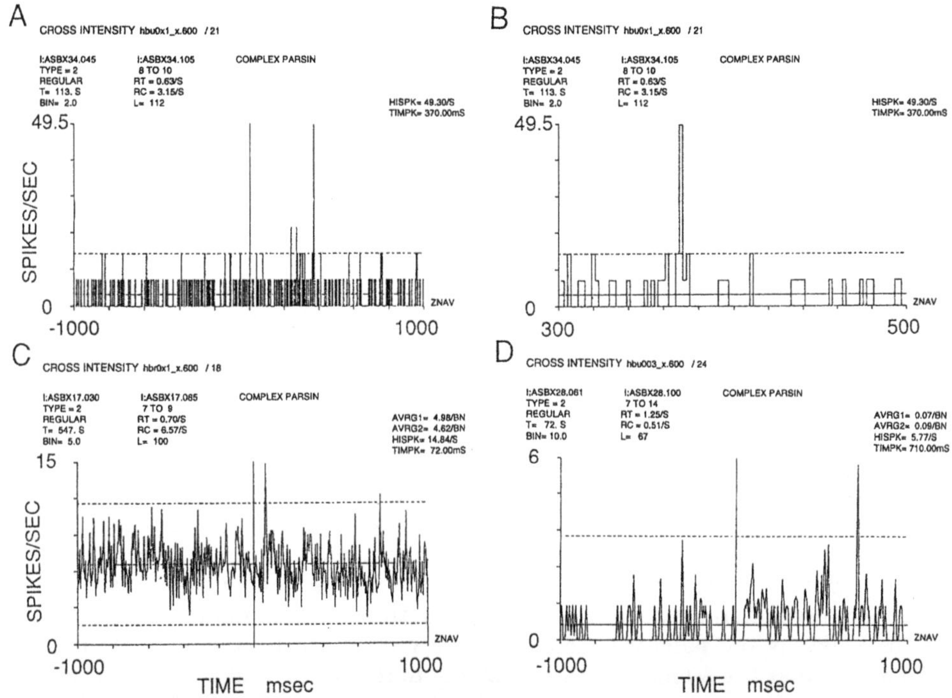

Figure 10. Delayed peaks in cross-correlations between units in frontal cortex of the behaving monkey. A. C. and D. for 3 pairs of units in a compressed time scale, B. same as A. in an expanded time scale.

in fact all the synapses on the somata of the cortical pyramidal cells are inhibitory [Peters and Kara, 1985]. Excitatory synapses are limited to the dendrites and they are all weak [Abeles 1991 chp. 3]. The model of variable Poissonian firing rates cannot support such tight synchronization, since in this model the time constants of the cross-correlogram cannot be shorter than those of the auto-correlograms of the single neurons [Sompolinsky private communication], and one does not observe 1-3 ms wide peaks in the auto-correlograms.

A phenomenon which seems to contradict the known physiology of the cerebral cortex neurons should have attracted a lot of attention. Yet, reports on this type of tight synchronization after a long delay are quite rare. Why is that so? There are probably two main reasons for that dearth. First, when we observe a phenomenon for which we have no explanation, we often discard it as "strange", particularly if it is a rare phenomenon. Second, cross-correlograms of this type are indeed rare. The 7 occurrences of unit no. 10 following unit no. 8 after exactly 372 ms described in Fig. 10B happened during 113 seconds of recording. Had units 8 and 10 had higher firing rates the excess of 7 coincidences might have passed unnoticed. Alternatively, had we used larger bins for the cross-correlograms, the tight synchronization would be smeared out.

So much for pairwise cross-correlations. When research was aimed at events which are more complex than pairwise coupling, many such instances were invariably found [Klemm and Sherry 1981; Abeles 1982b, chp. 6; Abeles et al. 1983; Dayhoff and Gerstein 1983; Legendy and Salcman 1985; Abeles and Gerstein 1988; Villa and Abeles 1990; Bartanyan and Pyrogov 1991; Villa and Fuster 1992; Abeles et al. 1993].

Given the variety of laboratories, preparations and methods of analysis used, all of which yielded excess of compound time locking among several spikes, one cannot doubt the existence of this phenomenon. Fig. 11 is an illustration of tight time locking amongst 3 single units in the frontal cortex of a behaving rhesus monkey. In this figure the method for measuring the Joint Peri Stimulus Time Histogram (JPSTH) [Aertsen et al., 1989] was used to investigate the three-fold correlations amongst units number 12,13 and 2.

In Fig. 11 the firing times of unit number 12 were used as a *stimulus*. The delayed firings of unit number 13 after this pseudo-stimulus were plotted along the *x-axis*, the delayed firings of unit number 2 were plotted on the *y-axis*, while the joint occurrences of a spike from unit 13 and a spike of unit 2 were plotted in the *x-y plane*. Thus, the marginal histogram plotted under the *x-axis* is the pairwise correlation between units 12 and 13, the marginal histogram on the left of the *y-axis* is the pairwise correlation between units 12 and 2, while the *x-y plane* describes the 3-fold correlation among units 12, 13, and 2. Fig. 11A shows the number of compound coincidences that were actually found. Notice that we "zoomed" in on a region in the *x-y plane* which is relatively remote from the time origin. Counts in the neighborhood of the time origin reflects instances where all three events are near simultaneous. Hence, the picture there is dominated by various sources of common input (not shown here). The region shown in Fig. 11A, however, is quite far from that busy center, and hence, the counts are generally down to the baseline level. Indeed most of the bins in Fig. 11A contain few counts (5±4) and appear in dark grey shading, but one bin contains over 20 counts and is plotted white. The fine grain renders the visual inspection of the histogram difficult. A small section of the histogram is blown up and plotted in 3 dimensions in Fig. 11B. There the difference between the number of actual counts and the number that would be expected by chance (if there were no compound dependencies beyond what could be expected by the pairwise dependencies described by the marginal histograms) is plotted. Finally, Fig. 11C shows the probability of observing the data in Fig.11B by chance.

Note the following features: The event (12,2,13);(0,151±1,289±1), (that is, a spike from unit 12 followed after 151±1 ms by a spike from unit 2, and after 289±1 ms by a spike from unit 13) stands out in isolation. This means that we observe here time locking better then 3 ms even after a delay of 289 ms. The probability of seeing such an event by chance is *exp(-16)*, i.e. 1 in 10 millions! We encountered hundreds of such events in our recordings.

In summary, compound firing patterns, in which tight time locking is observed after long delays are abundant. These events have extremely low probability of

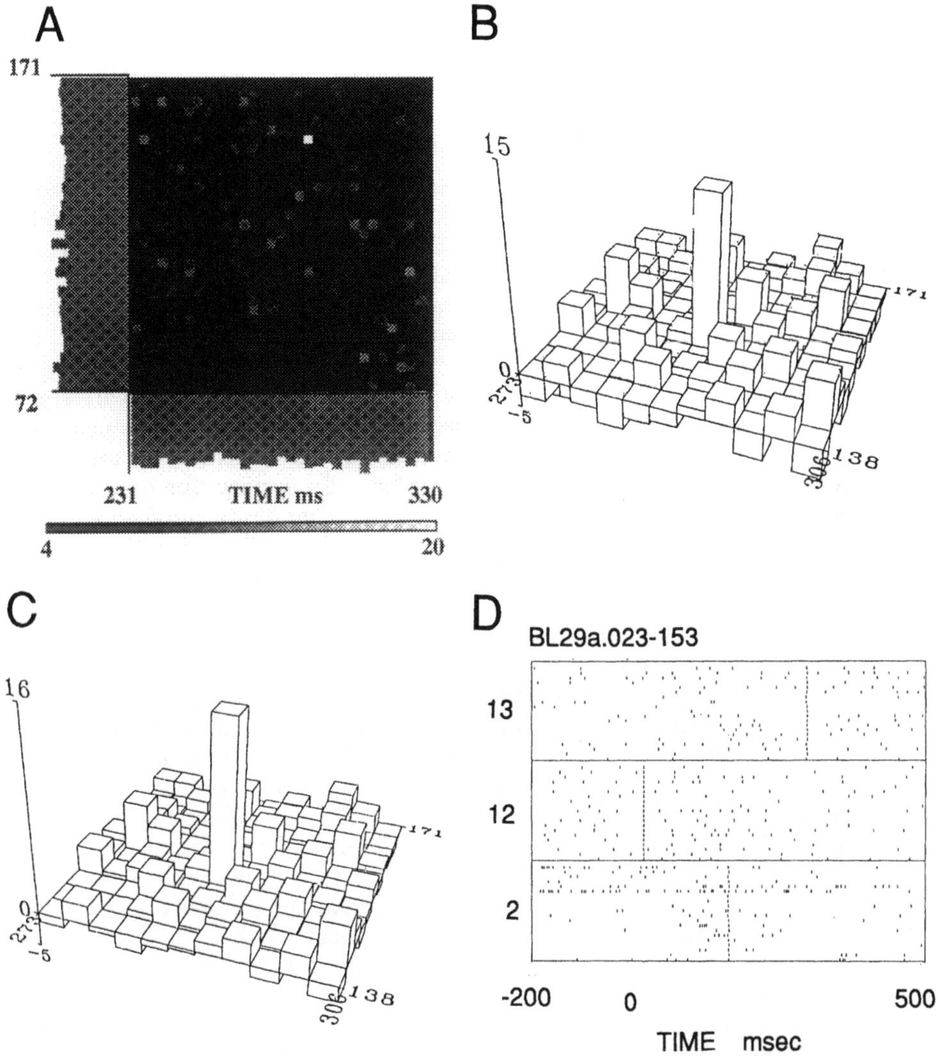

Figure 11. Triple correlation among units 12, 13, and 2 in frontal cortex of a behaving monkeys. A. Grey-level plot of the raw correlogram showing one bin in which 20 delayed coincidences occured. B. a 3-D plot of a small region around that peak, showing the difference between the counts found and the counts expected by chance. C. Surprise values of data in B. (i.e. *-ln{probability of getting by chance}*). D. Raster display of all the spikes around the 20 compound coincidences which contribute to the high bin in A, B, and C. The activity of each neuron is plotted in a different panel.

occurrence when considered as realizations of independent, time varying rate processes.

3.2. The Syn-Fire Chain.

The accurate spike timing which occurred even after delays of 500 ms must be generated by special mechanisms. Since the strength of the intra-cortical excitatory synapses is very small [Abeles, 1991, chp. 3] one must assume that any secure transmission takes the form of a convergence of several presynaptic cells on one post-synaptic cell. Similarly, since a single post-synaptic cell cannot have much of an effect on its own, there should be a group of post-synaptic cells receiving the same type of connections. This situation is illustrated in Fig.12, where two groups of cells are shown to be connected by multiple diverging-converging connections. Two qualifying points should be made when speaking of diverging-converging links: First, the drawing of Fig. 12 is not an anatomical plot, but rather a functional connectivity plot. In reality the neurons that comprise each set may be (and probably are) scattered and intermingled with each other and with many other neurons. They also make and receive thousands of additional synapses beyond those plotted in the figure. Second, it is not necessary to have every cell in the presynaptic set excite every cell in the post-synaptic set. In fact, it is sufficient to have a situation where every presynaptic cell excites only a fraction of the post synaptic cells. The rich intra-cortical connectivity supports such diverging-converging connections without the need for any special growth processes. For example, if we examine a column of 20,000 neurons in which every neuron is connected to 5000 of the neurons at random, then for almost any set of 30 neurons picked at random there exists another set of 30 neurons each of which receives synaptic contacts from 16-17 neurons in the first set. [See chp. 6 of Abeles 1991, for details about how to compute plausible cortical connectivities.]

The average cortico-cortical excitatory synapse is very weak. In the motor cortex of behaving monkeys its peak depolarization at the soma is approximately 0.1 mV [Fetz et al. 1991]. Consequently, at this level 16-17 EPSP's would have only a small effect on the firing rates. Therefore the multitude of random diverging-converging links have only small effects. However, if by learning processes these synapses become 5-10 times stronger, as was observed by Ahissar et al. [1992], these diverging-converging links may become very effective.

There is, however, a more important effect that should be taken into account when considering the combined effect of EPSP's arriving at a post-synaptic neuron. When the neurons are firing at low rate, the individual EPSPs have only a small effect on the firing of the receiving neuron. However, when these same EPSP's arrive in a synchronous mode, they can provide a very strong drive [Abeles, 1982a; Abeles, 1991 chp. 7]. In fact diverging-converging connections of the type shown in Fig. 12 act as a spatio-temporal filter which preferably transmits synchronized volleys. Thus even if the first links of a chain of diverging-converging connections is excited by prolonged excitatory processes, if the connections are of sufficient strength they would convert such diffuse

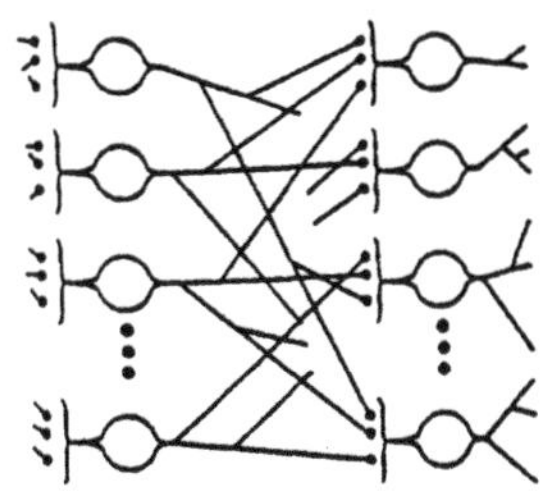

Figure 12. Multiple diverging converging connections between two sets of neurons.

excitation into a synchronous volley which is transmitted securely from link to link [Abeles, 1988; Abeles, 1991 chp. 7]. When activity is propagating in such a way within a chain of diverging-converging links, we refer to it as *syn-fire activity*. A chain that can transmit *syn-fire activity* is called a *syn-fire chain.*

We used neural network simulations to illustrate various properties of syn-fire chains. In the simulation there were 10% weak random connections among the cells but, in addition some of the neurons were connected by stronger diverging-converging connections to form a syn-fire chain. In addition, a large noisy current was delivered to each cell, to mimic the effects of many synaptic inputs from cells which are not in the simulated net. A net of this sort tends to explode by virtue of the positive feed-back provided by the excitatory connections [Abeles, 1991 chp. 5]. To prevent such "epileptic" explosion a global inhibitory mechanism was provided. The inhibitory cell population received excitatory synapses from all the neurons in the net, it responded by a firing rate that depended on the membrane potential of the inhibitory cells in a quadratic fashion [Amit and Treves, 1989]. The firing rate of the inhibitory cells affected all the excitatory cells in the net.

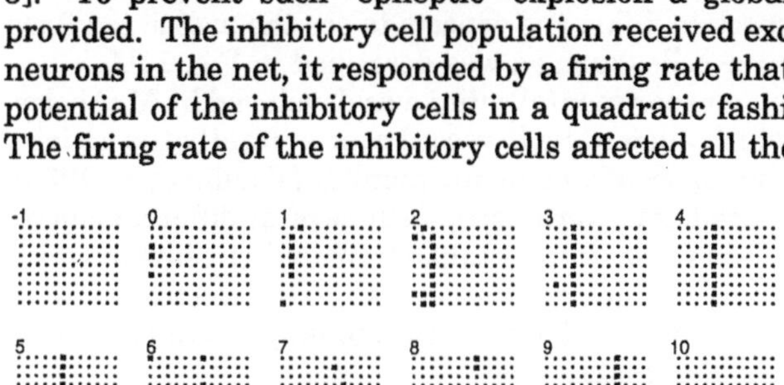

Figure 13. The initiation of syn-fire activity in a chain by constant current stimulation to the first (leftmost) three links. The stimulus was applied from time 0 to 9.

This process is illustrated in Fig. 13, where the activity of the first 10 links of a syn-fire chain with 26 links is shown. Each link was composed of 9 cells and each cell gave strong excitatory connections to 7 of the cells in the next link. The entire syn-fire chain was embedded in a neural-net of 1000 neurons. The neurons in the first 3 links were depolarized by a weak current (0.3 nA) for 10 ms. We see how this diffuse depolarization evokes some activity in time 0 which then builds up and within 3 ms produces a synchronized volley. From then on, this propagates in a secure manner along the chain. Note that despite the continued depolarizing current during time steps 0 to 9, the input neurons do not start another cycle of syn-fire activation. The reason is that, once activity is evoked, the inhibitory PSP's which are fed-back to the entire network prevent the input neurons from regenerating a second wave.

Once a chain of diverging-converging links starts to operate in the syn-fire mode, activity propagates in a secure and time-wise accurate manner from link to link. Provided the chain is long enough, the spikes from two neurons, one taking part in an early stage, the other in a late stage, are expected to exhibit a tight, delayed time locking (such as was shown in Figs. 10 and 11).

Thus, in our view syn-fire chains are a simple and both anatomically and physiologically plausible mechanism for generating such accurate timing as was found in cortical activity.

3.3 Experimental Evidence.

The present experimental techniques do not allow us to observe directly the syn-fire activity. If each link contains, say, 30 neurons and the chain is embedded in a column containing 20,000 neurons, the probability of recording simultaneously from 3 or even 2 neurons in the same link is vanishingly small. The best that one can hope for, is to obtain a simultaneous recording from 3 neurons, each of which takes part in a different link of the same syn-fire chain. If such a recording is obtained, then whenever the syn-fire chain is activated and the synchronized volley passes through the respective links, one expects to see the associated fixed firing pattern amongst the three cells, such as was described in Fig. 11D.

In order to obtain such data an experimental setup was developed by M. Abeles, H. Bergman and E. Vaadia, which allowed us to record concomitantly the activity of up to 16 single units from the cortex of a behaving monkey [Vaadia et al. 1993]. Several different behavioral paradigms were used, with several different micro-electrode arrangements. The data shown here is taken from experiments in which 6 micro-electrodes were introduced into frontal areas. The electrodes were arranged in a circle having a radius of 0.3 mm. The monkeys were trained to perform light and sound localization tasks.

To obtain data about excessively repeating spatio-temporal firing patterns, a pattern searching algorithm, which detects any repeating pattern regardless its single unit composition or duration [Abeles and Gerstein 1988], was applied. While applying this algorithm to multi spike data as recorded in these experiments, many instances in which excessively repeating patterns appear were indeed found. The details of this finding and a detailed description of their properties can be found in Abeles et al. [1993]. Here we summarize the main results which are relevant for our deliberations.

The patterns are very often associated with some external, i.e. stimulus or behavioral, event. Often, but not always, the patterns appeared in periods which one would characterize as having *event related activity* i.e. in which the firing rate of the units were elevated in relation to the time of some stimulus or movement.

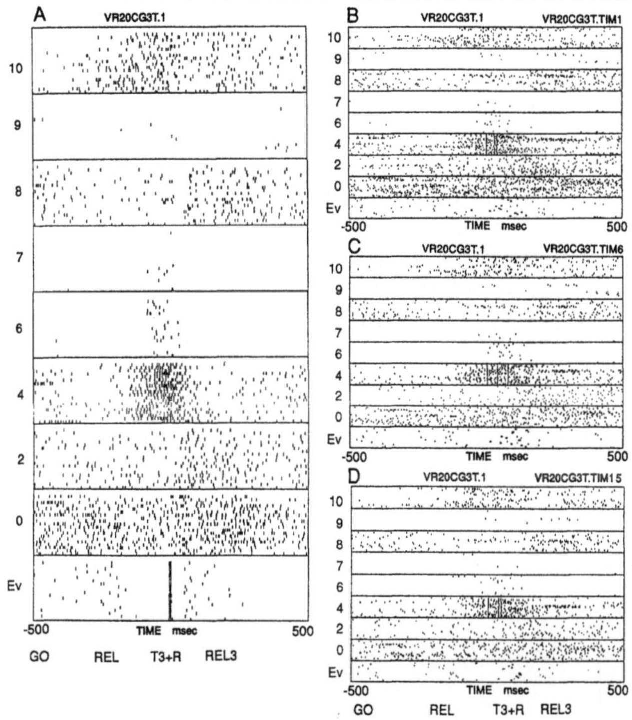

Figure 14. Event related activity and spatio-temporal firing patterns in the frontal cortex of a behaving monkey.
Raster displays of the activity of 8 neurons and the external events. A. The display is aligned around the time that the monkey touched target no. 3 (*T3*) and obtained a drop of juice as reinforcement (*R*). The other events are the moment at which the monkey was instructed to move (*GO*), at which the movement started (*REL*), and at which the monkey released the target (*REL3*). B., C., and D. Displays aligned around the first spike of (4,4,4);(0,30±1,39±1), (4,4,4);(0,39±1,78±1), and (4,4,4);(0,39±1,48±1) respectively.

Fig. 14 illustrates this point. In this case the activity of 8 neurons was recorded simultaneously. All the sections from 0.5 seconds before the monkey was instructed to move, towards target no. 3, up to 0.5 seconds after he completed the movements , were collated together. The responses of these units is displayed in Fig. 14A, where all the traces are aligned around the time that the monkey touched the target and obtained a drop of juice. We see that unit 4 elevated its firing rate markedly around this time, units 6,7, and 10 showed smaller activation, while the activity of units 2 and 8 was slightly supressed. On this data a search for excessively repeating patterns was made. In this search all patterns that lasted up to 100 ms with a timing accuracy of 3 ms were examined. The most prominent excess was found for unit no. 4, which generated 19 different patterns, each repeating 14 or more times (Note that one would expect to see 0.443 such patterns of unit 4 by chance, which means that the actual count presents an event that had a probability of less than 1 in 10^{24} to happen by chance).

Fig. 14B shows the pattern (4,4,4);(0,30±1,39±1) which repeated 21 times. All the traces are aligned around the time of the first spike in this pattern. Below the firing times of unit 0 the behavioral events are marked. We see that the pattern tends to appear shortly before the monkey touched the target. This pattern came intermixed with various other patterns of the type (4,4,4). The 16 repetitions of one of them, (4,4,4);(0,39±1,78±1), are plotted in Fig. 14C, the 14 repetitions of another pattern, (4,4,4);(0,39±1,48±1), are shown in Fig. 14D, and finally there were 16 other patterns of the type (4,4,4);(some other delays). In order to study the relations between the classical event related activity (Fig. 14A) and the patterns, we marked all the spikes of unit 4 that took part in one of the excessively repeating patterns and studied their relation to the external events.

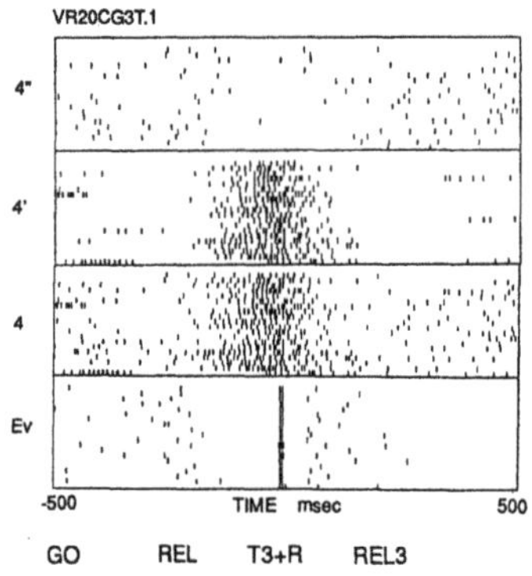

Figure 15. Event-related activity and spatio-temporal firing patterns. The same data as in Fig. 14. The panel marked by 4, displays all the spikes of unit 4, 4' displays only those spike of unit 4 which participated in excessively repeating spatio-temporal patterns, and 4" displays the remaining spikes of unit 4.

Fig. 15 illustrates our results. The activity is aligned around the time that the monkey touched target 3 and obtained a drop of juice. The behavioral events are shown in the bottom panel. The firing times of unit 4 are shown on the next panel (these two panels are a repeated plot of Fig. 14A.) The next panel shows all the spikes that were associated with one of the excessively repeating patterns, while the top panel shows the spike of unit 4 which were not associated with such patterns. Clearly *all* the event related spike were associated with excessively repeating patterns!

We studied in detail 20 cases for which a single unit showed a marked elevation of firing rate in association of an external event and had also an excess of repeating spatio-temporal patterns. In 13 of those, all (or almost all) the spikes within the event-related activity were forming excessively repeating patterns.

In summary, a large excess of repeating spatio-temporal patterns is often found. These patterns are frequently associated with event-related elevations of firing rates. Often the patterns embrace most, if not all, of the spikes within the event related activity. Hence, one must conclude that the neuronal mechanism which causes event-related elevations of firing rates has the capacity to generate a surplus of very accurate spatio-temporal firing patterns.

3.4. Syn-Fire Reverberations

While studying the abundance of repeated spatio-temporal patterns, it became clear that the findings do not fit completely the predictions derived from the simple syn-fire chain model. For a simple syn-fire chain the patterns are expected to be composed of spikes form 3 different cells (like 12,2,13 as in Fig. 11), whereas in the experimental data most of the patterns were composed of 3 spikes from the same single neuron (like 4,4,4 as in Fig. 14). The simple syn-fire hypothesis suggests that whenever the same process repeats (say, the monkey reaches for target 3), the same spatio-temporal firing patterns should appear. While we found that typically there are several patterns associated with each process, each individual pattern repeats only part of the times (as illustrated by Fig. 14). By combing all the different patterns, however, the entire activity related to the process could be accounted for (as illustrated in Fig. 15). See Abeles et al. [1993] for a more detailed description of this situation.

One may account for repetitive appearances of spikes from the same neuron in a single pattern by assuming that this neuron participates in several links of the

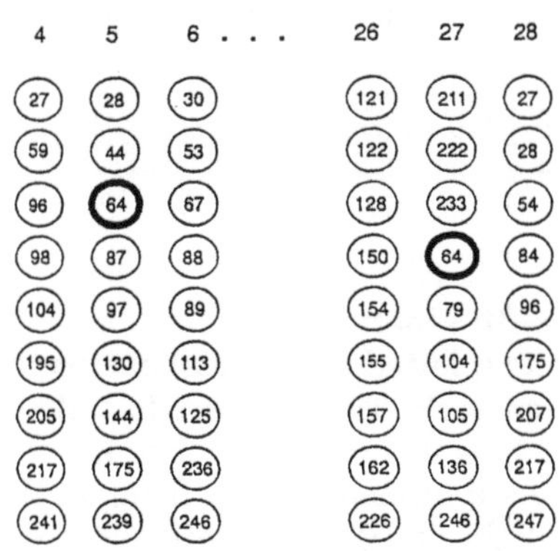

Figure 16. Syn-fire chain with repetitions.
The neurons in each column compose one link in a chain of diverging-converging connections. Unit 64 takes part in both link 5 and link 27.

syn-fire chain. Fig. 16 illustrates how that may happen. It shows two regions from a simulated syn-fire chain: a section around link 5 and a section around link 27. The syn-fire chain is similar in its structure to the one in Fig. 13. Unit 64 is part of the 5-th link, since it receives multiple connections from neurons 27,59, ...241, and excites neurons 30,53, ...246 Moreover, it is also part of the 27-th link, since it also receives multiple connections from neurons 121,122, ...226, and excites neurons 27,28, ...247. Whenever activity sweeps through the chain, neuron 64 is likely to fire twice with an interval of 22 (=27-5) synaptic delays in between.

If repetitions such as described in Fig. 16 occur often, the syn-fire chain ceases to be a straight feed-forward (multi-layered) structure, but starts to contain multiple feed-back loops. Such a structure may exhibit attractors in which a group of neurons excite themselves and maintain elevated firing rates for a long period [Hopfield 1982, 1984; see also Rubin and Sompolinsky, 1989; Amit and

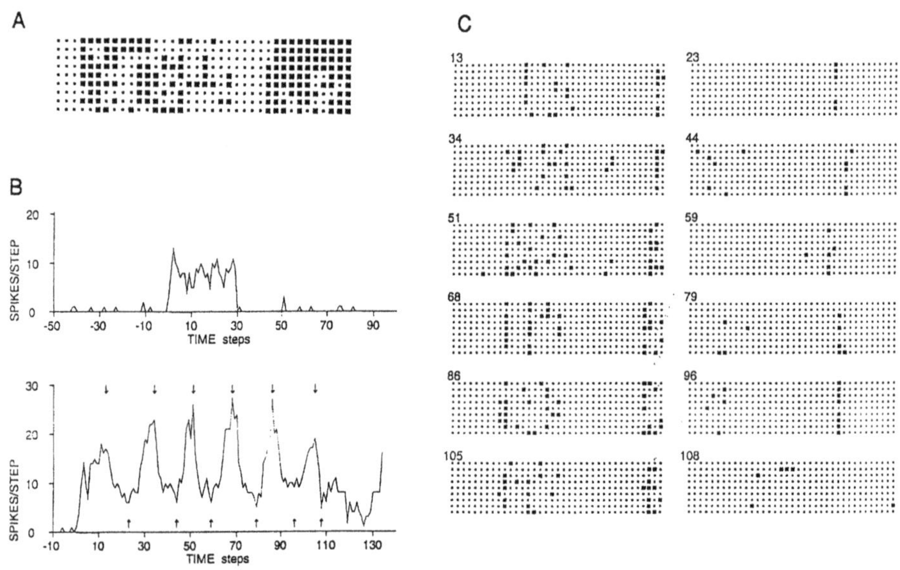

Figure 17. Reverberations in a syn-fire chain with repetitions.
A. The structure of the chain. 82 out of 256 units (plotted as fat dots) participate in two links, the others (thin dots) appear only once. B. The response to a steady stimulation of the first three links for 10 ms (steps 0 through 9). The abscissa describes the simulation time, the ordinate the total number of cells firing at that time. Top panel, the response of the syn-fire chain of Fig. 13, where every unit appeared only once. Bottom panel, the response of the syn-fire chain described in A. The arrows mark the peaks and troughs of activity levels. C. The firing constellations at the peaks (left column) and troughs (right column) of the response in B (bottom trace). A firing cell is marked by a fat dot.

Treves, 1989; Amit et al., 1990; Erb and Aertsen,1992, for feed-back networks whose neurons do not fire at their maximal rates when the network reaches an attractor].

In order to examine the behavior of syn-fire chains in which the same neuron participates more than once, we extended the simulation described in section 3.2 in the following way: 12 more links were added to the syn-fire chain of Fig. 13. The neurons in the new links were selected at random from 256 neurons, 234 of which were already taking part in the chain. A neuron was not allowed to participate in two links which are closer then 9 steps apart. Fig. 17A gives an overview of this new chain. Small dots represent neurons that appear only once in the chain, large dots represent neurons that appear more often. Approximately 33% of the neurons appeared twice in the chain.

The chain was again activated by depolarizing its input neurons (first three links) continuously for 10 ms. Activity started in the same way as in the simple syn-fire chain, but then exhibited more complex propagation patterns. These settled down to reverberations, in which activity continued to cycle within the chain for a very long time. This process is illustrated in Fig. 17. Fig. 17B shows the total number of firing neurons in each step (1 ms) of the simulation. The top trace shows the level of activity for the simple syn-fire chain described in section 3.2 (Fig. 13), while the bottom trace shows the level of activity in the chain with repetitions. The firing constellations at the arrows are shown in Fig. 17C. Clearly, even when as little as 33% of the cells participate twice in the chain, complex activation patterns appear, which in this case took the form of periodic oscillations.

Activation patterns become even more complex as the extent of repetitions in the links increases. When the syn-fire chain of Fig. 17 was further elongated by adding 15 more links, the activity pattern as seen in Fig. 18 was obtained. The level of activity and examples of the activation constellations are shown in Fig. 18B and C. In this syn-fire chain one obtains reverberations which can maintain elevated activity for very long time. The activity levels show large undulations, however these are not periodic anymore but exhibit very complex shapes. Successive peaks and troughs of the activity levels are associated with different firing constellations. The large undulations of the activity level may tell the observer that something is being processed by this network, however what actually goes on can be specified only in terms of the spatio-temporal firing patterns.

Although activity does not propagate just between successive links, the number of different propagation trajectories is limited. Thus, if the syn-fire chain is activated many times, there would be a tendency for the same spatio-temporal firing patterns to repeat at levels higher then chance. Suppose we could insert micro-electrodes into this network of the 1000 simulated cells and pick 3 of them at random. Our chances of getting 3 different cells from the same syn-fire chain are small. As only a quarter of the 1000 neurons take part in the syn-fire chain, the chances of getting 3 members is 0.016 (1/4 to the third power). The chances of seeing exactly one neuron which participates in the syn-fire chain is 0.42 (3•1/4•3/4•3/4). Thus we expect to see many more cases in which one neuron

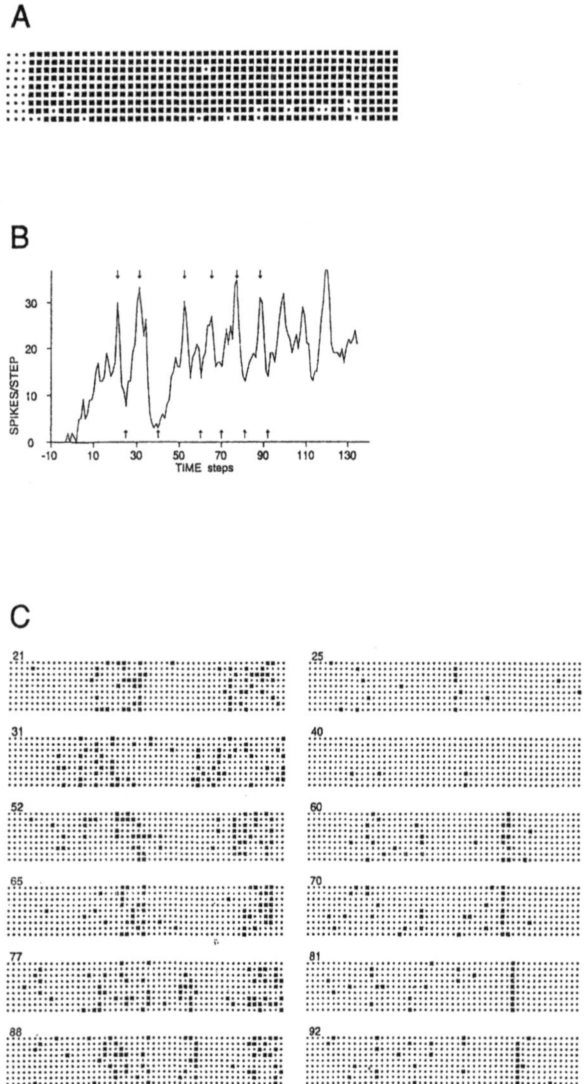

Figure 18. Reverberations in a syn-fire chain with many repetitions. A, B, and C. as in Fig. 17.

shows internal time locking amongst its spike, then cases in which three neurons show across-neuron patterns. Moreover, a single participating neuron in a syn-fire chain such as in Fig. 18, would show several repeating patterns, each associated with a different trajectory of reverberations.

3.5. Significance of Spatio-Temporal Patterns

In our data of cortical activity as recorded through micro-electrodes we found abundance of delayed synchronization. The delays can be as long as 500 ms, but a high time accuracy of ±1 ms is maintained throughout this interval. There is close association between the appearance of such accurately repeating spatio-temporal patterns and event related activity. We argued that these findings are a sign of the neural mechanism which generates the event related activity in the cortex. Up to this point we treated the spatio-temporal patterns merely as indications for the underlying neural mechanisms, but not as containing information from the brain's point of view.

This attitude was radically changed when one of us (Y.P.) discovered that units which responded in a similar fashion to external stimuli showed different profiles of spatio-temporal firing patterns, according to the meaning that the stimuli had. This phenomenon is illustrated in the following example.

A rhesus monkey was trained to perform a delayed-response task in the following manner: The monkey sat in a restraining chair, with an arena of touch buttons and light sources in front of it. Below, in the center of the arena, an additional touch button (the *start-trial* button) was provided. The monkey was trained to behave according to two alternating paradigms (*go* and *no-go*). In the *go* paradigm the monkey started a trial by touching the *start-trial* button. After a variable delay a red LED, referred to as the *ready LED*, was turned on in the

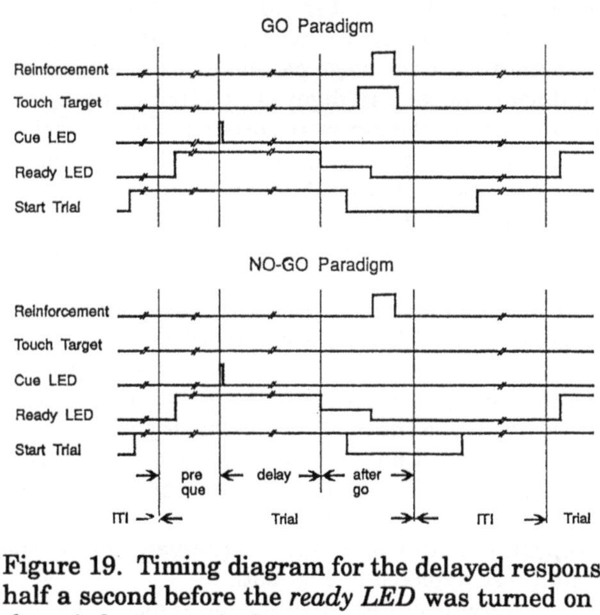

Figure 19. Timing diagram for the delayed response paradigm. The trial started half a second before the *ready LED* was turned on and ended half a second after the *reinforcement* (a drop of juice) was delivered. Breaks in the traces represent random delays introduced by the computer which controlled and monitored the behavioral experiment.

center of the arena. After another variable delay (3-6 s), a 100 ms long yellow light blink was flashed either from the left or from the right. We refer to this light blink as the *cue*. After an additional delay of 1,2,4,8,16 or 32 seconds the *ready* LED changed its color from red to orange, this served as a *GO* signal, after which the monkey had to touch the target from which the *cue* was delivered within 1.2 s. Correct responses were reinforced by a drop of juice. After 4 correct responses in the *go* paradigm, all the lights in the arena were switched on for 4 seconds; this instructed the monkey to switch its behavior to the *no-go paradigm*. In the *no-go* paradigm the same sequence of events proceeded, but the monkey had to keep its hand on the *start-trial* button after the *GO* signal. After 4 correct trials in the *no-go* paradigm, the LEDs of the arena were turned on again for 4 seconds and the monkey switched its behavior back to the *go* paradigm.

These event sequences are displayed graphically in Fig. 19. The figure also shows the division of the time into 4 periods. The *ITI* period covers the time between successive trials, the *pre-cue* period spans the time from turning on the *ready LED* until the *cue* was blinked, the *delay* period covers the time from the *cue* up to the *GO* signal, while the time from then on until the end of the trial was called the *after go* period.

On 5-May-1991 unit 6 had 80 excessively repeating patterns, however they did not occur with the same likelihood at the different time sections. The patterns (6,6,6); (0,52±1,338±1) and (6,6,6); (0,32±1,152±1) appeared most frequently during the *delay* period which followed the *cue* on the left. Fig. 20A shows the responses of unit 6 to the *cue* on the left. The responses are sorted such that the responses while the monkey was behaving according to the *no-go* paradigm are shown in the upper panel, while those made during the *go* paradigm are plotted in the lower panel. Moreover, responses within each paradigm were sorted so that the trials in which the *delay* was 1 s are displayed at the bottom, those with 2 s *delay* above them, and so on, up to *delays* of 32 seconds. Observe that unit 6 responds to the blink from the left by elevating its firing rate strongly for approximately 2 s in both paradigms. In addition, it also elevated its firing rate after the *GO* signal in the go paradigm .

The frequency of occurrence of the two patterns mentioned above is plotted in Fig. 20B. The occurrences of (6,6,6); (0,56±1,338±1) are marked by a dot, those of (6,6,6); (0,52±1,152±1) by a triangle. Notice that the first pattern tends to appear selectively when the monkey behaves according to the *go* paradigm, while the second pattern is the preferred one while the monkey is behaving in the *no-go* paradigm. Thus, the elevated activity of unit 6 tells us which stimulus (cue from the left) was given; in addition, the pattern of firing tells us what the "mental set" (*go* vs. *no-go*) of the monkey was. Such differential preference of patterns was found for all the 80 excesively repeating patterns of unit 6. Hiwever, usually the profiles were more complex than those depicted in Fig. 20B.

The differential preference of different patterns with the same single-cell composition (say 6,6,6) to appear in different states raises two questions: (1) What is the neural mechanism responsible for this phenomenon, and (2) Is this phenomenon significant for the brain, or only for the neurophysiologist studying it.

The appearance of different patterns within similar responses of the same neuron suggests that a group of neurons may reverberate in different modes. This situation is easily realized with the syn-fire concept. Just like the same neuron may participate in several different links of a single syn-fire chain, it may also

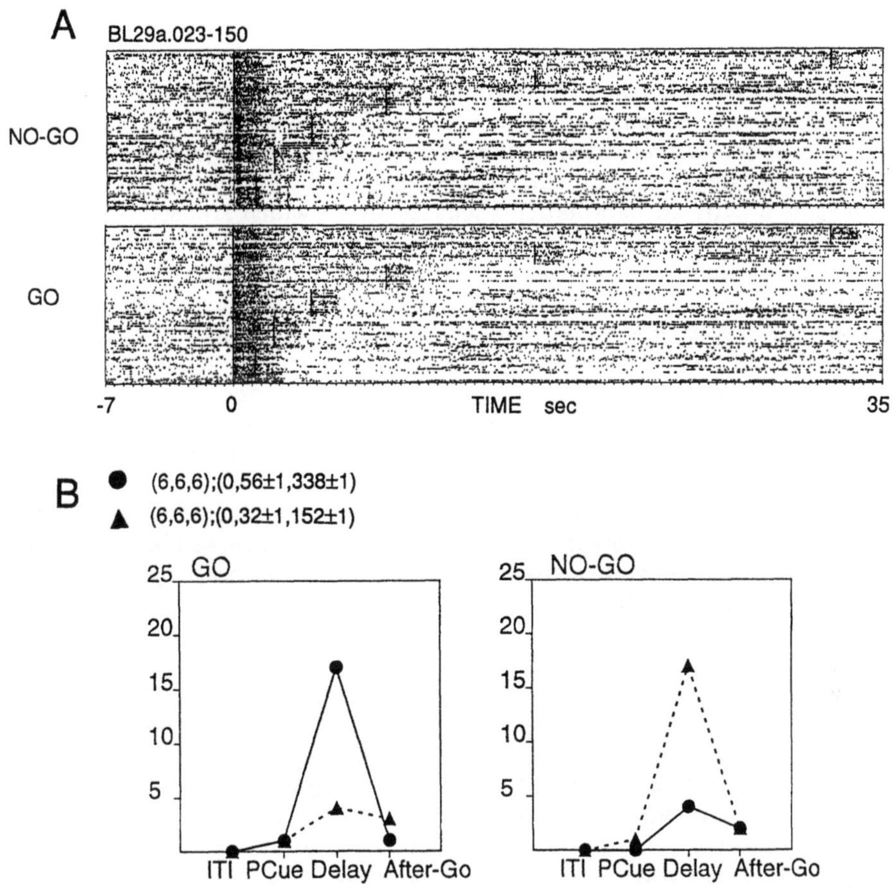

Figure 20. Profiles of spatio-temporal firing patterns of unit 6 (see text).
A. The activity of unit 6, from 7 seconds before the *left cue* up to 35 seconds after it. The heavy vertical bars at times 1,2,4,... seconds mark the times of the *go* signal. Top panel: activity while the monkey was performing according to the *NO-GO* paradigm. Bottom panel: activity while the monkey was performing according to the *GO* paradigm. B. The profiles of recurrence of two patterns in the different stages of the trial. • and the solid line, for the pattern (6,6,6); (0,56±1,338±1), ▲ and the broken line, for the pattern (6,6,6); (0,32±1,152±1).

participate in two different syn-fire chains. To test this idea we extended the simulation described in section 3.4 to a situation where 600 neurons were connected so as to form two chains of diverging-converging connections. The 3 input stages of the first chain had 25% overlap with the three input stages of the second one. Except for the input neurons, all other neurons participated in both chains, and most of them appeared twice in each chain. When the input links of chain 1 were stimulated by a constant depolarizing current for 10 ms, syn-fire activity started in this chain and continued to propagate securely through the chain until it reached the last (75-th) link. When the input links of chain 2 were stimulated, activity propagated in syn-fire mode according to the sequence of connections of the second chain. Thus, it is plausible that syn-fire chains may indeed reverberate in several different modes. The reverberation mode can contain some additional information about the event which aroused the activity.

Can the brain use this information? It was suggested by E. Bienenstock [1991; Doursat, 1991] that two syn-fire chains may be cross connected by a few connections. If the activity in these connections arrives at the appropriate times, the two syn-fire chains may in fact enhance each other and generate a more stable network. We have tested this idea by way of simulation and found this to be the case. Moreover, with the appropriate strength of cross connections, we observed that, when the activity in the two syn-fire chains becomes engaged, the overall activity levels (like those shown in Fig. 18B) of the two chains undulate in a coherent way.

In summary, syn-fire reverberations are not just a way to generate and "shape" the spatio-temporal structure of the cortical activity; they also can be used to bind the activity of multiple syn-fire chains in different cortical locations into a single coherent mental experience.

Acknowledgements

The authors are indebted to E. Ahissar and I. Nelken who helped in developing the experimental setup, to E. Margalit, I. Halman and H. Slovin who helped in carrying out the experiments, to B. Karmon for developing an automated analysis of correlations and organizing the results in a data base, and to V. Sharkansky for help in the artwork. This research was supported in part by grants from the United-States Israeli Binational Science Foundation (BSF), the Basic Research Fund administered by the Israel Academy of Sciences and Humanities, and the G.I.F. the German-Israeli Foundation for Scientific Reasearch and Development.

References

Abeles M. (1982a). Role of the cortical neuron: Integrator or coincidence detector? Isr. J. Med. Sci. 18:83-92.

Abeles M. (1982b). Local cortical circuits: An electrophysiological study. Springer, Berlin.

Abeles M. (1988). Neural codes for higher brain functions. In H.J. Markowitsch (ed.) Information processing by the brain. Hans Huber Pub. Toronto, pp. 225-238.

Abeles M. (1991). Corticonics: Neural circuits of the cerebral cortex. Cambridge University Press, New York.

Abeles M., Bergman H., Margalit E. and Vaadia E. (1993). Spatio-temporal firing patterns in the frontal cortex of behaving monkeys. submitted.

Abeles M. and Gerstein G.L. (1988). Detecting spatiotemporal firing patterns among simultaneously recorded single neurons. J. Neurophysiol. 60:909-924.

Abeles M., de Ribaupierre F. and de Ribaupierre Y. (1983). Detection of single unit responses which are loosely time-locked to a stimulus. IEEE Trans. SMC 13:683-691.

Aertsen A.M.H.J., Gerstein G.L., Habib M.K. and Palm G. (1989). Dynamics of neuronal firing correlation: modulation of "effective connectivity". J Neurophysiol 61: 900-917.

Ahissar E. and Vaadia E. (1990). Single cell cortical oscillators in a somatosensory cortex of awake monkey. Proc. Natl. Acad. Sci. 87. pp. 8935-8939.

Ahissar E., Vaddia E., Galun M., Bergman H., Arieli A. and Abeles M. (1992). Dependence of cortical plasticity on correlated activity of single neurons and on behavioral context. Science 257: 1412-1415.

Allum J.H.J., Hepp-Raymond M.C. and Gysin R. (1982). Cross-correlation analysis of interneuronal connectivity in the motor cortex of the monkey. Brain Res. 231:325-334.

Amit D.J., Evans M.R. and Abeles M. (1990). Attractor neural networks with biological probe records. Network 1:381-405.

Amit D.J. and Treves A. (1989). Associative memory neural networks with low temporal spiking rates. Proc. Natl Acad. Sci. U.S.A. 86:7871-7875.

Arieli A. (1992). Novel strategies to unravel mechanisms of cortical function: From macro- to micro-electrophysiological recordings. In: Aertsen A. and Braitenberg V. (Eds.) Information Processing in the cortex. Springer, Berlin. pp 123-137.

Bartanyan G.A. and Pyrogov A.A. (1991). Neurobiological foundation of higher brain functions. Nauka, Leningrad (in Russian).

Berger H. (1929). Ueber das Elektrenkephalogram des Menschen I. Arch. Psychiat. Nervenkr. 87:527-570.

Bienenstock E. (1991). Notes on the growth of a "composition machine". In: Contributions to Interdisciplinary workshop on Compositionality in Cognition and Neural Models. Abbaye de Royaumont, May 27-28.

Burns D.B. (1958), The Mammalian Cerebral Cortex. Arnold, London.

Burns D.B. and Webb A.C. (1979). The correlation between discharge times of neighboring neurons in isolated cerebral cortex, Proc. R. Soc. London B203:347-360.

Calvin W.H. (1983). A stone's throw and its launch window: timing precision and its implications. J. Theor. Biol. 104:121-135.

Calvin W.H. and Stevens C.F. (1968). Synaptic noise and other sources of randomness in motoneuron interspike intervals. J. Neurophysiol. 31:574-587.

Cope T.C. and Mendell L.M. (1982). Parallel fluctuations of EPSP amplitude and rise time with latency at a single Ia fiber-motoneuron connection in the cat. J. Neurophysiol. 47:455-468.

Cox D.R. (1962). Renewal Theory. Methuen, London.

Dayhoff J. and Gerstein, G.L. (1983). Favored patterns in spike trains. II. Application. J. Neurophysiol. 49:1349-1363.

Douglas R.J., Martin K.A.C. and Whitteridge D. (1991). An intracellular analysis of the visual responses of neurones in cat visual cortex. J. Physiol (London) 440:659-696.

Doursat R. (1991). Contribution a l'étude des représentations dans le système nerveux et dans les réseaux de neurones formels. Ph.D. Thesis, Université Paris VI.

Eccles J.C. (1957). The Physiology of the nerve Cell. Johns Hopkins Univ. Press, Baltimore.

Eckhorn R., Bauer R., Jorden W., Brosch M., Kruse W., Munk M. and Reitboeck H.J. (1988). Coherent oscillations: A mechanism for feature linking in the visual cortex? Biol. Cybern. 60:121-130.

Edelman G.M., Gall W.E., Cowan W.M. (1984). Dynamic Aspects of Neocortical Function. Wiley, New York.

Engel A.K., Konig P., Kreiter A.K., Schillen T.B. and Singer W. (1992). Temporal coding in the visual cortex: new vistas on integration in the nervous system. TINS 15:218-226.

Erb M. and Aertsen A. (1992). Dynamics of activity in biology-oriented neural network models: stability at low firing rates. In: Aertsen A. and Braitenberg V. (eds) Information Processing in the Cortex: Experiments and Theory. Springer, Berlin. pp 201-223.

Fetz E., Toyama K. and Smith W. (1991). Synaptic interactions between cortical neurons. In: Peters A. (ed.) Cerebral Cortex, Vol. 9. Plenum Publ. Corp.

Frostig R., Frostig Z., Frysinger R. and Schechtman V. (1985). Multineuron analysis reveals complex patterns of interaction among neurons. Soc. Neurosci. Abstr. 11:1020.

Gerstein G.L. and Kiang N.Y-S. (1960). An approach to the quantitative analysis of electrophysiological data from single neurons. Biophys. J. 1:15-18.

Gray C.M. and Singer W. (1989). Stimulus-specific neuronal oscillations in orientation columns of cat visual cortex. PNAS USA 86:1698-1702.

Hanzel D. and Sompolinsky H. (1992). Synchronization and computation in a chaotic neural network, Phys. Rev Let. 68:718-724.

Hochstein S. and Shapley R.M. (1976). Quantitative analysis of retinal ganglion cell classifications. J. Physiol. (London), 262:237-264.

Hopfield J.J. (1982). Neural networks and physical systems with emergent collective computational abilities. Proc. Natl. Acad. Sci. U.S.A. 79:2554-2558.

Hopfield J.J. (1984). Neurons with gradual response have collective properties like those of two-states neurons. Proc. Natl. Acad. Sci. U.S.A. 81:3088-3092.

Kiang N. Y-S.,Watanabe M.D., Thomas E.C., Louise F. and Clark A.B. (1962). Stimulus coding in the Cat's auditory nerve. Ann. Oto. Rhin. Laryng. 71:1009-1026.

Klemm, W.R., and Sherry, C.J. (1981). Serial ordering in spike trains: what's it "trying to tell us"? Int. J. Neurosci. 14:15-23.

Landolt J., Reinis S. and Weiss D. (1985). Identification of local neuronal circuits in the visual cortex of the cat. Soc. Neurosci. Abstr. 11:1010.

Lass Y. and Abeles M. (1975). Transmission of information by the axon. I. Noise and memory in the myelinated nerve fiber of the frog. Biol. Cybern. 19:61-67.

Legendy C., and Salcman M. (1985). Bursts and recurrences of bursts in the spike trains of spontaneously active striate cortex neurons. J. Neurophysiol. 53:926-939.

MacGregor R.J., Lewis E.R. (1977). Neural Modeling. Electrical signal processing in the nervous system. Plenum Press, New York.

McCormick D.A., Connors B.W., Lightall J.W.and Prince D.A. (1985). Comparative electrophysiology of pyramidal and sparsely spiny stellate neurons of the neocortex. J. Neurophys. 54:782-806.

Moore B.C.J. (1982). An introduction to the psychology of hearing. Academic Press, London, pp150-158.

Morest D.K. (1968). The collateral system of the medial nucleus of the trapezoid body of the Cat, its neuronal architecture and relation to the olivo-coclear bundle. Brain Research 9:288-311.

Morest D.K., Kiang N.Y-S., Kane E.C., Guinan J.J.Jr. and Godfrey D.A. (1973). Stimulus coding at caudal levels of the cat's auditory nervous system: II. Patterns of synaptic organization. In: Moller A.R. (ed.) Basic Mechanisms in Hearing. Academic Press, New York, pp 479-504.

Morison R.S. and Dempsy E.W. (1943). Mechanism of thalamocortical augmentation and repetition. J. Physiol (London), 138:297-308.

Murthy V.N. and Fetz E.E. (1992). Coherent 25-35 Hz oscillations in the sensorimotor cortex of the awake behaving monkey. Proc. Natl. Acad. Sci. U.S.A. 89:5670-5674.

Nelson J.I., Salin P.A., Munk M.H.J., Arzi M. and Bullier J. (1992). Spatial and temporal coherence in cortico-cortical connections: A cross-correlation study in areas 17 and 18 in the cat. Visual Neurosci. 9: 21-37.

Peters A. and Kara D.A. (1985). The neuronal composition of area 17 of the rat visual cortex. I. The pyramidal cells. J. Comp. Neurol. 234:218-241.

Rubin N. and Sompolinsky H. (1989). Neural networks with low local firing rates. Europhys. Lett. 10:465-470.

Sheer D.E. (1989). Focused arousal and the cognitive 40-Hz event-related potentials: Differential diagnosis of Alzheimer disease. Prog. Clin. Biol. Res. 317:79-94.

Softky W.R. and Koch C. (1993). The highly irregular firing of cortical cells is inconsistent with temporal integration of random EPSP's. J. Neurosci. 13:334-350.

Sompolinsky H., Crisanty A. and Sommers H.J. (1988). Chaos in random neural networks. Phys. Rev. Lett. 61:259-262.

Toyama K., Kimura M. and Tanaka K. (1981), Organization of cat visual cortex as investigated by cross-correlation technique. J. Neurophys. 46:202-214.

Vaadia E., Abeles M., Bergman H., Lavner Y., Margalit E. and Karmon B. (1993). Simultaneous recording of single units activity in the frontal and posterior parietal cortices of rhesus monkey during performance of a localizing behavioral task. I: Properties of single neurons. In preparation.

Villa A.E.P. and Abeles M. (1990). Evidence for spatiotemporal firing patterns within the auditory thalamus of the cat. Brain Research 509:325-327.

Villa A.E.P and Fuster J.M. (1992). Temporal correlates of information processing during visual short-term memory. NeuroReport 3:113-116.

Brain Theory / A. Aertsen (Editor)
1993 Elsevier Science Publishers B.V.

Bifurcation analysis of continuous time dynamics of oscillatory neural networks

Walter J. Freeman and Sven Jakubith

Division of Neurobiology, LSA 129, Department of Molecular & Cell Biology
University of California, Berkeley CA 94720 USA

Abstract

Two models are constructed to represent the dynamics of neuron populations. Each component population is described by a linear 2nd order ordinary differential equation and an asymmetric sigmoid input/output curve. Networks of components having these properties are predisposed to instability and bifurcations that depend on input and internal parameters. The simpler model restricted to negative feedback interactions has only point and limit cycle attractors. The model that allows positive feedback between excitatory populations and inhibitory populations revealed these attractors and also a chaotic domain in its phase space. The model having the three types of nonlinear feedback is proposed as a useful starting point for bifurcation analysis of neurodynamics in cerebral cortex.

1. INTRODUCTION

The goal of this contribution is to sketch how the architecture and the dynamical properties of a model neural network can be motivated by the findings from neurobiological research. A bifurcation analysis of two simple models is done with tools provided by dynamical systems theory.

Neural networks have been developed in three historical stages. The first models proposed by McCulloch & Pitts (1943) treated the neuron as a logical switch with two possible states. Networks of these units did Boolean algebra and led to the development of automata theory and programmable digital computers. In the second stage simple "add, multiply" units were connected massively in parallel and in multiple layers having modifiable connection strengths, with implementation by discrete time steps and with a variety of learning algorithms (Rumelhart & McClellan, 1990). In the third stage elements corresponding either to single neurons or to neural populations were embodied with nonlinear differential equations and explored with embodiments in analog (Mead, 1989; Eisenberg et al, 1989) or digital computers (Freeman, 1987).

In all cases the neural networks are characterized more by their architectures than by the dynamical properties of their components. The architectures describe which elements are connected (fully or sparsely) and the signs of the connections between the elements (excitatory + or inhibitory -) with or without fixed and distributed time delays. The dynamics of the components of a neural network are described by their input / output characteristics. Each operation is specified by equations that give a rule for transforming a known input into a measured output. Typically the dynamics of a component is expressed in a linear equation that describes temporal and spatial summation, and a nonlinear operator that imposes unilateral or bilateral saturation on the sum. There are different input /output functions in use, ranging from a step function through smooth nonlinear sigmoidal shaped functions in a wide variety, to a linear ramp. The interactions of biological neurons can be plastic, so that all models which are capable of learning must incorporate some functions by which the connection strengths (weights) between processing elements (models of neurons or of pools of neurons) are changed during learning.

The spatial and temporal behavior of a neural network can be characterized with tools provided by dynamical systems theory, thus finding fixed points, limit cycles or chaotic attractors and their basins of attraction in phase space. The behavior of a specific neural network as a function of input values and internal parameter settings can be summarized in a bifurcation diagram. In spite of the fact that bifurcation analysis is applied to a wide variety of systems, problems ranging from engineering chemical reactions to systems in surface and solid state physics (Krischer, 1990) and Hamiltonian mechanics (Abraham and Marsden, 1978), only a few bifurcation analyses have been performed for neural networks (Khiblik, Borisyuk, 1992). Example of such diagrams will be given here. These diagrams show immediately which dynamic properties can be expected for a given range of input or parameter values.

Formulating a neural network which is closely motivated by brain architecture and dynamics presents formidable problems, owing to the characteristics of brain circuits having multiple nonlinear elements, which are both excitatory and inhibitory, and which are connected in multiple types of feedback architectures having both fixed and distributed feedback delays. A carefully staged approach is developed here. First, a neural component is constructed whose characteristics conform to those refined by nature during evolution, and second, a network is constructed to act as coupled nonlinear oscillators, thus building a bridge to fields in engineering, where interactions of forced nonlinear oscillators constitute a wide field of investigations for example, in electronics and materials science (Awrejecewicz, 1991).

2. CONSTRAINTS FROM NEUROBIOLOGY FOR A MODEL

2.1. Clarifying remarks and nomenclature

The model described below reflects spatial arrangements and features of *neural masses*. Since the step from neurons to neural masses provides a major source of confusion, this issue will be outlined here first. A motivation for considering neural masses instead of single neurons will be given in section 2.2.).

A neuron (Fig. 1) can be divided into three parts reflecting its functions. The dendrites provide the input surface of the neuron with the synapses as sources or sinks for the dendritic current. The cell body directs a wide variety of biochemical activities that support but are not part of the transactions involving neural information processing. The trigger zone at the origin of the axon converts the den-

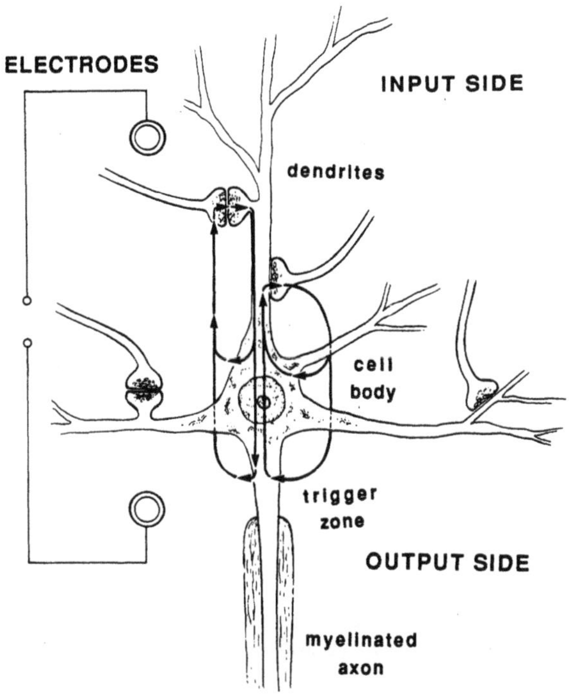

Figure 1.
(1a) Schematic diagram of a typical mammalian neuron. Connected to the dendritic tree are thousands of synapses, each releasing an electromotive force for a loop current. Excitatory current flow is inwardly at an excitatory synapse and outwardly at the trigger zone. Inhibitory loop current flows in the opposite direction. The sum of the currents at the trigger zone determines the output pulse rate. Also shown are extra cellular electrodes which record the averaged field potential (EEG) that is established by the flow of dendritic current in its return path across tissue resistance by a pool of neighboring neurons and a sample of the pulse density of the pool. In general there are hundreds to thousands of neurons located between two electrodes, so the EEG recording manifests the macroscopic activities of neural masses.

dritic current into a spike train, which propagates along the axon, the usual output pathway of the neuron. The input to the dendrites is provided by synaptic junctions with the axons of connected neurons. Each synaptic junction converts a spike rate into a current wave amplitude. These two different encoding schemes of the information (see Fig. 1b); current amplitude on the dendrites and spike frequency on the axon, reflect the different tasks of the input and output functions of the neuron. The dendrites perform a spatiotemporal integration of

Neural State Variables

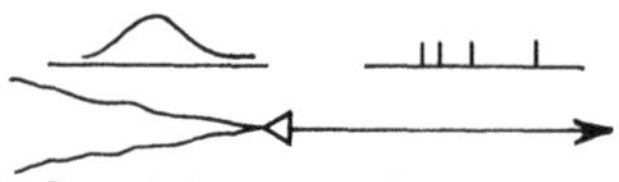

Form			Scales	
Structure	Dendrite	Axon	Time	Distance
Neuron:				
activity	membrane polarization	action potential frequency	msec	μm (point)
observable	postsynaptic potential	unit train		
Ensemble:				
activity	current density	pulse density	sec	mm (area)
observable	EEG wave	unit sum		

Figure 1.
(1b) Neuroelectrical activity occurs in two modes, the impulse action characteristics of the axon and the wave action characteristics of dendrites. From Freeman 1983.

the electric currents. The current flow from excitatory synapses is inward at the synapse and outward at the trigger zone, whereas the direction of current at inhibitory synapses is the reverse. The cost of this dendritic processing is an attenuation of the current wave with distance along the dendrites. The input to remote synapses is more strongly attenuated than is proximal input, but the remote dendritic surface area is much larger, so that in population dynamics there may be no significant loss of remote input.

The axons on the other hand distribute the information encoded in the spike density without attenuation over long distances. They perform this task by generating a traveling pulse along the axon. The cost of this stratagem is distance-dependent delay in transmission, and both upper and lower limits on the transmission band width. The minimum spike rate is zero at threshold. The maximum rate of single neurons is restricted by the absolute refractory period to an upper limit of about 1 kHz, above which cathodal block occurs (again, zero firing rate). A summary diagram of experimental findings for neurons is shown in Fig.2. The pulse-wave conversion at the synapses is time varying and non-linear, and the wave-pulse conversion at the trigger zone is also time varying, but linear between threshold and cathodal block.

A population of neighboring neurons forming a *neural mass* is characterized by the type of its internal synaptic connections. A neural mass without internal connections is called a K0 set. An example is the set of olfactory receptors. A mass with mutual synaptic connections all having the same sign is referred to as

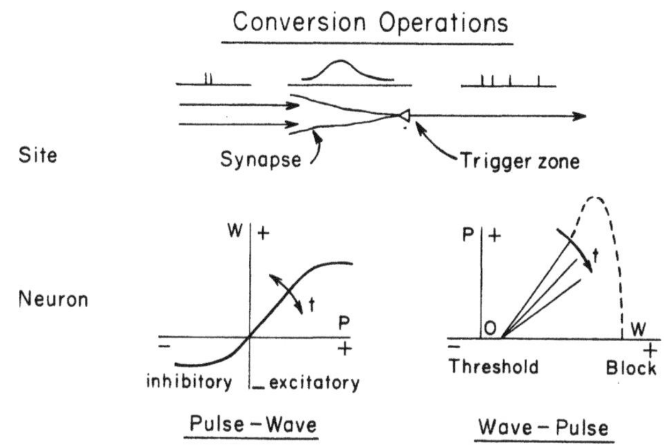

Figure 2.
The conversion characteristics from spike rate to a current wave at the axo-dendritic synapse (lower left) and the conversion of a current wave to a spike rate at the trigger zone (lower right) for a single neuron.

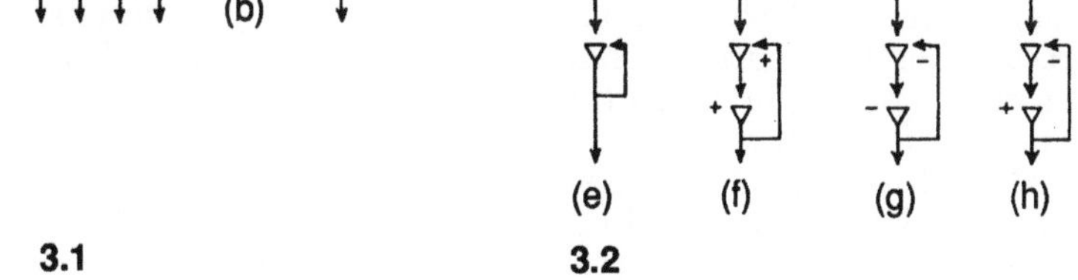

Figure 3.
(3.1) a) KO set with one input channel and one output channel, b) KI set with one input channel and one output channel.
(3.2) The topological conventions represent patterns of connection among neurons. a) convergence, b) divergence, c) series, d) parallel, e) auto feedback, f) positive excitatory feedback, g) positive inhibitory feedback, h) negative feedback. From Freeman 1975.

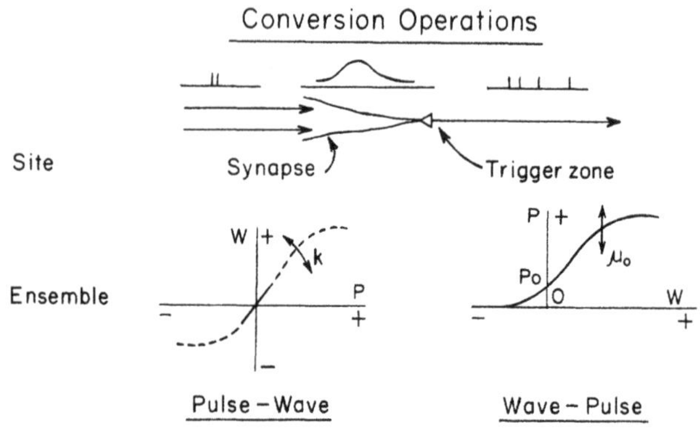

Figure 4.
Diagram of the input/output characteristics within a mutually connected neural mass. The pulsewave conversion is constrained to a small-signal linear range, and the slope is changed with learning. The wave pulse conversion is static and nonlinear and is changed with arousal.

a KI set. The connections are mutual, because each neuron transmits to thousands of others in its surround, and it also receives from thousands of others, but no neuron can transmit to or receive directly from itself. The connections are either excitatory or inhibitory, thus giving excitatory KI_e and inhibitory KI_i neural masses respectively. A KI_e mass that is reciprocally connected to a KI_i mass forms a KII set and a negative feedback loop. Negative feedback loops can give rise to sustained oscillations and are found widely throughout the cerebrum and spinal cord. These feedback loops are the focus of the analysis in this report. Fig.3 summarizes the nomenclature of the types of connectivity.

The characteristics of the wave-pulse conversion by a KI ensemble of neurons differ notably from the characteristics for the case of single neurons. This is due to the presence of the mutual connections in the ensemble. The output is a continuous pulse density, in which the actual firing is rotated randomly over the set at a low rate for each member neuron. The refractory periods are uncorrelated over the set, so that the sigmoid curve is static rather than time varying. Fig.4 shows the experimental findings. The equation for describing the sigmoid characteristic of neural masses motivated by experimental findings is given in section 2.3). The pulse - wave conversion at synapses in neural masses is maintained within a narrow small-signal linear range, because its domain of input is bounded by the preceding wave-pulse conversion in feedback loops (Freeman, 1975).

2.2. Spatial properties of neurons and neural masses

This section gives a brief introduction to the anatomy of the olfactory bulb (OB) and the olfactory cortex (OC). Also the motivation is given why to use neural masses instead of neurons as processing elements in neural networks.

When looking at a stained sample of the OB (Fig. 5), the striking features are the immense numbers of neurons and the spherical nests at the input layer called glomeruli. Each glomerulus confines a collection of synapses between axons and dendrites of neurons from the same and other layers. The schematic drawing (Fig. 5.2) shows the connectivity of the layers in the mammalian olfactory bulb. The axons from the olfactory receptors in the nasal cavity form the primary olfactory nerve (PON). Each axon projects without branching or interconnection into the glomeruli layer (GL) and terminates in a glomerulus, where it branches and forms excitatory axodendritic connections to the apical dendrites of a mitral cell. The ratio of receptors to mitral cells is thought to be about 1,000 to 1 in the glomeruli. The mitral cells have axons that transmit from the OB to the OC by the lateral olfactory tract (LOT). They also have branches called collaterals that excite other mitral cells, so that they form a KI_e set. A few exited OSN can stimulate over this pathway hundreds of mitral and granule cells in a process called "avalanche conduction".

There are two main types of interneurons. The external interneurons are called periglomerular cells (PG), and they form a KI_e set (not shown). The internal interneurons are called granule cells, and they form a KI_i set. The connection from mitral to granule cells is excitatory, whereas the connection from granule to mitral cells is inhibitory. Thus the interconnected populations of granule and mitral cells form a negative feedback loop, which can give rise to sustained oscillations. Synaptic transmission is divergent in both directions , involving many thousands of interconnected neurons for each glomerulus. From these findings it is evident why it is necessary to consider neural masses instead of single neurons when simulating brain dynamics.

2.3. Dynamical operations of neural masses

The input / output function (Sigmoid function)

An analytic expression for the input / output function (sigmoid function) is derived from two premisses. The first is that in a neural mass most neurons most of the time have membrane potentials that fluctuate just below the threshold for firing. In this subthreshold range the probability of firing increases exponentially with depolarization. This property contributes the concave upward lower part of the curve. The second premiss is that the actual firing rate in a population asymptotically approaches an upper limit determined by the hyperpolarizing afterpotentials of spikes. This property forms the convex upward part of the curve. A detailed derivation has already been published (Freeman, 1979). Here we give here only the equation for the sigmoid function, which has a single parameter Q_m defined as the upper asymptote in normalized coordinates $Q = (p - p_0) / p_0$, where p_0 is the mean pulse density.

$$Q(x, Q_m) = \begin{cases} Q_m \cdot \left[1 - e^{-\frac{(e^x - 1)}{Q_m}} \right], & x > -u_0 \\ -1, & \text{else} \end{cases} \qquad u_0 = -\ln\left\{ 1 - Q_m \cdot \ln\left\{ 1 + \frac{1}{Q_m} \right\} \right\}$$

Q_m is an adjustable parameter, which is used for fitting the function to the experimental data. It is decreased by sleep and by anesthetics, going to zero for very deep anesthesia, and it is increased by waking and emotional arousal. Fig. 6 shows a) the sigmoid function for different values of Q_m , b) the gain (defined as $\partial/dt\ Q[x,Qm]$) and c) the position of the maximum gain as a function of the parameter Q_m . The main features of this sigmoid function are:

1. A near-linear range for the domain of small values of x about the origin.
2. A constant value for large values of x and -x (wave amplitude later), thus reflecting the observed fact of bilateral saturation.
3. Asymmetry of the asymptotes about the origin.
4. Diminished gain approaching zero at both extremes of wave amplitude (Fig. 6b).
5. Increasing gain with increasing positive wave amplitude from the origin.
6. Position of the maximum gain shift to the right for increasing values of Q_m (Fig. 6b&c).

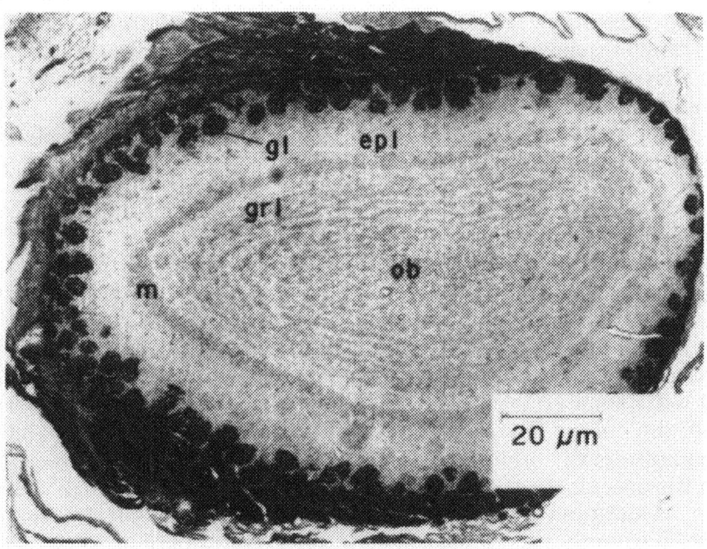

Figure 5.
(5.1) Frontal section through the forehead of a young mouse (10 days postnatal) at the level of the olfactory bulbs. The sensory fibers from the sensory neurons reach ipsilaterally the olfactory bulbs (ob), where they form the glomeruli (gl). A slight photographic contrast enhances the other layers of the bulb. epl, external plexiform layer; m, mitral cell layer; grl, granule cell layer. Courtesy of Dr. P.P.C. Graziadei, Florida State University (Graziadei, 1990).

In an experimental study (Eeckman & Freeman, 1991) of the relationship between the pulses of a local cluster of neurons and the dendritic potentials recorded in spatial proximity gave the findings of Fig. 7, which shows examples of the experimental data and the curves fitted to the data. The only adjustable parameter was Q_m in the sigmoid function. Fig. 8 summarizes the experimental

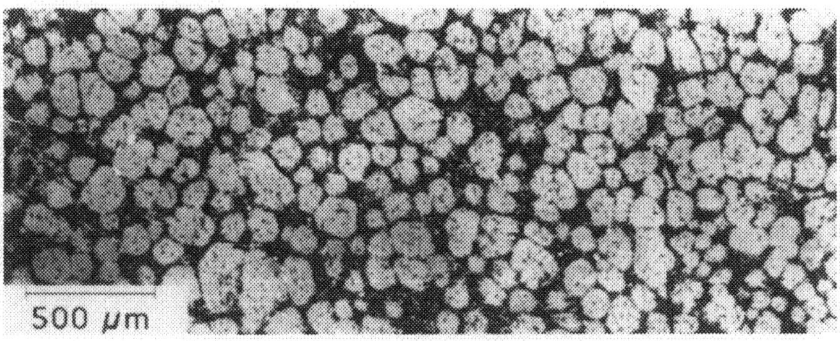

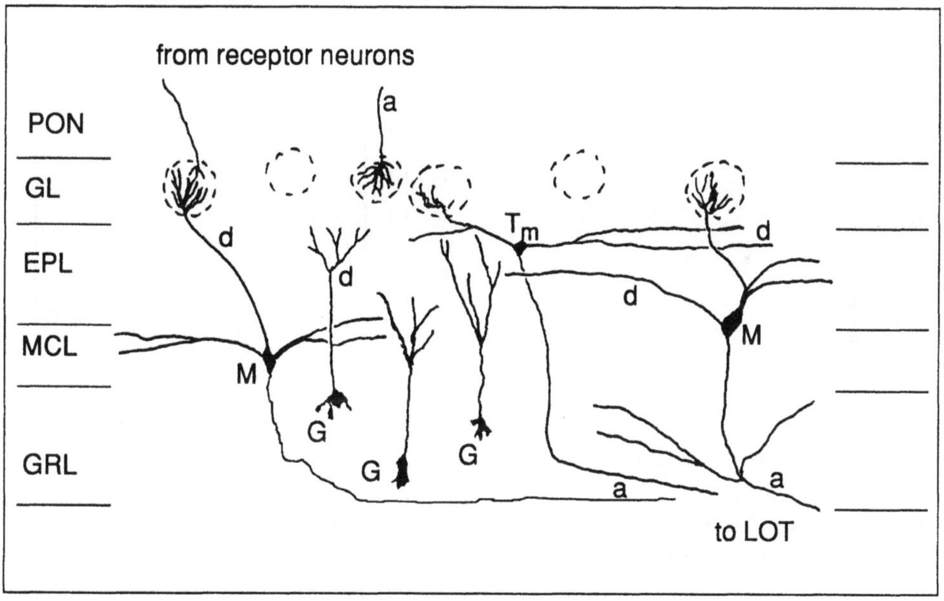

Figure 5.
(5.2) Composite photomicrograph of a section in the plane of the glomerular layer. From Freeman (1975).
(5.3) The neural elements of the mammalian olfactory bulb in a simplified diagram. Abbrevations for layers. PON, primary olfactory nerve; GL, glomerular layer; EPL, external plexiform layer; MCL, mitral cell layer; GRL, granule cell layer; Abbrevations for cells. M, mitral cell; Tm, middle tufted cell; G, granule cells; a, axon; d, dendrite; LOT, lateral olfactory tract;

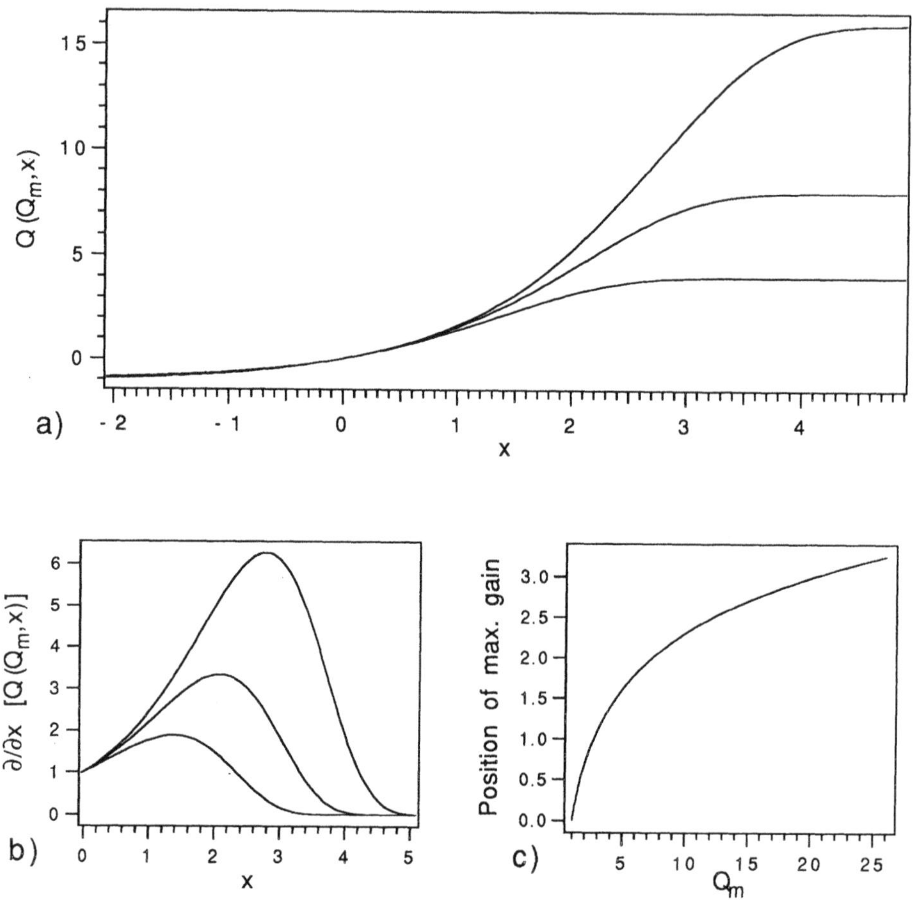

Figure 6.
(a) The sigmoid function for three different values of Q_m . 4, 8, 16, from bottom curve to top curve.
(b) The gain of the sigmoid function for the three curves in a).
(c) Value of wave amplitude for the maximum gain vs. the parameter Q_m .

findings. Two features were remarkable. First, the Q_m value was much higher in the OC than in the OB. This fact suggests that different tasks are performed by the OB and the OC. The second was that the distributions were skewed to the side of higher Q_m values. When the same experiment was carried out with mildly anesthetized animals the Q_m values were found overall to be smaller. This indicates that a change in the state of arousal can be expressed by a change in the corresponding Q_m values. This remarkable fact should be incorporated in a fully developed neural network, which simulates brain dynamics. This dynamical aspect should not be confused with learning, which changes the synaptic strengths between selected neurons and thus the overall feedback gains within neural masses.

Derivation of the linear ordinary differential equations, which describe the time dependent dynamical behavior of neural masses

Experimental data of the temporal impulse responses of a neural mass after an electrical stimulus were obtained with deeply and lightly anesthetized animals (Freeman, 1975). In the first case Q_m was reduced to zero and an open loop response was measured with the feedback gain at zero to suppress synaptic

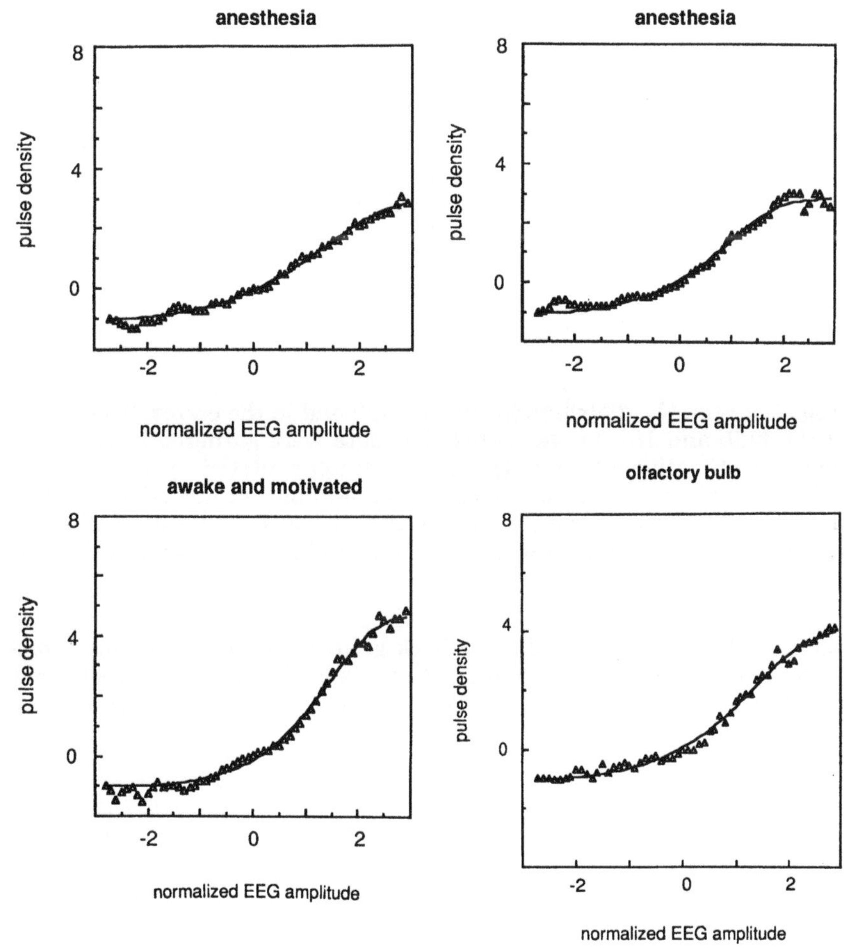

Figure 7.
Sigmoid curves fitted to the normalized conditional pulse probability density. The curves are steeper in awake and motivated animals (bottom two illustrations). The vertical scales are identical. From Eeckman 1991.

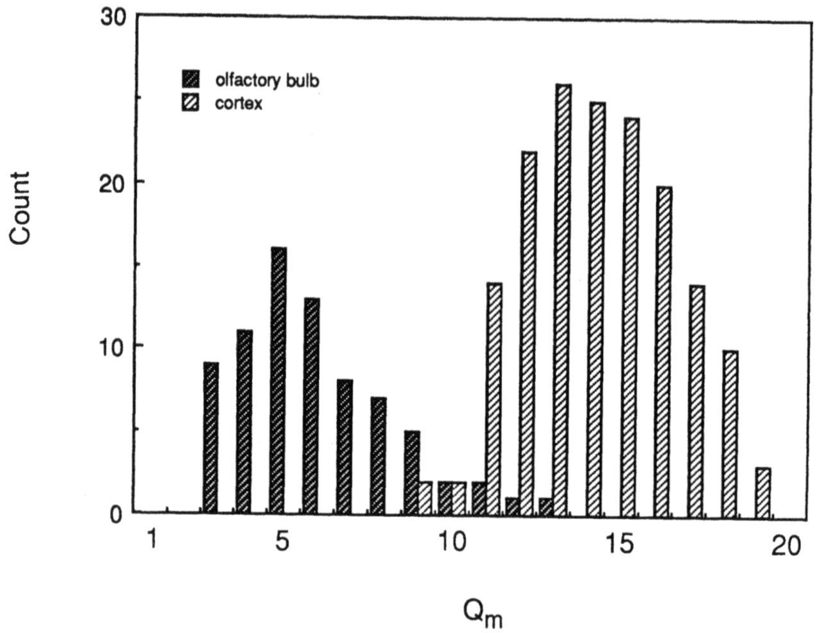

Figure 8.
Comparison between Q_m distributions in the bulb and in the cortex. There are 75 cases for the bulb and 162 for the cortex. The data were gathered in 76 animals in the awake and motivated state. Q_m is a parameter related to the height and slope of the sigmoid curve that describes the wave-to-pulse conversion in a population of cells. It can be seen from the graph that there is a significant difference between the distributions of values from the bulb and cortex. From Eeckman and Freeman (1991).

interactions. In the second case the feedback gains were non-zero, and damped oscillations occurred with impulse input. Fig. 9 and Fig. 10 show experimental data and fitted curves for both cases. The fitted curves are the solution of a differential equation of the form:

$$(1) \qquad \frac{1}{B}\left(\frac{d^2}{dt^2}x(t)+A\frac{d}{dt}x(t)+Bx(t)\right)=P(t) \quad , \quad A=a+b \; , \; B=ab \; , \; a=0.22\,/\,ms \quad , \; b=0.72\,/\,ms$$

a and *b* are time-independent parameters, P(t) is a time dependent input (the forcing term) and x(t) is the variable. *a* and *b* are fitted to the experimental data and reflect rate constants of the biological system, specifically the lumped synaptic and dendritic cable delay (b) and the passive membrane decay rate manifested in single shock postsynaptic potentials (PSPs). P(t) is in the experiment a brief (10 microsecond) electrical stimulus and a delta function in the calculated case.

For the calculated curves in Fig. 10 the feedback gain, K_{mg}*K_{gm}, also has to be taken into account, and the equations have the form:

$$\frac{1}{B}\left(\frac{d^2}{dt^2}m(t)+A\frac{d}{dt}m(t)+Bm(t)\right)=P(t)+K_{gm}\cdot Q(g(t)) \quad , \quad K_{gm}<0$$

(2)

$$\frac{1}{B}\left(\frac{d^2}{dt^2}g(t)+A\frac{d}{dt}g(t)+Bg(t)\right) = K_{mg}\cdot Q(m(t)) \quad , \quad K_{mg}>0$$

K_{gm} and K_{mg} are the coupling constants between the neural masses and Q() is the above defined sigmoid function. x(t) is replaced by m(t) and g(t) denoting the neural masses of the mitral and granule cells respectively. Note that the input is only to the mitral cells, and that it is excitatory.

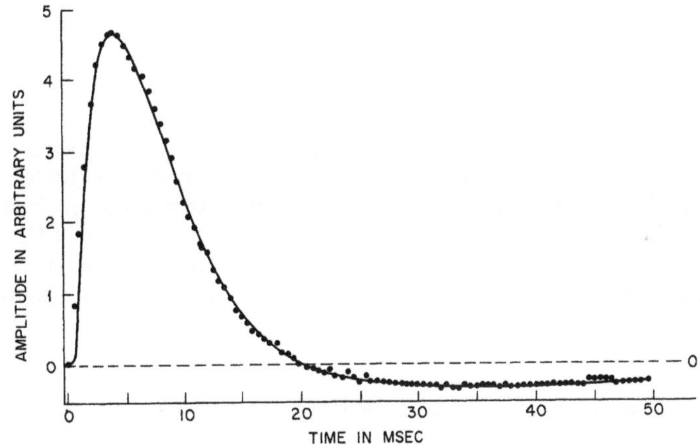

Figure 9.

Impulse response of the prepyriform cortex in the open loop state. Points, observed values; curve, calculated values. Fitted function is $v_e(t) = \sum_{j=1}^{4} B_j e^{-a_j t}$
From Freeman (1964).

3 MATHEMATICAL FORMULATION OF TWO MODELS

3.1 Coupling of two neural masses in a negative feedback loop

This model is restricted to negative feedback connections between neural masses of granule and mitral cells. The connectivity of a network described by

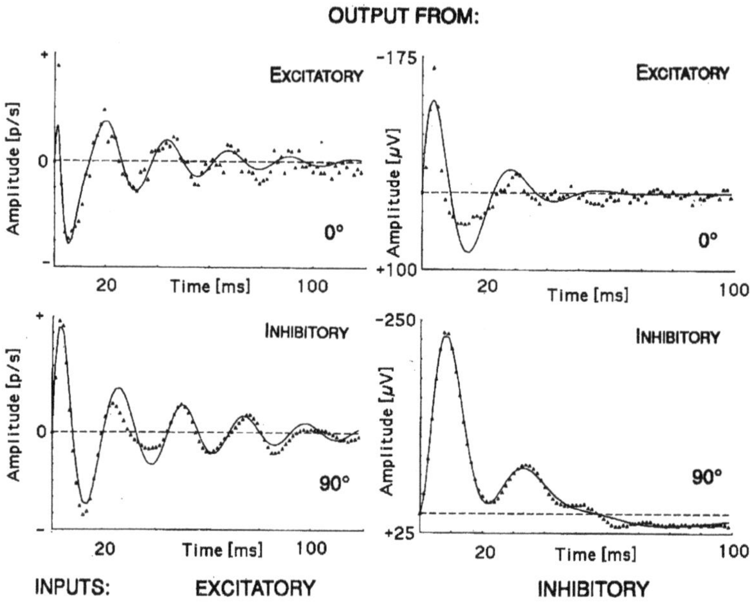

Figure 10.
Oscillatory impulse responses that were simultaneously taken from the excitatory neurons (upper frames) and the inhibitory neurons (lower frames) of the olfactory bulb show ed that the two populations had a common frequency and decay rate, but the output of the inhibitory neurons lagged that of the excitatory neurons by approximately ninety degrees, whether the input was to the excitatory cells (left frames) or to the inhibitory cells (right frames) From Freeman (1972).

equation (2) is visualized in Fig. 11a. A time independent input is given to the mitral cell populations, P(t) = constant. The system of coupled differential equations is:

$$\frac{1}{B}\left(\frac{d^2}{dt^2}m(t)+A\frac{d}{dt}m(t)+Bm(t)\right)=P+K_{gm}\cdot Q(g(t)) \quad , \quad P=\text{constant}$$

(2a)

$$\frac{1}{B}\left(\frac{d^2}{dt^2}g(t)+A\frac{d}{dt}g(t)+Bg(t)\right) = K_{mg}\cdot Q(m(t))$$

3.2. Coupling of six neural masses

An extended network (Fig. 11b) includes excitatory-excitatory and inhibitory-inhibitory connections, forming positive feedback loops between the mitral cells and positive feedback loops between the granule cells (a recurrent inhibitory -

inhibitory connection is also called a positive feedback loop). Input, $P_1(t)$, is only to one mitral cell pool (M_1).

$$\frac{1}{B}\left(\frac{d^2}{dt^2}m_i(t)+A\frac{d}{dt}m_i(t)+Bm_i(t)\right)=P_i+K_{gm}\cdot Q(g_i(t))+\sum_{\substack{j=1\\ j\neq i}}^{3}K_{mm}\cdot Q(m_j(t))\ ,\quad P_i=\begin{cases}P=const\ ;\ i=1\\ P=0\ ;\ else\end{cases}$$

(3)

$$\frac{1}{B}\left(\frac{d^2}{dt^2}g_i(t)+A\frac{d}{dt}g_i(t)+Bg_i(t)\right)=K_{mg}\cdot Q(m_i(t))+\sum_{\substack{j=1\\ j\neq i}}^{3}K_{gg}\cdot Q(g_j(t))\ ,\quad i=1,2,3$$

The differential equations of equation (2a) and equation (3) were rewritten into a system of first order differential equations and numerically integrated with a NETLIB integration routine (LSODE, available via anonymous ftp server: research.att.com) in double precision on a SUN Sparkstation SLC.

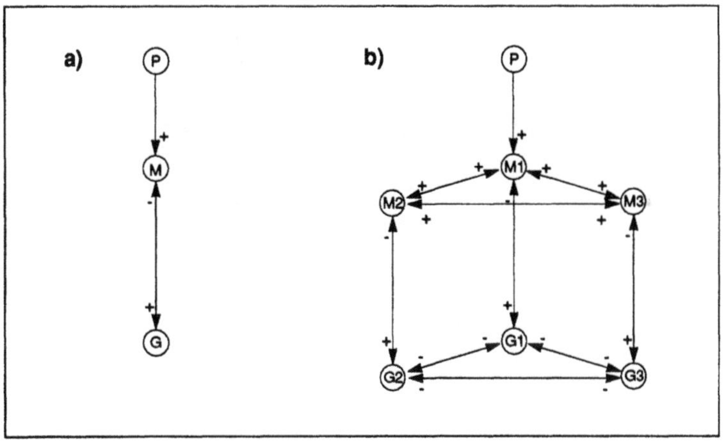

Figure 11.
Schematic diagram of two networks of neural masses.
(a) One mitral cell and one granule cell population in a negative feedback loop. Excitatory input is only to the mitral cell population.
(b) Extended network with lateral connections between mitral cell groups and lateral connections between granule cell groups. Excitatory input only to one mitral cell population.

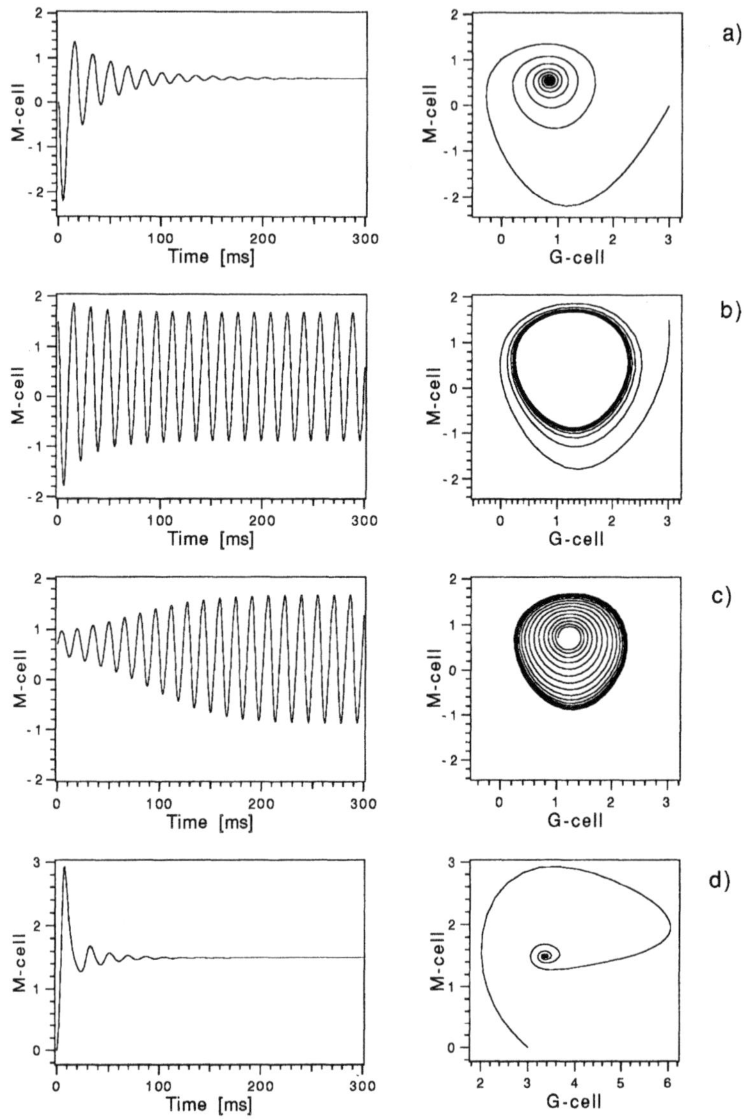

Figure 12.
Numerical integration of equation. (2a) for different parameter values and initial conditions.

$K_{mg} = -K_{gm} = 1.12$, $Q_m = 8$

a) $P = 1.99$, $m(0) = 0$, $g(0) = 0$, $m'(0) = 0$, $g'(0) = 0$

b) $P = 3.19$, $m(0) = 1.5$, $g(0) = 3$, $m'(0) = 0$, $g'(0) = 0$

c) $P = 3.19$, $m(0) = 0.5$, $g(0) = 1.1$, $m'(0) = 0$, $g'(0) = 0$

d) $P = 10.76$, $m(0) = 0$, $g(0) = 3$, $m'(0) = 0$, $g'(0) = 0$

4. RESULTS

4.1. Numerical integration of the nonlinear differential equations

Fig. 12 shows the temporal development of the system described in equation (2a) for a range of different parameter values and initial conditions. The parameters are in this case the time independent input P to the M cell and the coupling coefficients $K_{mg} = -K_{gm}$, which are symmetric and can thus be treated as a single parameter value. The integration starts at t=0, where P gets a positive constant value for t>0. For this reason the damped oscillations of Fig. 12 a) & d) are only transient effects. The system settles into a fixed point in phase-space after a short time.

In the cases of Fig. 12 b) & c) the system settles into sustained oscillations. The parameter settings are the same, only the initial conditions are different, in the first case the system spirals "inwardly" to the limit cycle, in the second case "outwardly".

For a set of certain parameter values between the first two and the last two

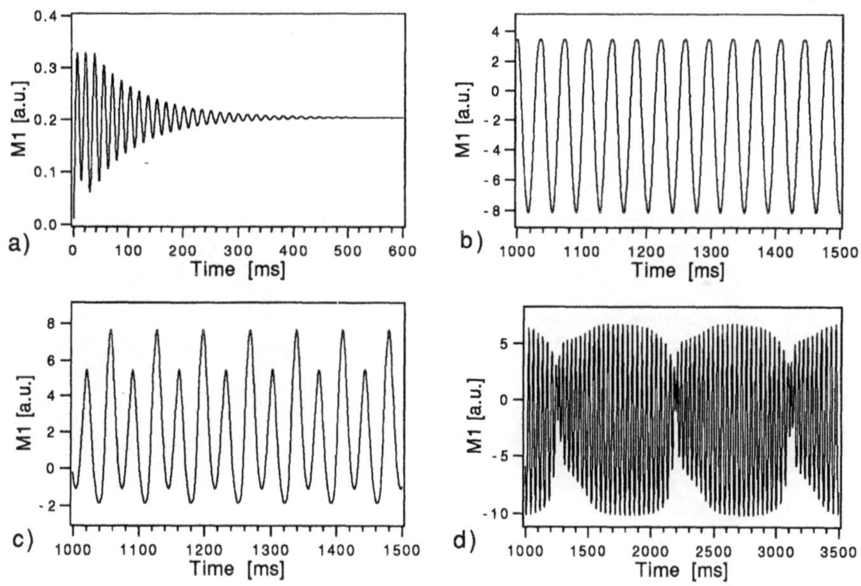

Figure 13.
Numerical integration of equation. (3), shown is the temporal development of the M1-cell. b) to d) after the transients have died out.
$m_i(0) = 0$, $g_i(0) = 0$, $m_i'(0) = 0$, $g_i'(0) = 0$, i = 1,2,3
a) $K_{mg} = -K_{gm} = 1.99$, $K_{mm} = -K_{gg} = 0.80$, P = 0.60
b) $K_{mg} = -K_{gm} = 1.99$, $K_{mm} = -K_{gg} = 0.80$, P = 2.39
c) $K_{mg} = -K_{gm} = 2.39$, $K_{mm} = -K_{gg} = 0.80$, P = 15.94
d) $K_{mg} = -K_{gm} = 2.79$, $K_{mm} = -K_{gg} = 0.80$, P = 7.97

diagrams the system changes its dynamical behavior by a Hopf bifurcation: a stable fixed point becomes unstable and is surrounded by a stable limit cycle (Hopf, 1942). The dynamic behavior of the network defined by equation (3) is more complex. Some, but not all, temporal behaviors of this system are shown in Fig. 13 and Fig. 14. Between Fig. 13 a) & b) again a Hopf bifurcation occurs, giving rise to oscillations of a single period. Between b) & c) the system

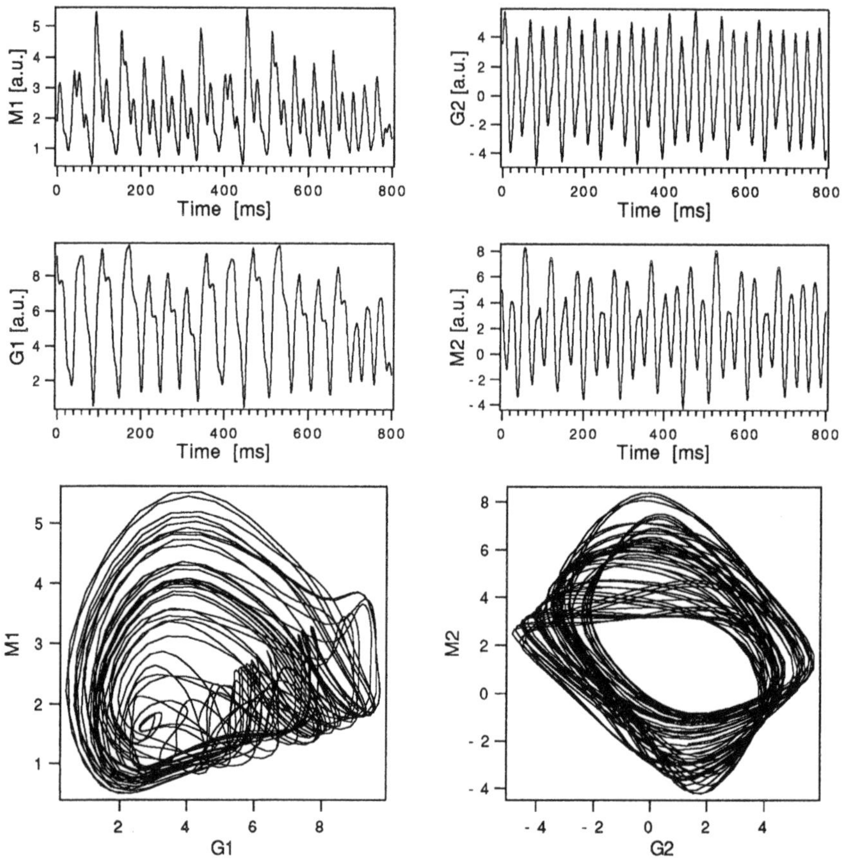

Figure 14.
Numerical integration of equation (3), shown is the temporal development of the M1, G1, M2, and G2-cells (upper four diagrams).
$m_i(0) = 0$, $g_i(0) = 0$, $m_i'(0) = 0$, $g_i'(0) = 0$, $i = 1,2,3$
$K_{mg} = -K_{gm} = 1.99$, $K_{mm} = -K_{gg} = 0.80$, $P = 10.02$
In the lower two diagrams the M1 (M2) cell is plotted against G1 (G2) cell. The M cells have a phase lag of $\pi/2$ against the G cell, therefore these diagrams represent a phase space reconstruction in 2-dim.

undergoes a period doubling bifurcation, whereas in d) two slightly different frequencies co-exist, giving rise to beat phenomena. This network is not restricted to oscillations of finite period, as is shown in Fig. 14. Here the system is in a chaotic state and the curves in phase space close after infinite time. Note that the outputs of the M1 and G1 cells are more irregular than those of the M2 and G2 cells. We attribute this to the fact, that only the M1 cell receives an input in this model.

4.2. Bifurcation diagrams

The dynamical behavior of the two networks are summarized in bifurcation diagrams (Fig. 15 and Fig 16). The bifurcation analysis was carried out using the package AUTO (Doedl and Kernévez, 1986). The diagram for the network defined by equation (2) is calculated for three different values of Q_m. Within the enclosed area, which is bounded by the Hopf bifurcation curve, sustained oscillations occur. This areas is shifted to larger values of the coupling constants for smaller values of Q_m. The parameter domain P where oscillations occur is more extended for larger values of Q_m.

The bifurcation diagram for the network of equation (3) and Fig. 16 shows more curves and also a new type of bifurcation: a saddle node. The temporal behavior of this system is now not only dependent on the choice of the parameter

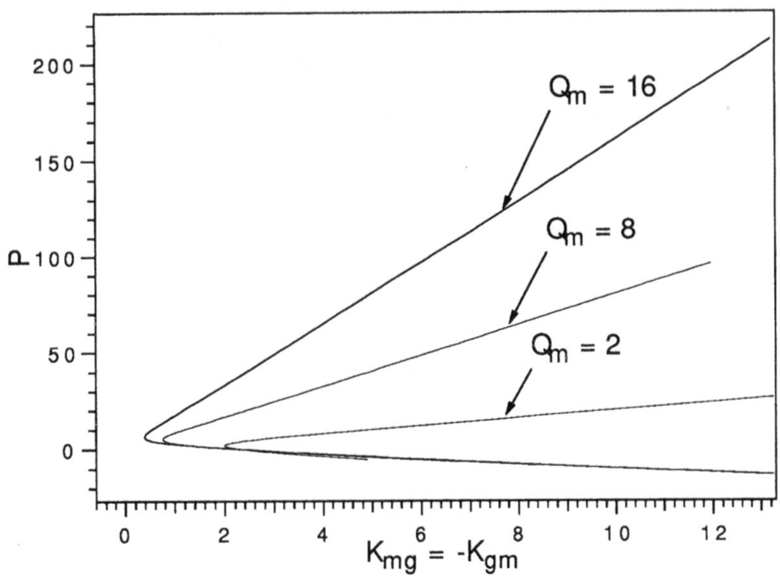

Figure 15.
Bifurcation diagram for equation (2) for three different values of Q_m. The curves show for which parameter values a Hopf bifurcation occurs.
Compare with Fig. 12.

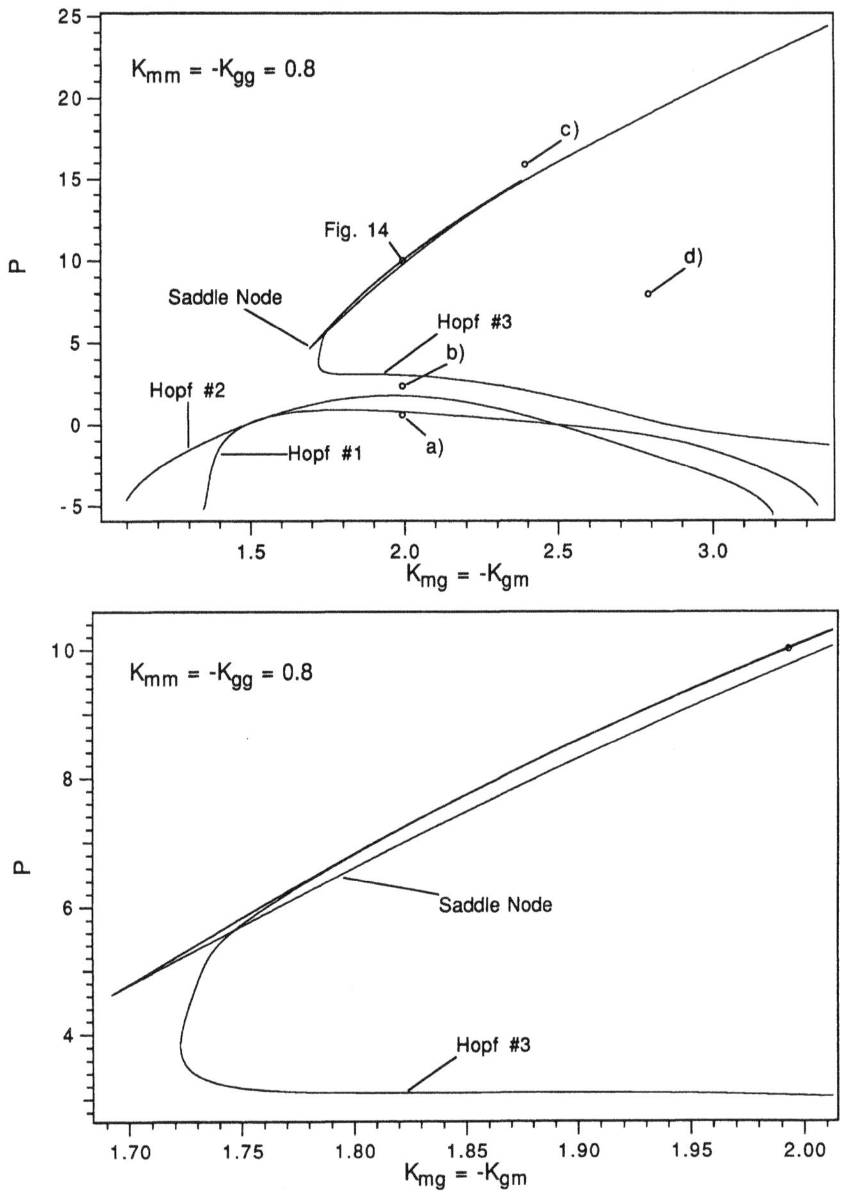

Figure 16.
Bifurcation diagram for equation (3). There exist three Hopf bifurcations and one saddle node bifurcation, the letters in the upper diagram refer to Fig. 13.

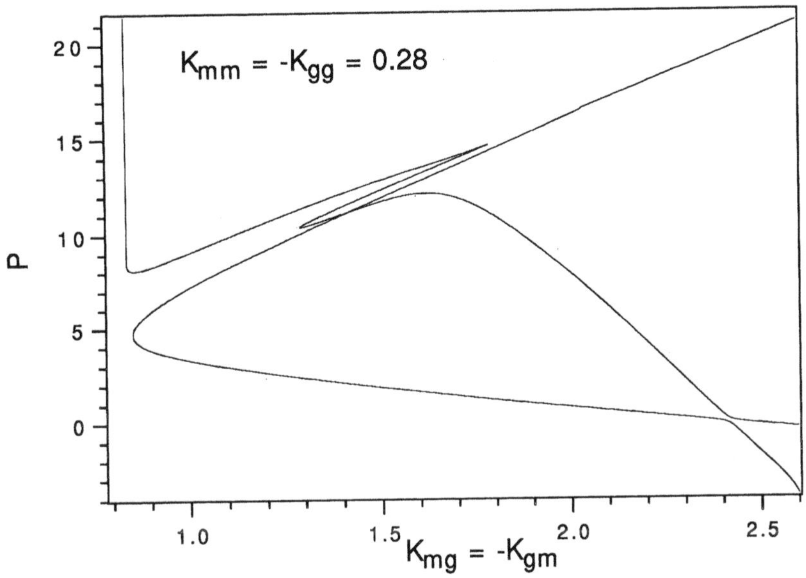

Figure 17.
Bifurcation diagram for equation (3) , only two Hopf bifurcations are shown; note the folded form of the upper one.

values, but also on the initial conditions. Stable limit cycles and saddles (points in phase space which are attracting in one direction and repelling in an another direction) can exist for a fixed set of parameter values. Basins of attraction for these two points are divided by a separatrix. The dynamical behavior of the network depends on which side of the separatrix the system starts (initial conditions).

These dynamical properties call for a deeper analysis, which is beyond the scope of this contribution.

This network has actually three free parameters and the bifurcation curves are surfaces in the parameter space. This is illustrated in the last three figures. Fig. 17 shows a bifurcation diagram for smaller values of Kmm = -Kgg . This case shows, that bifurcation curves can be folded, so that small steps through the parameter space can yield a variety of network behaviors . In Fig. 18 the tip of the lower curve of Fig. 17 is traced out for different fixed values of P, in this case the coupling coefficients are varied. The latter diagram can also be visualized as a surface, this is shown in Fig.19.

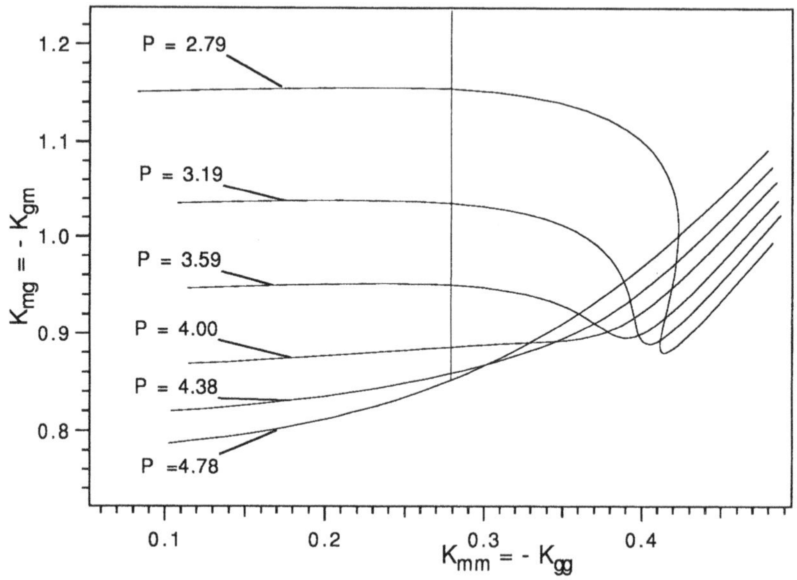

Figure 18.
Bifurcation diagram for equation (3) in this case P is fixed and $K_{mg} = -K_{gm}$ and $K_{mm} = -K_{gg}$ are varied and the Hopf bifurcation curve is traced out. The vertical line is the left tip of the lower bifurcation curve in Figure 17.

5. DISCUSSION

The basic component of the networks described here represent a local population rather than a single neuron, so that its state variables for dendritic current and pulse density can both be represented as continuous in time, and the sigmoid function is static instead of time-varying as in the case for single neurons. Two properties deriving from the physiological analysis are especially important. One is the use of a second order ODE to represent the kernel of linear dendritic integration, instead of the more commonly used first order ODE with self-excitation. The other is the asymmetry of the sigmoid, in which excitatory input increases both the output and the forward gain of the population. Both of these properties predispose the networks to destabilization by input, which is an undesirable feature of artificial neural networks, but which figures prominently in brain function (Freeman, 1991). The present results are consistent with those from other models (cited above), and also with those from preceding studies of the KII model (Freeman, 1975) with the notable exception of the verification of the existence of at least one domain of chaotic behavior with parameter values not previously selected. This finding indicates that we are far from having a complete description of this network, which is the simplest that can suffice to describe the dynamics incorporating the three types of feedback that interconnected populations of excitatory and inhibitory neurons can sustain. Further studies are clearly needed to establish this network as a kernel in models for cortical neurodynamics.

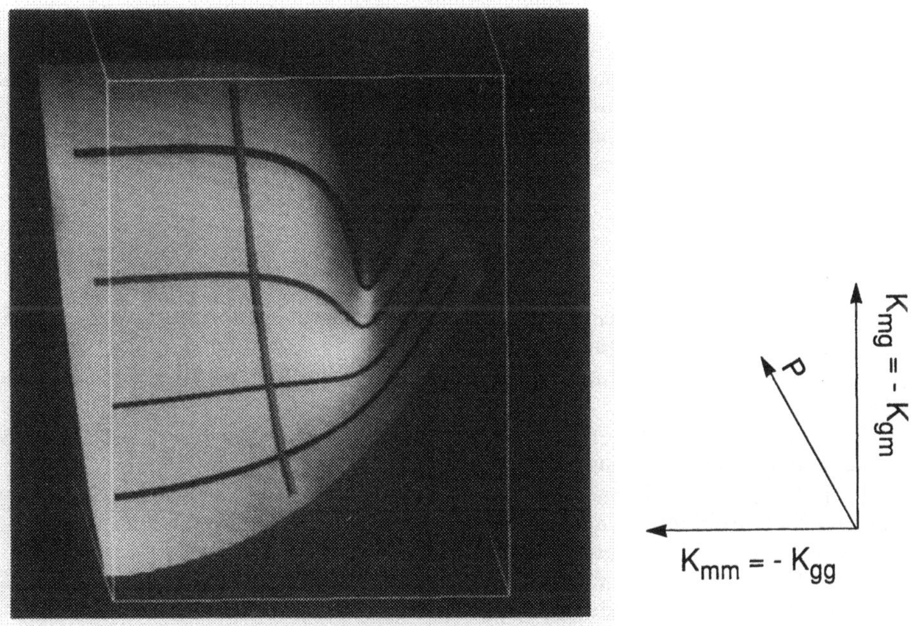

Figure 19.
The bifurcation curves of Fig. 18 form a surface in the three dimensional parameter space. If the system parameters are chosen so that they lie under the surface, the system settles to a fixed point. In the case where the system parameters lie above the surface a limit cycle with sustained oscillations occurs.

6. SUMMARY

We have shown a footpath here, which is favorable to take, when constructing a neural network, which is closely related to brain architecture and dynamics. Emphasis on the architecture was given to the fact, that masses of neurons instead of single neurons should be used as processing elements. Therefore only the input / output functions of masses were discussed in detail.

Two simple networks, consisting of excitatory and inhibitory neural masses were constructed and their temporal behavior discussed. In the next step bifurcation diagrams for a range of parameter values were constructed. These diagrams summarize the dynamical properties of the networks. In future investigations larger networks will be considered and emphasis should be made on time-varying inputs which can drive the network from one state into another, for example from periodic solutions to chaos, or from one chaotic attractor to another. Investigating these dynamical processes will be helpful in understanding how the brain processes information rapidly and with a great fault tolerance.

Acknowledgments

This research was supported by a grant from the Office of Naval Research ONR - N63373 - Nonlinear Neurodynamics of Biological Pattern Recognition. The authors gratefully acknowledge assistance in computer programming from Michael Greenspon and in computer graphics from Dr. Tim Robinson.

7. REFERENCES

Abraham R. and J.E. Marsden (1978) Foundations of mechanics, Reading MA., Benjamin/Cummings Publ.

Awrejecewicz J. ,(1991) Bifurcation and Chaos in Coupled Oscillators, Singapore, World Scientific

Borisyuk R.M. and Kirillov A.B., (1992) Bifurcation analysis of a neural network model Biological Cybernetics 66 , 319 - 325

Doedl, E.J. and J.P. Kernévez, (1986) AUTO,: Software for continuation and bifurcation problems in ordinar differential equations, Applied Mathematics Report, California Institute of Technology, 226 pages

Eeckman F.H. and Freeman WJ (1991) Asymmetric sigmoid nonlinearity in the rat olfactory system. Brain Research 557: 13-21.

Eisenberg J., Freeman W.J., Burke B. (1989) Hardware architecture of a neural network model simulating pattern recognition by the olfactory bulb. Neural Networks 2: 315-325.

Freeman W.J. (1964) Use of digital adaptive filters for measuring prepyriform evoked potentials from cats. Experimental. Neurology 10: 475-492.

Freeman W.J. (1972) Measurement of oscillatory responses to electrical stimulation in olfactory bulb of cat. Journal of Neurophysiology 35: 762-779.

Freeman W.J. (1975) Mass Action in the Nervous System. New York, Academic Press.

Freeman W.J. (1979) Nonlinear gain mediating cortical stimulus-response relations. Biological Cybernetics 33:237-247.

Freeman W.J. (1983) Dynamics of image formation by nerve cell assemblies. In: Synergetics of the Brain. Basar E, Haken H, Mandell AJ (eds) Berlin, Springer-Verlag. pp 102 - 121.

Freeman W.J. (1987) Simulation of chaotic EEG patterns with a dynamic model of the olfactory system. Biological Cybernetics 56: 139-150.

Freeman W.J. (1991) Colligation of coupled cortical oscillators.by the collapse of the distributions of amplitude-dependent characteristic frequencies Chapter 5 in: Gear, C.W. (ed.) First NEC Research Symposium. Philadelphia, SIAM, pages 69-103.

Graziadei P. (1990), In: Coleman, JR (edt.), Development of sensory systems in mammals, p. 519 - 566, John Wiley & sons

Guckenheimer J. and Holmes P. (1983) Nonlinear Oscillations, Dynamical Systems, and Bifurcations of Vector Fields. Berlin, Springer-Verlag

Hopf, W. (1942) Abzweigung einer periodischen Loesung von einer stationaeren Lösung eines Differentialsystems. Berlin Math.-Phys. Klasse Sachs. Akad. Wiss. Leipzig 94: 1-22.

Jakubith S., Rotermund H.H., Engel W., von Oertzen A. and Ertl G. (1990) Spatiotemporal Concentration Patterns in a Surface Reaction: Propagating and Standing Waves, Rotating Spirals, and Turbulence, Phys. Rev. Lett. 65: 3013 - 3016

Khiblik A.I., Borisyuk R.M. and Roose D. (1992), Numerical bifurcation Analysis of a model of coupled neural oscillators, International Series of Numerical Mathematics, Vol. 104 (1992), 215 - 228

Krischer K. (1990), Nichtlineare Dynamik zweier Grtenzflächenreaktionen, Thesis (in german) , Freie Universität Berlin, Germany

McCulloch W.S. and Pitts W. (1943) A logical calculus of the ideas immanent in nervous activity. Bulletin of Mathematical Biophysics 5: 115 - 133

Mead C. (1989), Analog VLSI and neural systems , Reading, Mass. : Addison-Wesley, 1989

Rummelhart D.E. and McClelland J.L. (Eds.), (1990) Parallel distributed processing (Vol.1. Foundations), Cambridge, MA: MIT Press.

Brain Theory / A. Aertsen (Editor)

Dynamic Aspects of Cortical Function: Processing and Plasticity in different Sensory Modalities

Hubert R. Dinse, Friederike Spengler, Ben Godde and Bernt Hartfiel

Institut für Neuroinformatik - Lehrstuhl für Theoretische Biologie

Ruhr-University Bochum, Germany

Abstract

Based on the assumptions that the entire temporal structure of neuron responses carries significant information and that single cell receptive fields (RFs) and representational maps (RMs) typify representative stages of cortical processing, their dynamic properties were investigated in three modalities. The resulting time dependent receptive fields and representational maps are interpreted as specific adaptations to processing of inherently time-variant signals. An also time-based concept of temporal distributed processing (TDP) across subcortical and cortical substrates is presented that accounts for time-dependent RFs and RMs due to feedforward, lateral interactions and feedback mechanisms. Post-ontogenetic plastic adaptive processes are supposed to act within this scenario to provide sufficient flexibility to slow changing conditions in the environment and individual performance requirements. In our view, the stimulus-triggered emergence of complex spatio-temporal activity patterns are a key feature of sensory information processing. A hypothesis is presented that attempts to map such spatio-temporal activity onto behavioral states.

1. General assumptions

1.1. Observables

The article deals with dynamic aspects of cortical functions. This venture poses several constraints upon the possible observables and substrates. Accordingly, the studies will focus on cortical areas with special emphasis on subcortical, thalamic structures.

There is a substantial variety of processing elements that are present at different subcellular and cellular levels, i.e. channels, synapses, single neurons and cortical areas. However, the logic behind the transformational steps that lead from one element to another is in most cases puzzling. In search of basic elements which can provide the operational basis for a description of informational processing, we choose single neurons and as a next higher, macroscopic level, cortical areas. At an early processing level, they contain topologically ordered representations of sensory environments, so-called representational maps.

To rely mainly on RFs is mostly because they provide a link between the outside world and neuron activity that can be assessed by available recording techniques. It is therefore mainly a pragmatic and methodological motivated measurement instruction. In this view, the term RF is used in a very broad meaning: RFs contain rules for mapping physical events onto representations based on neural activity (see also chapter RF concepts).

The advantage of using maps as an additional basic component of processing comes from the unambiguous relationship between RFs and maps. It is the topology of RFs, their neighborhood relationship that determines the map. Detailed knowledge of RF properties appears therefore essential for understanding representational maps and the networks they contain. Also, due to new recording techniques like optical monitoring, maps can now be measured directly, providing an additional operational motivation.

Clearly, there is a wide gap between single neurons on the one hand and cortical areas on the other. There are many names for concepts that lie in between, such as neuron assemblies, neuron groups, neuron populations or population coding, and subrepresentations. It appears fair to attribute the conceptual difficulties primarily to a lack of experimental procedures to measure them as direct and straightforward as single neuron RFs. Though in the beginning reports on multi-neuron studies were mostly anecdotal, recent progress in routinely performed multi-electrode experiments developed in many laboratories provides first insights into that processing level, leaving many open questions, related to the feasibility of such experiments, data analysis and interpretation (Abeles 1977, Krüger 1983, Reitboeck 1983, Gerstein 1988, Vaadia & Aertsen 1992). Accordingly, there is a need to diminish this gap by combined efforts based on new methods, experimental data and conceptual frameworks.

1.2. Time scales

Neural systems organize behavior in naturally occurring environments. However, environments and the constraints they impose change on a variety of time scales. Therefore, each system operating in such an environment must possess considerable adaptive capacities. Under the assumption that the ontogenetic development is completed, these adaptational capacities of adult systems must be twofold: First, in order to deal appropriately with the dynamic dimensions of the environment, there must be a match between processing dynamics with that of the dynamics in the stimulus space.

Secondly, as these requirements are subject to slow changes, there must be additional plastic-adaptive processes. In fact, during the last years considerable post-ontogenetic plastic-adaptational capacities have been demonstrated for cortical systems. Besides long-term changes due to lesion- and learning-induced reorganization (Merzenich et al. 1984,

1988, Eysel 1982, Kaas et al. 1990, Pascal-Leone et al. 1992, Recanzone et al. 1992a), numerous manipulations have been shown to act at a minute-scale on cortical processing such as classical conditioning, a few hours of paired peripheral tactile stimulation (PPTS) or intracortical microstimulation (ICMS) and even peripheral lesions (Weinberger et al. 1990, Ohl et al. 1992, Recanzone et al. 1992b, Spengler & Dinse 1992ab, Godde & Dinse 1993, Calford & Tweedale 1988). However, in spite of the overwhelming evidence for plastic-adaptive capacities, systems must possess some generic stability. With this we mean the requirement of the entire system to perform stable operations despite continuous changes in the environment. One can expect a trade-off between modifiability on the one hand that warrants sufficient flexibility, and stability on the other that enables the system to achieve a state of invariance. One possibility to meet both requirements would be to hold them separated along a learning-related time scale from that determined by the dynamics of the stimulus space. In all neocortical modalities, the latter is in the millisecond-range, sufficient to allow separation from the minute to hour and longer time scales. This separation, however, is supposed not to prevent interactions between both which are necessary to allow learning related plasticity to interfere with the normal, ongoing processing.

Based on these considerations, the following time scales will be studied:

1. a ten to several hundred millisecond scale, during which intrinsic RF dynamics become apparent
2. a minute to hour scale that deals with plastic-adaptive processes that overlay and interact with the faster scale.

1.3. Modalities

Experiments were performed in anesthetized rats and cats with recordings made in striate and extrastriate visual areas as well as in primary somatosensory and primary auditory cortices.

Comparison of differences and similarities found in different modalities offer a unique way of differentiating modality specific modes of processing from those that are biased by constraints of cortical architecture and cortical processing principles.

2. Time dependence of receptive fields

2.1.1. General methodology

Time dependence of receptive field organization was investigated by applying the so-called time slice technique. The main idea behind this protocol is to avoid average or maximal measures of neuron responses, but to utilize and to rely on the entire response episodes. First, descriptors of receptive field properties were conventionally measured and calculated (Dinse et al. 1990a, 1991). Descriptors used in the following were response planes (visual and somatosensory system), orientation tuning and length tuning (visual system), frequency-intensity planes and frequency tuning (auditory system). Next, time windows or time slices of constant width were chosen and the respective descriptors were recalculated for each time slice. By this, a temporal sequence of descriptors is obtained which provides information about the instantaneous and actual content of the descriptor. It is worth mentioning that this approach can be traced back to Podvigin who used it for investigation of spatio-temporal RFs of neurons recorded in the lateral geniculate nucleus (LGN) in the visual portion of the thalamus (1974). For striate cortex, Shevelev reported changes of preferred orientation over time (1987). Using the reverse correlation technique as introduced by de Boer and Kuyper (1968, see also this volume), similar results on time dependence of RF organization have been reported for neurons of the frog and cat auditory system (Aertsen et al. 1979, Aertsen & Johannesma 1981, for review Eggermont 1991) and of the cat visual cortex (Krause & Eckhorn 1983, Jones & Palmer 1987).

2.1.2. Temporal isolation

All experiments were performed using so-called stationary, on-off type stimulation in order to relate the resulting temporal structure of neuron responses unequivocally to the highly reliable and controllable time course of sensory stimulation which is not possible when moving stimuli are used. Also, the described findings are from an experimental protocol called *temporal stimulus isolation*. Under these conditions, cells are studied by stimulating them from a *zero state* condition. Actually, this is the way most studies are done, which requires long interstimulus-intervals (ISIs) of more than 1/2 second. This protocol prohibits interaction between successive stimuli thus warranting that the cell has reached zero state. Therefore, the observed dynamics are interpreted as the manifestation of the intrinsic dynamics of the system, undisturbed by additional dynamics caused by timing constraints of the stimulation (Dinse et al. 1990a). So far, little is known about possible differences between zero state conditions and different steady states which can be regarded as perturbations of the system at different base levels.

Available data suggest that sufficient stimulus isolation is only warranted for long ISIs which, however, can be different in different modalities (Watanabe et al. 1989, Selpien & Dinse 1992, Dinse & Schreiner unpublished). First experiments indicate that the above defined intrinsic dynamics are subject to considerable changes when cells are no longer stimulated from zero state but from various degrees of a steady state introduced by additional timing effects due to short ISIs.

2.2.1. Visual system

Details of RF dynamics have been described elsewhere (Best et al. 1989, Dinse et al. 1990a, 1991). They consist of substantial changes of RF organization indicating non-separability of space and time as well as time dependence of tuning characteristics.

Most interesting are observations that cortical selectivity such as orientation selectivity or length preference evolves over time preceded by early, unspecific states in which the cell appears invariant against changes of the stimulus. This gives rise to dynamic orientation selectivity and dynamic length preference. Most notable is the match between the emergence of anisotropic RF organization over time and the development of orientation selectivity which both evolve after a state of isotropic organization during which no selectivity can be found.

2.2.2. Auditory system

Time dependence of auditory RF in AI of cats are under investigation in collaboration with Dr. Schreiner from UCSF (Dinse & Schreiner 1992). When RFs were analyzed with the time slice technique, complex diagonal bands of activity across the frequency-intensity planes became visible which were found in equal proportions along the low or high frequency side of the tuning curves. As a rule, these bands were concealed in conventional, time-avaraged tuning characteristics. When the data were re-plotted as frequency-time plots, the time-dependent frequency characteristics revealed frequency transitions and time-dependent multi-peaked tuning properties. In the extensive studies of the grass frogs auditory system by the Nijmwegen group, time dependent frequency tunings have been reported (Aertsen & Johannesma 1981, Eggermont 1990).

2.2.3. Somatosensory system

2.2.3.1. Single cell RF dynamics

Response plane techniques combined with the time slice technique were recently also applied to study RF dynamics of neurons in the hindpaw representation of rat primary somatosensory cortex (SI). Up to 28 stimulations sites on the hindpaw were selected for computer controlled tactile stimulation (Dinse et al. 1993). Preliminary inspection of the data revealed a time dependent spread of activity resulting in complex spatio-temporal pattern. Most notable was the gradual shift of activity starting at the distal part of the digit representation which then emerged further proximal within the skin representation of the pads or the emergence of focal zones of activity that then spread towards the peripheral portions of the time-averaged RFs, resembling very much the overall spatio-temporal behavior described for the visual and auditory system (Fig. 1 A).

A similar approach was taken in the whisker barrel representation in the ventral posterior (VP) complex of the thalamus of awake rats (Nicolelis et al. 1992). These authors observed time dependent spatial shifts of RF centers over 50 ms. Inspecting latencies encountered in different cortical layers following vibrissa stimulation, Armstrong-James et al. (1992) described complex patterns of depth-dependent temporal sequences of activation of laminar subunits.

2.2.3.2. Representational map dynamics

Given the above described relationship between RFs and representational maps and, also taken into account the observed dynamic nature of RF organization found in all modalities

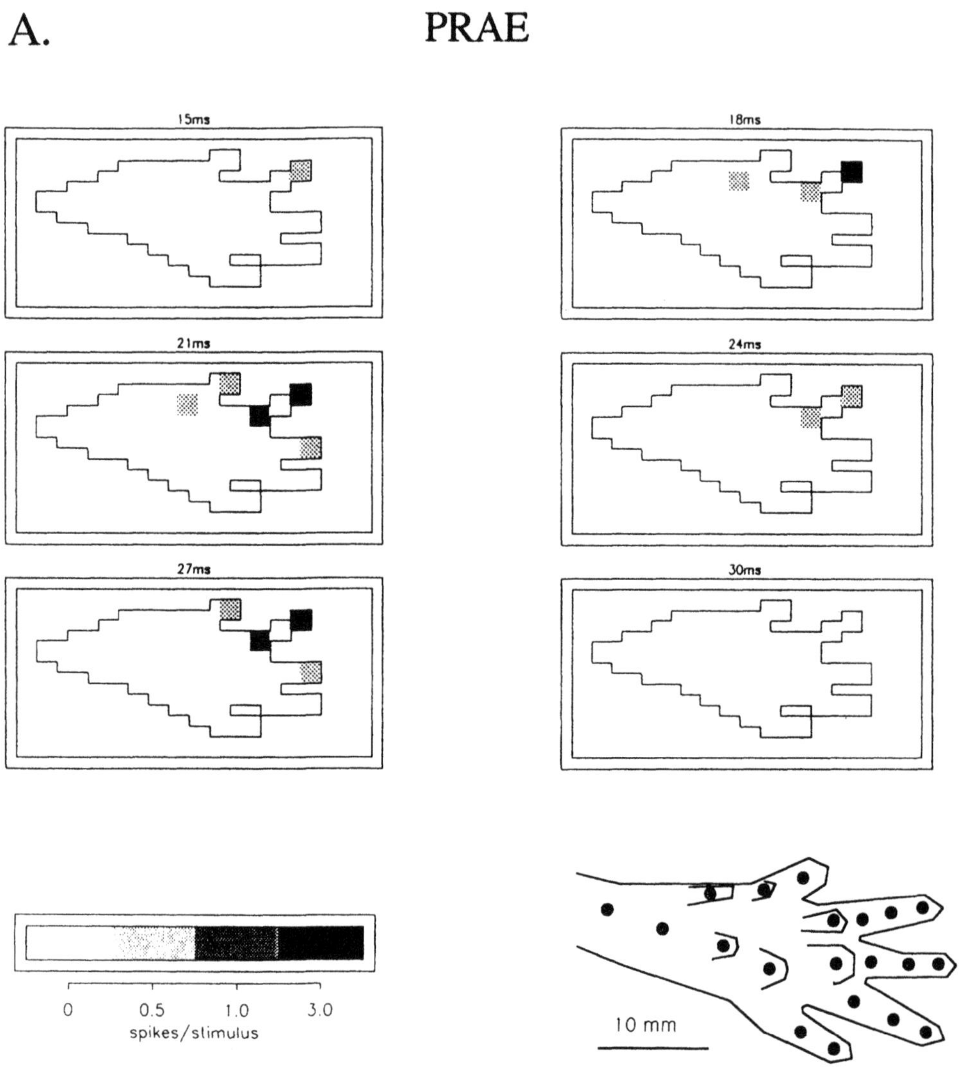

Fig. 1. Time dependence of RF organization recorded in the hindpaw representation of rat somatosensory cortex (SI). Temporal sequence of response planes in steps of 3 ms illustrating the spatio-temporal spread and shift of activity following tactile stimulation. The locations of 20 selected stimulation sites on the hindpaw are marked on the figurine at the bottom. Each pixel represents activity evoked by a given site. Neuron activity is given in different gray levels. A. Control (PRE) and B. after 5 h of PPTS-induced plasticity (POST). Paired peripheral tactile stimulation (PPTS) increases RF size and elongates the temporal structure of the neural response leading to an overall alteration of RF dynamics.

B. POST

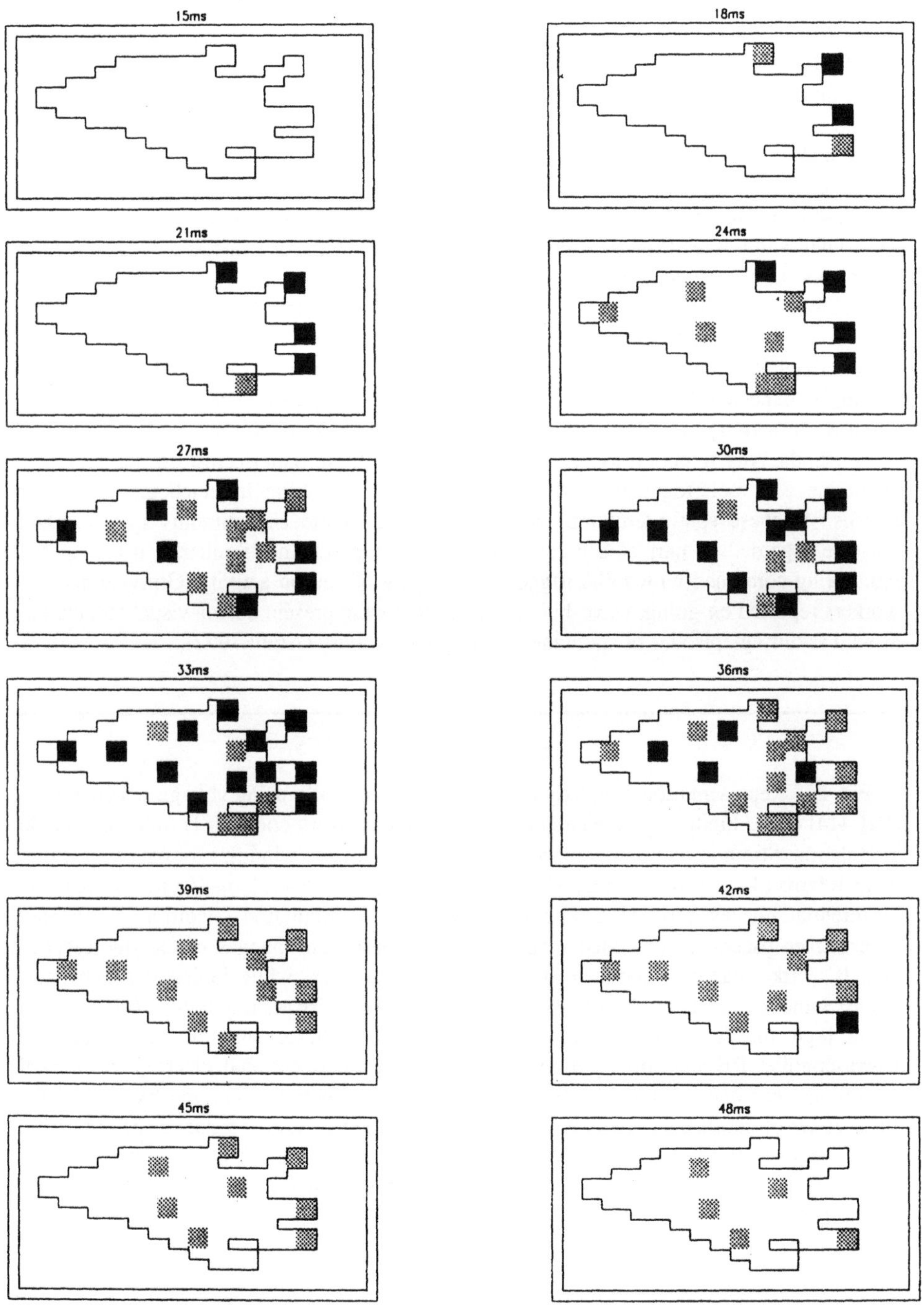

tested so far, it appears conceivable that comparable dynamics act at the macroscopic level of cortical maps. To unravel those dynamics we initiated studies based on optical monitoring using voltages sensitive dyes (RH 795, RH 414) in the somatosensory cortex of the hindpaw representation. This method allows real-time monitoring of two-dimensional, spatio-temporal activity distributions (Grinvald et al. 1988). Data acquisition windows were set in such a way that a coarse temporal resolution of the time course of the activity distributions could be obtained (Dinse et al. 1992).

The spatial extent of the representational zones following point-like stimulation at selected skin fields was different when recorded at different times after stimulation. Also, the zone of cortex being activated was considerably larger when longer integration times were used. In this case, the activated area covered the different zones mapped with the short integration times (Fig. 2), suggesting similar dynamic properties as described for the organization of single cell RFs.

Preliminary results in the somatosensory and auditory cortex (Fukunishi et al. 1992) in fact revealed the existence of considerable dynamics that are compatible with what is known from RF dynamics. However, they imply much more complicated patterns than would be expected from single cell recordings. The available data suggest wave-like patterns of activity, implying that cortical maps change on a time scale of tens of milliseconds. Any temporal averaging cannot resolve these pattern. Instead, the temporal averaging will melt together small zones of activity, resulting in much larger activity pattens. As these spatio-temporal patterns are reproducible and stimulus specific, they must be regarded as part of a macroscopic processing scheme. Similarly, using optical monitoring combined with spike triggered averaging of neuron activity, Grinvald and co-workers reported on-going, wave-like processes that were present during visual stimulation as well as during episodes of spontaneous, on-going activity (Arieli 1992).

Fig. 2. Representational map dynamics recorded optically using voltage sensitive dyes (RH 414) in the hindpaw representation of rat somatosensory cortex (SI) following tactile stimulation. Stimulation point is marked by an asterisk. Time windows for data acquisition were 100 ms (30 to 130 ms) in a., 35 ms (30 to 65 ms) in b., and 55 ms (65 to 120 ms) in c. Differential images between a stimulation and a non-stimulation condition. Left: Gray copies from pseudo color coded maps of activity based on changes of the fluorescence signals. Dark areas correspond to decrease of fluorescence, white indicates no changes. Right: Same images as on the left side. Focal zones of activity are highlighted (black areas) using high thresholds for pseudo color coding to indicate shift of their locations. Corresponding RF sites on figurines as measured by conventional electrophysiological recording techniques. There are clear differences in the spatial extent and location of the focal zones of activity recorded at different time windows. Note the shift of activity from distal parts of the digit to its more proximal portion and parts of the pad representation over time and a much larger zone after longer integration times, which seems to melt together the smaller ones.

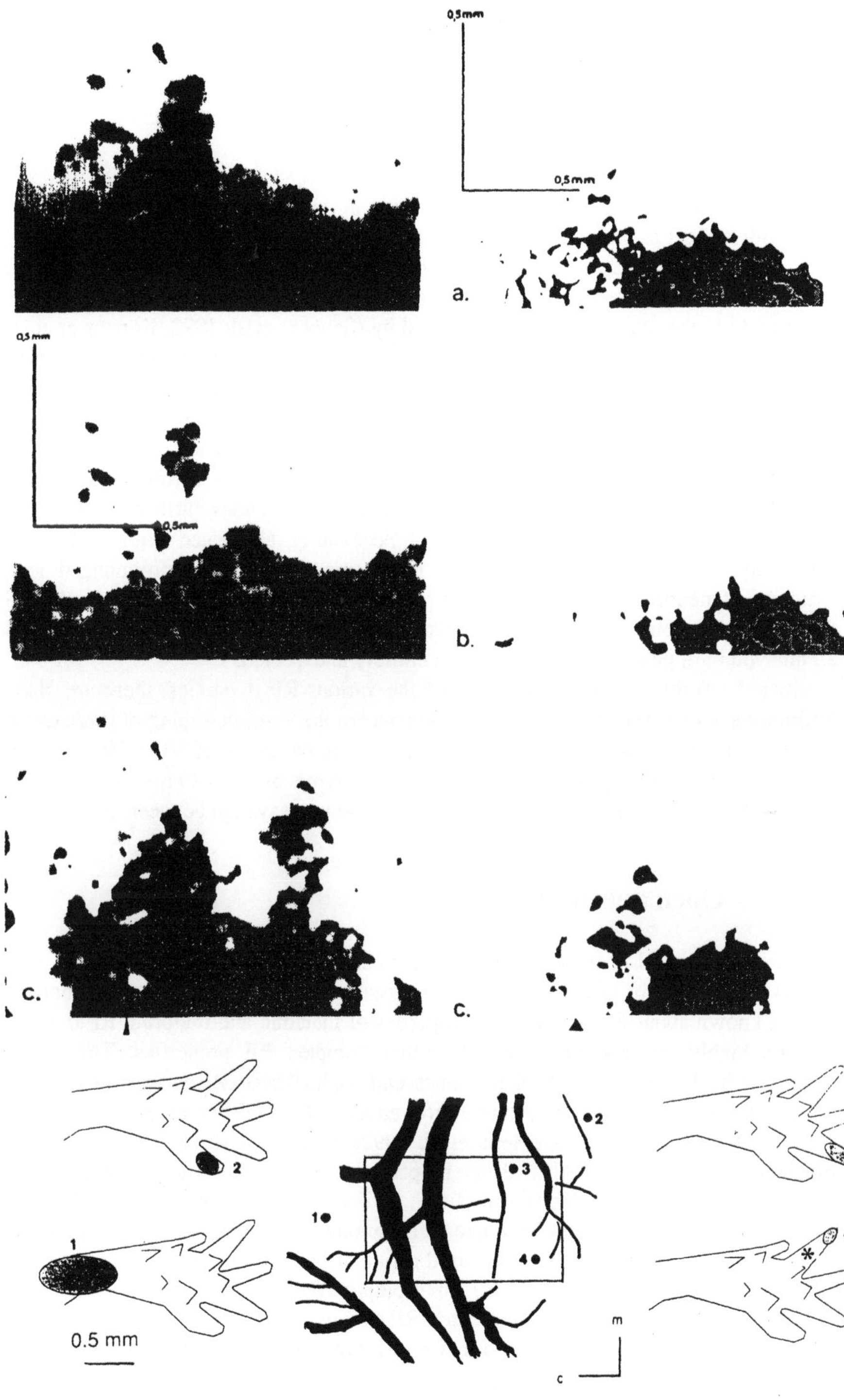
0,5mm
0,5mm
a.
0,5 mm
0,5mm
b.
c.
c.
1
2
3
4
0.5 mm
m
c

2.3. Common features of time-dependence and absolute timing

Time dependence of RFs was now investigated in three modalities whose organization principles in terms of neocortical architecture suggest a common mode of processing (see also chapter 3.2).

As a rule, two response components could be distinguished which reflect two different states of responsiveness. It should be noted that in this article the terms *early* and *late* are used with respect to the response following sensory stimulation (see also chapter 3.1.2). It does therefore not include putative low-frequency oscillatory components (Dinse et al. 1990a, 1991). The response components denoted here as *early* and *late* therefore correspond to the *latency peak* as introduced by (Schöner et al. 1992, Kopecz et al. 1992). According with this definition, the early component exhibits an unspecific state which is basically not tuned at all and which is paralleled by a simple, isotropic RF organization. In contrast, the late succeeding state displays a sharp tuning which, in many instances shifts the maximal tuning, and which is characterized by the emergence of an anisotropic RF organization. However, in terms of neural activity, the spike rate during the early state can be much higher than during the late state. This fact provides further evidence for the argument that high spike rates must not necessarily associated with high levels of selectivity (Dinse et al. 1990a). Under the assumption that the pronounced temporal structure of neuron responses is in fact used as a time code (Perkel & Bullock 1968, Dayhoff & Gerstein 1983, Freeman & Skarda 1985, Optican & Richmond 1987, Abeles et al. this volume), these findings appear not entirely unexpected.

Apart from the obvious similarities of the various RF dynamics, there are also clear differences across modalities which mainly concern the absolute timing of these dynamics. While the time constants of the visual system are in the range of 80 to 200 ms, the time constants of the auditory are much shorter and in a range of 10 to 20 ms, while preliminary data indicate an intermediate timing in the somatosensory system between 10 and 50 ms.

2.4. Open questions

There are several open questions related to single cell RF dynamics. So far, our own results focus entirely on cortical areas dismissing subcortical substrates. Although very little is known about time-dependent properties of thalamic and midbrain RF dynamics, it appears highly conceivable to expect rather complex RF properties. This is in fact suggested by findings on time-dependence and sophisticated RF properties in the visual (Podvigin 1974, Stevens & Gerstein 1976), auditory (van Gisbergen et al. 1975, Wagner 1990) and somatosensory (Nicolelis et al. 1992) subcortical structures. However, at the momenı iı is unclear whether these properties are due to cortico-thalamic feedback, properties generic to subcortical structures, or both. The temporal distributed processing scheme (TDP) described later in chapter 3.1.4. predicts time-dependent RFs at all levels of processing due to the primarily time-based nature of processing along sensory pathways.

The layer-specific arrangement of input-output connections found in all modalities is also mostly ignored in current research on RF dynamics. This is even more unsatisfactory, as this arrangement is supposed to give rise to complex depth-dependent patterns over time

(Best et al. 1986, Armstrong-Jones et al. 1992). Including such depth-dependent dynamics will certainly help to understand time-dependent RFs.

As to time-dependent representational maps as described by Dinse et al. 1992 and Fukunishi et al. 1992 using voltage sensitive dyes, several difficulties were encountered. Most impressive, the resulting two dimensional activity distributions were only partly compatible with what is known from microelectrode recordings. For example, the well-established iso-frequency lines of AI could not be visualized (Fukunishi et al. 1992). These difficulties can partly be attributed to the fact that micro-electrode-based control measurements are temporal averages. Besides this, concepts like iso-frequency organization are obviously a too schematic image of cortical organization to be true. Careful mapping experiments (Schreiner & Mendelson 1990) based on a parametric analysis revealed in fact a much more complicated, patchy-like topography in AI which resembles the patchy character described by Fukunishi et al. (1992). It remains to be seen whether similar difficulties will be encountered in the other modalities.

2.5. Modality-specific adaptations to inherently time-variant signals

The available data suggest that time dependence of RFs is a general feature of cortical information processing. Its functional significance, however, await further exploration. Time dependence of RFs is interpreted as modality-specific adaptation to the processing of inherently time-variant signals. This idea together with the implications of a time-based nature of thalamocortical and intra- and interareal processing will be discussed in the following chapters.

3. Mechanisms vs. Functions

The above described findings can be discussed under two main aspects. The first one deals with the mechanisms that might be responsible for rendering these dynamics, the second one speculates about their possible functional implications.

3.1. Mechanisms

3.1.1. Feedforward vs. feedback

In a first attempt it seems conceivable to attribute the observed RF dynamics to purely *feedforward* mechanisms. In addition, differences in the time course of facilitatory and suppressive actions might play an important role, an aspect denoted in the following as *lateral interaction*. Finally, *feedback* mechanisms are likely to contribute substantially to the observed effects.

It might be tenable that within a pure feedforward system, RF dynamics can evolve due to specific distributions of latencies and thresholds. However, there is little doubt that cortical and subcortical systems are paragons for the dominating appearance of feedback (Braitenberg 1986, Braitenberg & Schüz 1990, Freeman & Skarda 1985, von Seelen et al. 1986, 1987). It therefore appears appropriate to incorporate feedback mechanisms from the very beginning.

We have recently proposed a very general feedback scheme based upon which the above described dynamics could be explained (Dinse et al. 1990a). Here, we suggest a more detailed framework with the advantage that it incorporates the contributions of feedforward, lateral interactions and feedback, with special emphasis on the latter. Here, feedback mechanisms are not treated in a formal, abstract sense but can be pinned down to identified anatomical substrates. This framework is essentially based on experimental data on absolute timing conditions (response latencies and response durations) observed in different processing stages of the visual system, but similar data for the auditory and somatosensory system available so far are fully compatible with this scheme. The entire concept was called *Temporal Distributed Processing* - TDP.

3.1.2. Timing across hierarchical stages

In order to find the absolute timing conditions of processing along different stages of the visual system, we inspected visual response latencies and response durations following on-off stimulation in the lateral geniculate nucleus (LGN), area 17, area 18, 19, area PMLS (visual area in the posteromedial portion of the lateral suprasylvian sulcus) and area 7. Details of the connectivity between these substrates are laid down in Dinse 1989, Dinse et al. 1990a). There is general agreement about the serio-parallel nature of the visual pathway. However, the clear hierarchical character as demonstrated for the primate visual system (van Essen 1985), though present also in cats, is not as obvious there. The main results, illustrated in Fig. 3 for LGN, area 17 and PMLS, are the following 1. There is a broad scatter of latencies over a range of 40 to more than 100 ms. 2. There is also a broad scatter and wide range of response durations between 50 to up to more than 200 ms. 3. Accordingly, there is a considerable overlap of responses suggesting that most of the neurons within different stages of a hierarchically organized sensory pathway are simultaneously active when stimulated.

3.1.3. No wait-states in the subcortical-cortical processing chain?

On the basis of the rich body of neuroanatomical findings we conclude that processing in the sensory systems in both cats and primates occurs along a heavily interconnected system of subcortical and cortical stages. In the light of the above described findings, this chain of processing seems to be activated in a temporally continuous fashion. The high percentage of overlap of response latencies between subsequent stages, which indicates a high percentage of simultaneously activated neurons, makes it rather unlikely that proceeding stages wait until the end of the processing of a previous one. What seems more plausible is a continuous interaction that links together all stages of the processing path including higher-order areas (Dinse et al. 1990a).

Reports concerning response latencies measured in monkey inferotemporal cortex (IT) following stimulation with faces (Perret et al. 1982), geometric patterns (Richmond et al. 1987) or objects (Tanaka et al. 1991) were inspected, an area which is regarded as an example of a sophisticated high-end substrate for processing of complex visual objects. According to their data, response latencies in the range of 80 to 100 ms are a common finding. This fits remarkably well into the outlined framework of a continuous, interareal interaction that is essentially time-based.

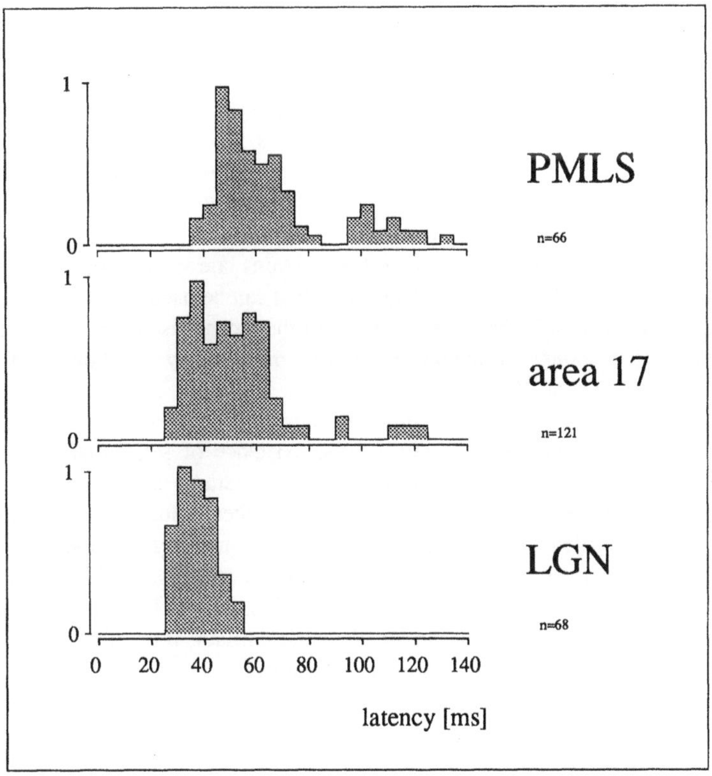

Fig. 3. Latency distribution of neurons recorded in the lateral geniculate nucleus (LGN), area 17 and area PMLS (visual area in the posteromedial portion of the lateral suprasylvian sulcus) following stationary flashed stimulation. Note the considerable overlap of the onsets of firing.

3.1.4. Temporal Distributed Processing - TDP

The basic idea of the TDP concept is illustrated in Fig. 4. The idea is grounded on the lack of wait-states which favor a temporally continuous mode of interactions. In detail, the profound overlap of response latencies due to only minor time differences between different stages suggests that the first available information about external stimulation is fed without delay through the entire pathway. This scheme represents the feedforward aspects of the response characteristics. During the continuation of the response, this behavior is repeated. However, after this very first response components, different types of lateral interactions and feedback mechanisms come into play. Among these are the thalamo-cortical feedback, the intracortical feedback and the feedback from other and higher areas. The time required for this action to impose impact might be variable for each of the described mechanisms and might depend on the type of stimulation and on the modality. Taken together, these temporally delayed influences can be assumed to modify to a considerable but variable degree the informational content being processed. It seems therefore conceivable that neuron responses are changed systematically over time making late components completely different from early ones. This is exactly what could be found experimentally: late response components differed decisively and systematically from early ones and were found to carry either high-level selectivity or modified selectivities, both not present in the early components.

In this view, TDP represents a scheme that explains late selectivity and time-dependent selectivity. Moreover, TDP provides a substrate that can be used for routing back memory-related information from higher areas. In view of the recent discussion of "memory-based" processing, late response components are the most likely candidates to carry such information (Poggio et al. 1992).

The lack of any types of wait-states bears interesting implications for brain models and concepts of cortical functions that rely on the existence of sequential processing in the sense that some information is extracted at a given stage and then used as input for processing in subsequent stages: in view of the described results such assumptions might turn out to be not sufficiently close to processing modes used by real brains.

As for parallel processing of aspects like form and motion (and color) in a system that is morphologically not entirely segregated, the idea of TDP would allow both to separate and to combine the different channels within the time domain, thus offering high functional flexibility independent of neuroanatomical constraints.

Fig. 4. Schematic illustration of the temporal distributed processing scheme (TDP). Schematic neuron responses are shown as post stimulus time histograms (PSTHs) for different stages of an assumed sensory pathway. Due to the time dependent action of feedforward components, lateral interactions, different types of feedback and temporally delayed and modified feedforward, the late response episodes are assumed to be significantly altered in their informational content compared to the early parts, independent of their lower firing rates.

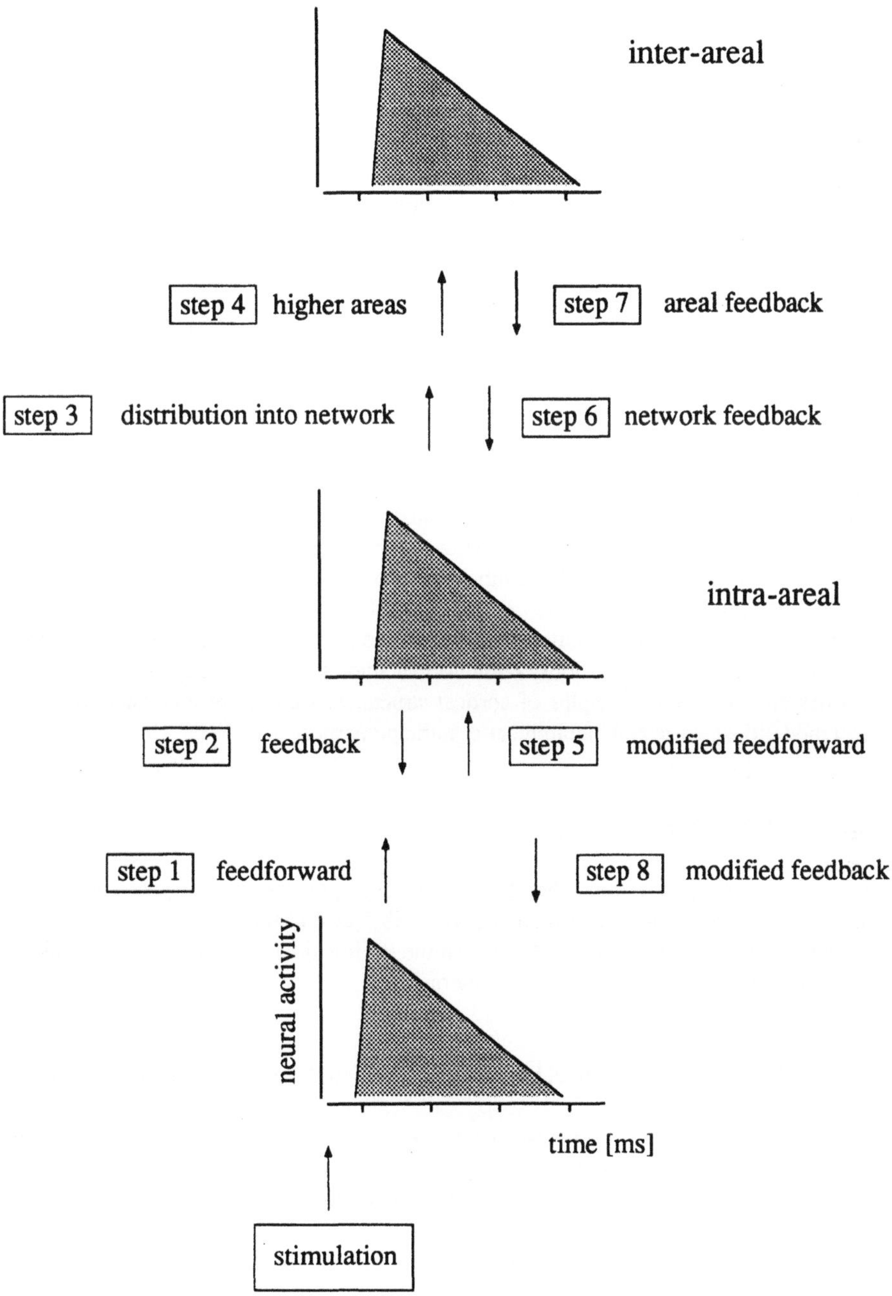
inter-areal
step 4
higher areas
step 7
areal feedback
step 3
distribution into network
step 6
network feedback
intra-areal
step 2
feedback
step 5
modified feedforward
step 1
feedforward
step 8
modified feedback
neural activity
time [ms]
stimulation

3.2. Anatomical constraints for modality-specific RF dynamics

While there are considerable similarities across modalities for the time-dependence of RF properties, the absolute timing is very different. One might have expected substantial differences along the periphery and subcortical pathways, but that as soon as the information enters cortical areas, a similar, now cortex-based time scale would be encountered. This is obviously not the case. However, the differences found are quite remarkable with respect to the cortical architecture.

Guided by the assumption of a structure-function relationship, many attempts were made during the last years to extract and identify common principles of neocortical neuroanatomy that are shared by all cortical areas across species and modalities. Due to their general properties, they can be assumed to be causally related to cortical functions. Especially, the work of Braitenberg and Schüz (1990) has contributed considerably to our understanding of cortical architecture. Based on their data and that of others, a "mean neuroanatomy" was defined and incorporated into computer simulations (details see von Seelen et al. 1986, Krone et al. 1986). These simulations predicted time-dependent RFs at time scales that were in fact observed in the visual system. Now, the interesting question arises what parameters might be responsible for the differences of the time courses observed experimentally in visual, auditory and somatosensory system. Candidates are thresholds, transmission times, time constants (excitatory-inhibitory), nonlinearities and the degree of feedback. Neuroanatomical analysis combined with computer simulations might offer insight in these questions. While former studies stressed the need for extracting similarities and common principles of cortical structures, studies are now necessary that unravel modality-specific and hierarchical-specific properties.

3.3. Functions

It is an interesting question whether informational processing strategies and coding properties of the brain are experimentally accessible at all (see also chapter 4). In the following, some attempts are made to outline a functional framework that relates information processing to neural activity pattern.

3.3.1. Time-dependent RFs as detectors for time-variant signals?

It appears tempting to regard time-dependent RFs as detectors for time variant signals, such as motion-detectors (Adelson & Bergen 1985) or FM-detectors (Eggermont et al. 1981). One requirement to be met to hold this assumption would be that their dynamics must be rather robust against further variations of the stimulus. So far, there are two lines of evidence suggesting that RF dynamics can be easily modified. Systematically evoked interference between stimulus timing and neural dynamics (Selpien & Dinse 1992, Dinse & Schreiner unpublished), revealed considerable alterations of RF dynamics. Similarly, the incorporation of further stimulus aspects such as binaural interaction were found to have considerable effects on the RF shape (Epping & Eggermont 1985). A second and even more important line of evidence comes from the investigation of cortical post-ontogenetic plasticity. Using intracortical microstimulation (ICMS) and paired peripheral tactile stimulation (PPTS) as a tool for inducing short-term plastic reorganization, many descriptors of RFs can be modified such as RF size, RF location, RF overlap and RF latencies. Also, time-dependence of RF organization is changed (see Fig. 1 B) as well as interneural interactions (Dinse et al. 1990b, Recanzone et al. 1992b, Spengler & Dinse 1992ab, Godde et al. 1993a). Finally, there are many studies suggesting a high degree of stimulus dependence of RF properties when a multi-dimensional stimulus approach is used which reveals a so-called mutual stimulus dependency (Spitzer & Hochstein 1985, Richmond et al. 1987, Dinse 1990, Dinse et al. 1990a, Stern et al. 1993). Although so far not explicitly tested, it appears rather likely that such an approach also modifies RF dynamics.

3.3.2. RF concept

From a conceptual point of view, the introduction of an additional though more complex type of detector seems counterproductive to attempts to develop a more general concept of RFs that incorporates all recent findings on plasticity, multi-unit RFs, population ideas and multi-dimensional stimulus approaches. How can the ideas about RF dynamics be incorporated into such a general concept of RFs? While more than 10 years ago, many studies were devoted to understand RFs, today many questions are still open. The findings described in this article on RF dynamics could be regarded as a simple extension to current concepts in that they simply add a temporal dimension. However, in view of the impacts imposed by these findings, a simple supplementary expansion seems inadequate. This was discussed already in more details by demonstrating that spatial and temporal behavior is not separable due to the multiplicative coupling between both factors (Dinse et al. 1991). According to the results on auditory RF dynamics, the same considerations hold for non-separability of spectral-temporal behavior (Eggermont et al. 1981, Dinse & Schreiner 1992).

While these considerations cover only two dimensions, the above mentioned multi-dimensional stimulus approach suggest the existence of at least a large number. In agreement with others, we believe that at the moment one has to include all possible aspects of RFs such as plastic-adaptive properties, multi-dimensional stimulus dependence, aspects of cooperativity and synchrony based on multi-unit recordings and oscillatory properties. Recently, the interesting aspect of destination in terms of projectional patterns was raised which certainly has to be included (Barlow 1992). The same holds for layer-specific and topology-related considerations (Schreiner & Mendelson 1990). In this view, RFs are the windows to the world and allow the mapping of physical events onto neural activity.

3.3.3. Adaptation to inherently time-variant signals

Independent of how the dynamics are generated, they seem to constitute an essential feature of cortical functions. At a first attempt to clarify their possible functional implications, it seems conceivable that they represent specific adaptations for processing of inherently time-variant signals. This notion can easily be counterchecked by comparing the range of RF dynamics with typical signal dynamics. In the case of the visual system, motion of either objects or of the system itself represents typical examples of signal dynamics. In the case of the auditory system, frequency transitions of second-order formants in human speech resemble the dynamics observed during the changes of RF organization (Liberman & Studdert-Kennedy 1978). Video analysis of the rat walking pattern revealed time scales of sensorimotor behavior that are fully compatible with that observed during somatosensory RF dynamics (Dinse et al. 1993). Of course, such matches are only the initial prerequisite for the above described idea of modality-specific adaptations. At the moment, they appear plausible, but they are certainly no proof. Further experiments as well as computer simulations might be helpful to elucidate this hypothesis.

3.3.4. Availability of information over time

From a general point of view, there are arguments in favor of time-variant information processing besides the assumed adaptation to time-variant signals. They all have in common that it is essential to keep information available over a longer period of time. This includes estimation and integration tasks, and the generation of continuity. Finally, all aspects related to short-term memory functions, though reaching somewhat beyond the time scales covered in this article, are of highest interest as they fill the gap between low-level processing dynamics and higher cognitive functions (Nakamura et al. 1992, Schöner et al. 1992, Villa & Fuster 1992).

3.3.5. Mapping spatio-temporal activity onto behavioral states

In contrast to the other modalities, the somatosensory system is unique in that it is inherently related to motor action and behavior. This intimate relationship can be utilized in search of interpretations of spatio-temporal activity pattern in terms of behavioral significance and relevance. The main problem is to identify such patterns. The constraints imposed by the somatosensory system on the patterns available are extremely valuable in providing clues to identify them.

The video-based analysis of walking patterns of rats and the time course of representational map dynamics observed for rat extremities match to a surprisingly high degree. Under normal conditions, walking consists of a certain sequence of steps, causally related to a fixed sequence of neural events. It is hypothesized that the late response components anticipate in their spatial and temporal spread of activity behaviorally relevant patterns, making the expected and anticipated pattern easier to evolve. It appears conceivable that this requires some act of learning to establish the sequence. Accordingly, complex spatio-temporal patterns are a manifestation of highly relevant behavior that are created either during the critical developmental period or during reorganization of the system following adult posy-ontogenetic plasticity, or both. In either way, RF dynamics link learning and memory to normal, continuously on-going processing. They represent highly adaptive states based on acquired knowledge that are crucial for facilitating sequences of neural events that can anticipate fixed sequences of behavior-related action. However, these dynamics are highly variable and subject to plastic alterations. From the point of view of this hypothesis, re-learning would be required to establish new spatio-temporal patterns whose temporal order and sequence would represent again behaviorally relevant adaptations to changes in the environment or changes in the significance of the environmental events.

This scheme appears at least, to some extent, also straightforward in the auditory system. In this case, speech - or less speculative and more general, complex sounds are assumed to elicit spectral distortions and frequency transitions whose temporal sequence are mapped into the dynamics of the time-dependent frequency tuning (Dinse & Schreiner 1992). According to the above described hypothesis, the temporal sequence represents a specific adaptation to the dynamics of the sounds the system has to deal with. Again, the considerable match makes it highly plausible that it has been acquired by learning. This learning-based dynamic processing might have genetic constraints and might have differential sensitivity during ontogenetic development and during post-ontogenetic plasticity.

Thus, this framework tries to link spatio-temporal activity distributions and RF dynamics to behavioral relevant states and learning-related plastic-adaptive behavior. Unfortunately, for the visual and to some extent also to the auditory system, this idea is much more difficult to apply. This is mainly, because the relevant parameters are less obvious to deduce. While motion-related parameters are certainly crucial, the naturally occurring range is much broader than the observed RF dynamics. Also, vision covers many more aspects of time-dependent action. Besides these *where-related* perspectives, the *what-aspects* are of similar importance, presenting even more difficulties due to their close relation to short-term memory functions (see chapter 3.3.4). More experiments are needed to unravel the specific constraints of the dynamics associated to visual processing. In analogy to the advantages of sensorimotor relationships of the somatosensory system, work on eye movements might be an experimentally feasible way to overcome these difficulties.

4. Outlook and conclusions

So far, time dependence of RFs and representational maps have been found in a variety of modalities and species, indicating that the stimulus triggered emergence of complex spatio-temporal activity patterns are a key feature of sensory information processing. However, their functional roles are mostly speculative. The main problem consists in assigning functional significance to neural response pattern. How can behavioral relevance be designated to neural phenomena? One possible approach is to study carefully the occurrences of such spatio-temporal patterns and their relation to on-going events in the outside world (see also Abeles et al. this volume). Another possible approach is the modelling of assumed functions whose feasibility would provide arguments in terms of plausibility and logical coherence (e.g. Neven & Aertsen 1992).

Recent work on short-term post-ontogenetic plastic reorganization of cortical functions in our laboratory has offered new insights in cortical organization that can turn out beneficial when approaching the above described problem. In this sense the investigation of cortical plasticity offers a unique tool: using various types of manipulations one can artificially create different sets of highly idiosyncratic layouts of cortical and subcortical organizations that utilize different types of processing strategies. Simultaneous assessment of the behavioral performance allows to establish a distinct relationship between the individual neural layout and its accompanying level of performance.

Using a protocol of a few hours paired peripheral tactile stimulation (PPTS) in both acute rat electrophysiological and in human psychophysical experiments enabled us to relate the considerable increase of RF overlap associated by a loss of fine grained topology to an increase of spatial discrimination performance found in humans (Godde et al. 1993b)). It is conceivable that RF overlap and thus the number of neurons activated are causally related to the change of performance. Though the first impression is the dramatic loss of topographic details preserved in small, non-overlapping RFs, the resulting large RFs with high overlap warrant the engagement of many more neurons than normal and thus an increase of neuron interaction which, taken together might be assumed to result in better performance. This fits remarkably well with recent findings about an increase in correlated activity in cortical regions that underwent plastic reorganization (Dinse et al. 1990b). According to this view, we expect further insight in the functional role of cortical dynamics by investigating them during a broad scale of plastic-induced alterations while monitoring simultaneously their behavioral impacts.

In recapitulation, these considerations try to link temporal aspects of informational processing strategies to plastic-adaptive processes. The ideas about a temporal distributed processing scheme (TDP) add a further time-based aspect. In this scenario, spatio-temporal patterns are generated as a synergetic process through the entire sensory pathways including very high processing stages, and are subject to changes due to plastic-adaptive processes along slower time scales.

The relation between neuron numbers and performance as discussed above, was often subject to speculations (Hoffmann & von Seelen 1983, Dinse & Krüger 1990), and again raises the questions for a multi-neuron approach. It therefore appears reasonable to apply the idea of a population coding scheme as successfully pursued in the motor system (Georgopoulos et al. 1989) now rigorously to sensory systems as well. We believe that such an approach might turn out extremely valuable in order to understand the gap between single cell RFs and maps in terms of dynamic cooperative processes.

Acknowledgements

The results presented in this paper are based on experiments performed in the Institut für Neuroinformatik at the Ruhr-University Bochum (RUB), Germany and in the Keck Center for Integrative Neuroscience at the University of California, San Francisco (UCSF), USA in collaboration with Dr. Chris Schreiner. We are thankful for extensive discussions to our colleagues and friends, especially to Ad Aertsen, Klaus Kopecz, Michael Merzenich, Gregor Schöner, Werner von Seelen and Chris Schreiner. We acknowledge the support of Michael Schäfer in the optical recording experiments and we are grateful to Dietmar Plenz for valuable advice and help. We wish to express our gratitude to Ad Aertsen for critical reading of various versions of this manuscript.

The research was supported by the Deutsche Forschungsgemeinschaft DFG (Di 334/4-1, 334/5-1, Se 251/31-1, the Forschergruppe at the RUB Neurovision Ey 8/17-1,2) and a NATO collaborative research grant CRG 910626. Exchange visits at the UCSF were supported by the DFG and NATO.

5. References

Abeles M (1977) Proc IEEE 65: 762-773

Adelson EH, Bergen JR (1985) J Opt Soc Am A2: 284-299

Aertsen A, Smolders JW, Johannesma P (1979) Biol Cybern 32: 175-185

Aertsen A, Johannesma P (1981) Biol Cybern 42: 133-143

Arieli A (1988) in: Aertsen A, Braitenberg V (eds) Information Processing in the Cortex. Experiments and Theory. Springer, pp 123-138

Armstrong-James M, Fox K, Das-Guypta A (1992) J Neurophysiol 68: 1345-1358

Barlow HB (1992) in: Aertsen A, Braitenberg V (eds) Information Processing in the Cortex. Experiments and Theory. Springer, pp 175-178

Best J, Reuss S, Dinse HR (1986) Brain Res 385: 356-360

Best J, Mallot HA, Krüger K, Dinse HR (1989) in: Personnaz L, Dreyfus G (eds) Neural Networks. From Models to Application. pp. 107-116, IDSET, Paris

Braitenberg V (1986) in: Palm G, Aertsen A (eds) Brain Theory (Springer) pp 81-96

Braitenberg V, Schüz A (1990) Anatomy of the Cortex, Statistics and Geometry. Studies of Brain Function. Springer

Calford MB, Tweedale R (1988) Nature 332, 446-447

Dayhoff JE, Gerstein G (1983) J Neurophysiol 49: 1334-1348

de Boer F, Kuyper P (1968) IEEE 15: 169-179

Dinse HR (1989) Informationsverarbeitung im visuellen System der Katze. Neuronale Grundlagen. Thieme Verlag, Stuttgart

Dinse HR (1990) in: Elsner N, Roth G (eds) Brain-Perception-Cognition, Thieme, pp 244

Dinse HR, Krüger K (1990) Exp Brain Res 82: 107-122

Dinse HR, Krüger K, Best J, (1990a) Concepts in Neuroscience (CINS) 1: 199-238

Dinse HR, Recanzone G, Merzenich MM (1990b) in: Eckmiller R, Hartmann G, Hauske G (eds) Parallel Processing in Neural Systems and Computers. Elsevier, pp 65-70

Dinse HR, Krüger K, Mallot HP, Best J (1991) in: Krüger J (ed) Neuronal Cooperativity. Springer, pp 67-104

Dinse HR, Hartfiel B, Schäfer M, Krüger K, von Seelen W (1992) in: Elsner N, Richter DD (eds) Rhythmogenesis in neurons and networks. Thieme, pp 109

Dinse HR, Schreiner CE (1992) Soc Neurosci Abstracts 18: 150

Dinse HR, Godde B, Spengler F (1993) in: Elsner N, Heisenberg M (eds) Gen - Gehirn - Verhalten, Thieme, in press

Eggermont JJ (1990) The Correlative Brain. Springer

Eggermont JJ, Aertsen A, Hermes DJ, Johannesma PIM (1981) Hearing Res 5: 109-121

Epping WJ, Eggermont JJ (1985) Hearing Res 18: 244-260

Eysel UT (1982) Nature 299: 442-444

Freeman WJ, Skarda CA (1985) Brain Res Rev 10: 147-175

Fukunishi K, Murai N, Uno H (1992) Biol Cybern 67: 501-509

Georgopoulos AP, Lurito JT, Petrides M, Schwartz AB, Massey JT (1989) Science 243: 234-236

Gerstein GL (1988) in: von Seelen W, Shaw G, Leinhos UM (eds) Organization of Neural Networks. VCH Verlag, pp 53-75

Godde B, Spengler F, Dinse HR (1993a) in: Elsner N, Heisenberg M (eds) Gen - Gehirn - Verhalten, Thieme, in press

Godde B, Spengler F, Dinse HR (1993b) in: Elsner N, Heisenberg M (eds) Gen - Gehirn - Verhalten, Thieme, in press

Grinvald A, Frostig RD, Lieke E, Hildesheim R (1988) Physiol Rev 68: 1285-1366

Hoffmann KP, von Seelen W (1984) Behav Brain Res 12: 101-120

Jones PP, Palmer LA (1987) J Neurophysiol 58: 1187-1211

Kaas JH, Krubitzer LA, Chino YM, Langston AL, Polley EH, Blair N (1990) Science 248: 229

Kopecz S, Kopecz K, Krüger J, Wolf M, Porada I, Schöner G, Dinse HR (1992) in: Elsner N, Richter DD (eds) Rhythmogenesis in neurons and networks. Thieme, pp 344
Krause F, Eckhorn R (1983) Neurosci Letters S14: 209
Krone G, Mallot HA, Palm G, Schüz A (1986) Proc Roy Soc Lond B 226: 421-444
Krüger J (1983) Rev Physiol Biochem Pharmacol 98: 177-233
Liberman AM, Studdert-Kennedy M (1978) in: Handbook of Sens. Physiol. Springer, pp 143-178
Merzenich MM, Nelson RJ, Stryker MP, Cynader MS, Schoppmann A, Zook JM (1984) J Comp Neurol 224: 591-605
Merzenich MM, Recanzone G, Jenkins WM, Allard T Nudo RJ (1988) in: Neurobiology of neocortex. Dahlem Konferenzen 1988, Wiley, pp 41
Nakamura K, Mikami A, Kubota K (1992) NeuroReport 3: 117-120
Neven H, Aertsen A (1992) Biol Cybern 67: 309-322
Nicolelis MAL, Lin RCS, Chapin JK (1992) Soc Neuroscience Abstr 18: 1392
Ohl F, Simonis C, Scheich H (1992) Soc Neuroscience Abstr 18: 841
Optican LM, Richmond BJ (1987) J Neurophysiol 57: 1162-178
Pascal-Leone A, Cohen LG, Hallet M (1992) TINS 15: 13-14
Perkel DH, Bullock TH (1968) Neural Coding. Neurosci Res Prog Bull 6: 221-348
Perret DI, Rolls ET, Caan W (1982) Exp Brain Res 47: 329-342
Podvigin NF, Kuperman AM, Chuyeva IV (1974) Biofizika 19: 341-346
Poggio T, Fahle N, Edelman F (1992) Science 256: 1018-1021
Recanzone GH, Merzenich MM, Jenkins WM, Grajski K, Dinse HR (1992a) J Neurophysiol 67: 1031-1056
Recanzone GH, Merzenich MM, Dinse HR (1992b) Cerebral Cortex 2: 181-196
Reitboeck HJ (1983) IEEE 13: 676-682
Richmond BJ, Optican LM, Podell M, Spitzer H (1987) J Neurophysiol 57: 132-146
Schöner G, Kopecz K, Spengler F, Dinse HR (1992) NeuroReport 3: 579-582
Schreiner CE, Mendelson J (1990) J Neurophysiol 65: 1445-1459
Selpien K, Dinse HR (1992) Soc Neurosci Abstracts 18: 1388
Shevelev IA (1987) Sensornje Sistemje 1, 73-82
Spengler F, Dinse HR (1992a) in: Elsner N, Richter DD (eds) Rhythmogenesis in neurons and networks. Thieme, pp 129
Spengler F, Dinse HR (1992b) Soc Neurosci Abstracts 18: 345
Spitzer H, Hochstein S (1985) J Neurophysiol 53: 1244-1265
Stern, E, Aertsen A, Vaadia E, Hochstein S (1993) in: Giles CL, Hanson SJ, Cowan JD (eds) Advances in Neural Information Processing Systems 5. Morgan Kaufmann Publishers.
Stevens JK, Gerstein G (1976) 39: 239-256
Tanaka K, Saito H, Fukada Y, Moriya M (1991) J Neurophysiol 66:170-189
Vaadia E, Aertsen A (1992) in: Aertsen A, Braitenberg V (eds) Information Processing in the Cortex. Experiments and Theory. Springer, pp 81-122
van Essen DC (1985) in: Peters A, Jones EG (eds) Cerebral Cortex Vol 3, Visual Cortex. Plenum Press, pp 259-329
van Gisbergen JA, Grashuis JL, Johannesma P, Vendrik AJ (1975) Exp Brain Res 23: 367-386
Villa AEP, Fuster JM (1992) NeuroReport 3: 113-116
von Seelen W, Mallot HP, Krone G, Dinse HR (1986) in: Palm G, Aertsen A (eds) Brain Theory. pp. 49-79, Springer
von Seelen W, Mallot HA, Giannakopoulos F, (1987) Biol Cybern 56: 37-49
Wagner H (1990) Europ J Neurosci 2: 949-959
Watanabe JM, Gawne TJ, Richmond BJ, Optican LM (1989) Soc Neurosci Abstr 15: 1054
Weinberger NM, Ashe JH, Metherate R, McKenna TM, Diamond DM, Bakin J (1990) Concepts Neurosci (CINS) 1, 91-132

Brain Theory / A. Aertsen (Editor)

Neuronal Interaction in the Cortex – Quantitative Characterization by Cross-Interval Statistics

Stefan Rotter[a] [1], Ad Aertsen[b] and Eilon Vaadia[c]

[a]Max-Planck-Institut für biologische Kybernetik, Spemannstraße 38, 7400 Tübingen, Germany

[b]Institut für Neuroinformatik, Ruhr-Universität, P.O. Box 10 21 48, 4630 Bochum, Germany

[c]Department of Physiology, Hadassah Medical School, P.O. Box 1172, Hebrew University, Jerusalem, Israel

Abstract

In view of the relatively poor temporal precision which is generally observed for the generation of action potentials in cortical neurons, we argue that a certain amount of "stochasticity" may be immanent to single neuron function in the cortex. This fact seems to hamper many attempts at a quantitative assessment of neuronal interaction by means of cross-correlation analysis, especially in higher cortical areas. We give arguments why it might be advantageous to look at cross-interval histograms of low order separately from each other. In fact, theoretical considerations within the framework of *interacting stochastical point processes* suggest how certain characteristics of neuronal interaction may be inferred from spike trains that were recorded extracellularly from several neurons simultaneously under circumstances where "fluctuations" seem to dominate the signals.

1. RANDOMNESS IS IMMANENT TO CORTICAL NERVE CELLS

The *temporal precision* in the generation of action potentials in single cortical neurons seems to be relatively poor. This may be concluded from electrophysiological recordings *in vivo* as well as from *in vitro* experiments, where parameters are better under control. For the *in vivo* case it appears to be a general rule that a neuron's response to repeated presentations of identical sensory stimuli shows a considerable amount of variability over different stimulus presentations (for examples, see [16, 30]). Moreover, even under the most controlled *in vitro* conditions the timing of action potentials is far from being "exact" and reproducible (see also the contribution by Heck et al., this Volume). In the simple experiment, where constant input into the cell under otherwise stationary conditions can be achieved by intracellular DC-current injection, the interval histograms are still

[1]Present address: Max-Planck-Institut für Entwicklungsbiologie, Spemannstraße 35/IV, 7400 Tübingen, Germany. Additional support for this study was received from the German Ministry for Science and Technology (BMFT), the German-Israeli Foundation for Scientific Research and Development (GIF), and the Graduiertenkolleg Neurobiologie, Tübingen.

remarkably broad. For example, in layer II/III neurons in an acute slice preparation of rat visual cortex, we found standard deviations for the interval histograms ranging up to tens of milliseconds [25] (see Fig. 1).

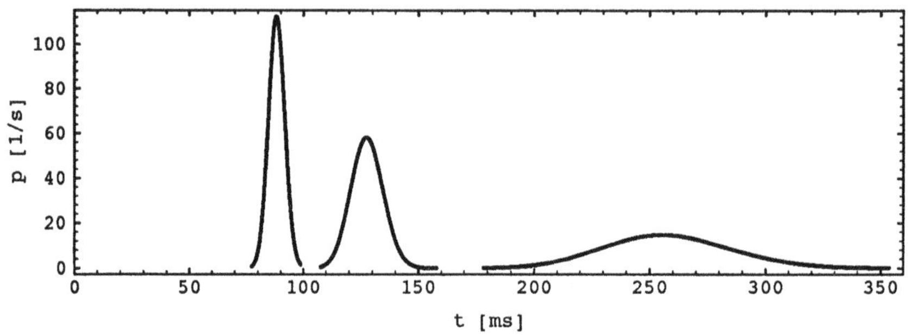

Figure 1. Fits to interval histograms for the repetitive discharge of a cortical neuron under intracellular injection of DC-currents using γ-distributions [8, 10]. We show examples for the respective probability density functions p (scaled in counts/second) from an acute slice preparation of rat visual cortex, the injected current was 320, 390 and 480 pA, respectively. Each sample contained about 800 action potentials, the mean intervals were 88, 128, 258 ms with standard deviations of 3.6, 6.7, 27.4 ms, respectively. A more detailed description of the experiments is given in [15, 25].

Such findings lead us to the even more drastic view that some sort of *random principle* might be a fundamental property of single cell function in the cortex [23, 24]. If this was true, the following questions would immediately arise (see also the contribution by Abeles et al., this Volume): How can the brain perform efficient information processing on the basis of devices whose output is so much "contaminated" by random fluctuations? How is it possible to represent infomation on a time scale down to the millisecond range with units being one order of magnitude less precise?

It is our feeling that this apparent randomness of cortical single neuron firing deserves specific attention from experimentalists. At the same time, it is of considerable interest to explore the functional implications of randomness as a theoretical concept [17, 19, 23, 24]. In this context, the mathematical theory of *interacting stochastical point processes* seems to provide a suitable framework for an approach to the questions stated above (see also [7]). Here, we discuss certain implications from this theory concerning the quantitative assessment of neuronal interaction in stochastic settings.

2. CROSS-INTERVALS AND CROSS-CORRELATION FUNCTION

Cross-correlation is the commonly used tool to assess the interaction between two simultaneously recorded neurons. It estimates the probability density for an action potential in neuron A as a function of the time lag to an action potential in neuron B [18, 22]. In fact, an evaluation of the relative timing of action potentials remains the only possibility to approach questions on neuronal interaction when intracellular signals are not available, as is most often the case in the *in vivo* situation. The experimental literature provides a large number of examples where cross-correlation analysis led to valuable insights into the functional organization of cortical structures [1–3, 10, 27–30]. In view of the fluctuations omnipresent in neuronal signalling, however, we assert that it may be advantageous to use a more refined tool for analysing the experimental data.

The method we propose is based on a selection of action potentials by their "order". Using a spike in neuron B as a point of reference, each spike in neuron A is attributed a number accounting for its proximity to the reference spike: the first spike following is of order $+1$, the second one following is of order $+2$, and so on. Likewise, the first spike in neuron A preceding the reference spike in neuron B is of order -1, the second one is of order -2, and so forth. Thus, using all action potentials in neuron B's spike train as a reference spike, one ends up having as many intervals of any order k as there were reference spikes. We consider their distribution by forming appropriate normalized histograms $p^k_{AB}(\tau)$ which have been called *cross-interval histograms* [22]. From this construction it is obvious that we get the usual cross-correlation function $c_{AB}(\tau)$ as the sum of all cross-interval histograms, if one takes care of the proper sign for the interval

$$c_{AB}(\tau) = \sum_{k=1}^{\infty} \left(p^{+k}_{AB}(+\tau) + p^{-k}_{AB}(-\tau) \right). \tag{1}$$

Separate inspection of the cross-interval histograms and comparison to the respective expectation for zero-interaction may reveal quite detailed information about range and dynamics of interaction between the two neurons. In fact, theory suggests a way to compute the approximate zero-interaction expectation for the cross-interval histogram of any order from the data. In the case of (total) cross-correlation this zero-interaction expectation is simply the "flat" correlation

$$c_{A0}(\tau) \equiv \frac{1}{\mu_A}, \tag{2}$$

where μ_A is the average length of the intervals in neuron A's spike train. Instead of giving the expectation for the cross-interval histograms analytically [18, 22], we process the data in a way which provides additional valuable information on the *significance* of any deviation from the zero-interaction case. To this end, we *shuffle* neuron B's intervals and force them into a pseudo-random order. Again, we compute the cross-interval histograms from the randomized spike train with neuron A's original spike train. Shuffling will destroy all the intricate timing relations which might have existed between neuron B and neuron A. By this procedure, neuron A is left untouched, whereas neuron B's interval distribution remains the same. In fact, if B's firing could be described as a renewal process [8, 10], which is often not too far from reality, shuffling would not change B's statistical characteristics at all. The shuffling procedure is performed a reasonable number of times (in our case, 20 times) to enable a reliable estimation of the statistical fluctuations occurring in the estimation of the cross-interval histograms. For each entry of the cross-interval histogram, we determine the average value and the standard deviation of the zero-interaction histogram estimates gained by shuffling. This is done by applying conventional formulas to each set of 20 values, respectively. Fig. 2 shows an example of the outcome of this procedure.

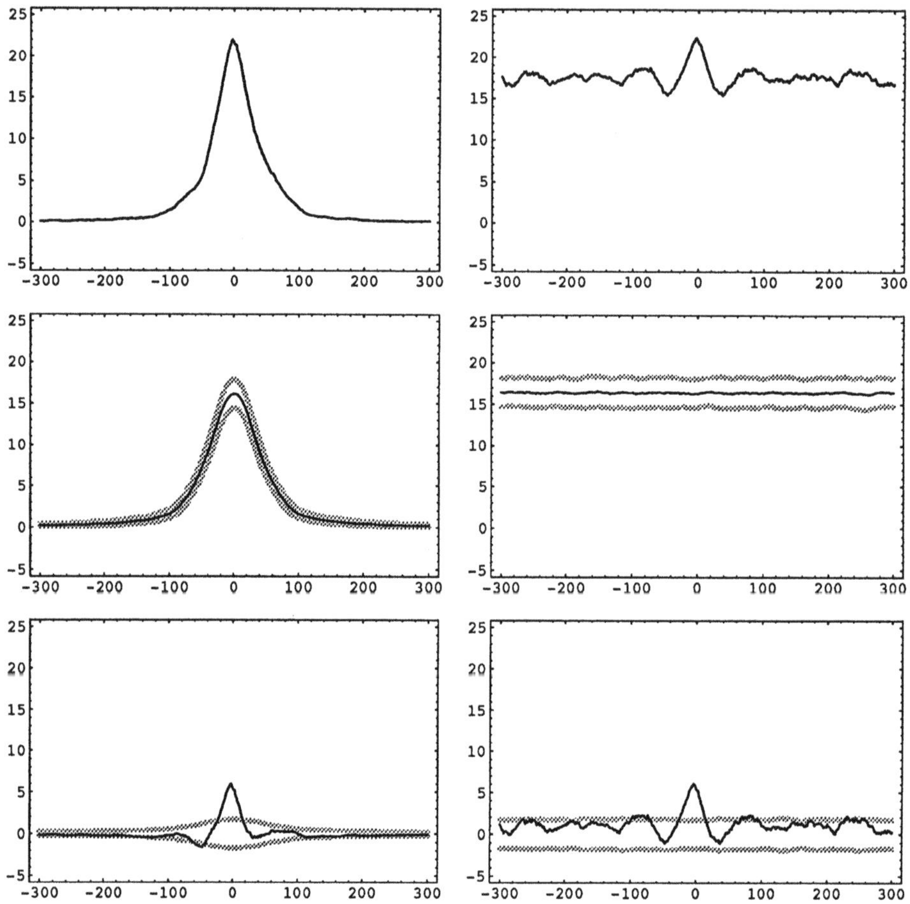

Figure 2. Example for the decomposition of correlation functions between two cortical neurons on the basis of the shuffling procedure described in the text. The data are from two simultaneusly recorded neurons in the prefrontal cortex of an awake, behaving monkey [30]. The plots show a first order cross-interval histogram (top left) and a cross-correlation function (top right), their zero-interaction parts (center, dark lines) and the residuals (bottom, dark lines), respectively. The vertical scaling is counts/second. The residual may be interpreted as the net effect of interaction due to direct synaptic interaction and/or more complex cooperative network effects. The plots also show one standard deviation from the mean, both in positive and in negative direction (grey lines), thus providing a measure for the "significance" of deviations from the zero-interaction case. All curves shown were smoothed by a moving average of width 20 ms.

3. INTERPRETATION OF CROSS-INTERVAL HISTOGRAMS

There is a great interest in detailed information on the relation between direct synaptic interaction and synchronized inputs into a cortical nerve cell. Especially the question what mechanism(s) of interaction contribute to the main features of the correlation structure in neuronal signals continues to give rise to considerable debate. Even partial answers to this question would enable substantial progress in the elaboration of more global ideas on the meaning of cortical spike activity, such as *neuronal assemblies* [13, 14], *synfire chains* [1, 2], *response synchronization* [9, 11, 20] and related concepts. It seems that the point process approach can indeed give new hints on the interpretation of observations made in simultaneous recordings from several neurons. As an example, we will discuss what can be inferred from the distribution of cross-intervals, rather than from the usual cross-correlation function.

Typical post-synaptic currents induced by transmitter release at cortical synapses from neuron B to neuron A in response to a presynaptic action potential have durations of up to some tens of milliseconds [12]. Hence, in view of the low firing rates in the cortex, leading to intervals of length 100 ms and more on average, it is generally only the generation of the first spike following the trigger spike which is eventually influenced by direct synaptic effects. Therefore, we expect that the action potentials in neuron A immediately following an action potential in neuron B are the ones carrying most information on direct synaptic interaction. All the successors are indirectly influenced in terms of a sustained time shift in A's spike train and, in addition, may reflect the effects of more complex forms of interaction between the two neurons due to ensemble synchronization effects.

The first action potential following a trigger spike is, of course, included in the cross-correlation function. However, there are more or less serious conceptual and statistical complications arising while adding the higher order spike events:

- The serial correlations *internal* to neuron A's spike train are also included and may mask the interaction effects. For neurons with a prolonged dead time there is no obvious way to correct the cross-correlation function for such effects [26].
- The effects on mutual timing induced by interaction are corrupted by fluctuations. The first spike in neuron A following the trigger spike is least affected by this, higher order spikes are subject to random displacements increasing more or less linearly with the order.

Analysis of neuronal interaction on the basis of first order cross-intervals can partially avoid such unwanted effects. In the presence of complex interactions between neurons, however, it is also of interest to examine higher order interval distributions. In fact, there is evidence that the time constants of ensemble synchronization effects may be in the range of hundreds of milliseconds or even seconds [1, 2, 4, 30] (see also Abeles et al., this Volume). Also, one would expect a quite different time structure for correlated activity as compared to the case of elementary synaptic interaction. Particularly the higher order interval distributions might exhibit a non-trivial structure reflecting these long-term interactions (see Fig. 3 for an illustration).

As a general rule, higher order cross-interval histograms from cortical recordings tend to be more and more corrupted by fluctuations. It is nevertheless instructive to successively

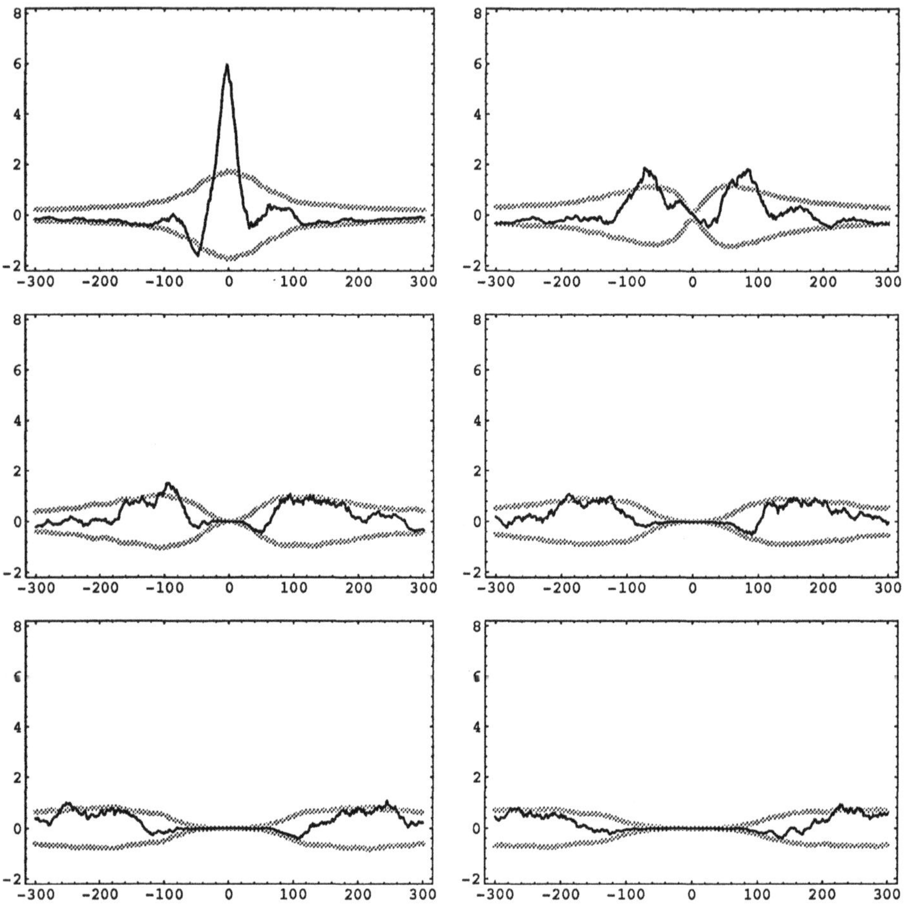

Figure 3. Residual cross-interval histograms up to order 6 from the same data as shown in Fig. 2, the zero-interaction expectation obtained by shuffling was subtracted. The grey lines indicate the respective plus/minus one standard deviation levels. Vertical scaling is again in counts/second. This is, in fact, a rather typical example, where the time structure in the cross-correlation function cannot be reduced to the first order cross-intervals alone (cf. Fig. 4). Generally, however, the "signal to noise ratio" appears to decrease rapidly with the order of the intervals.

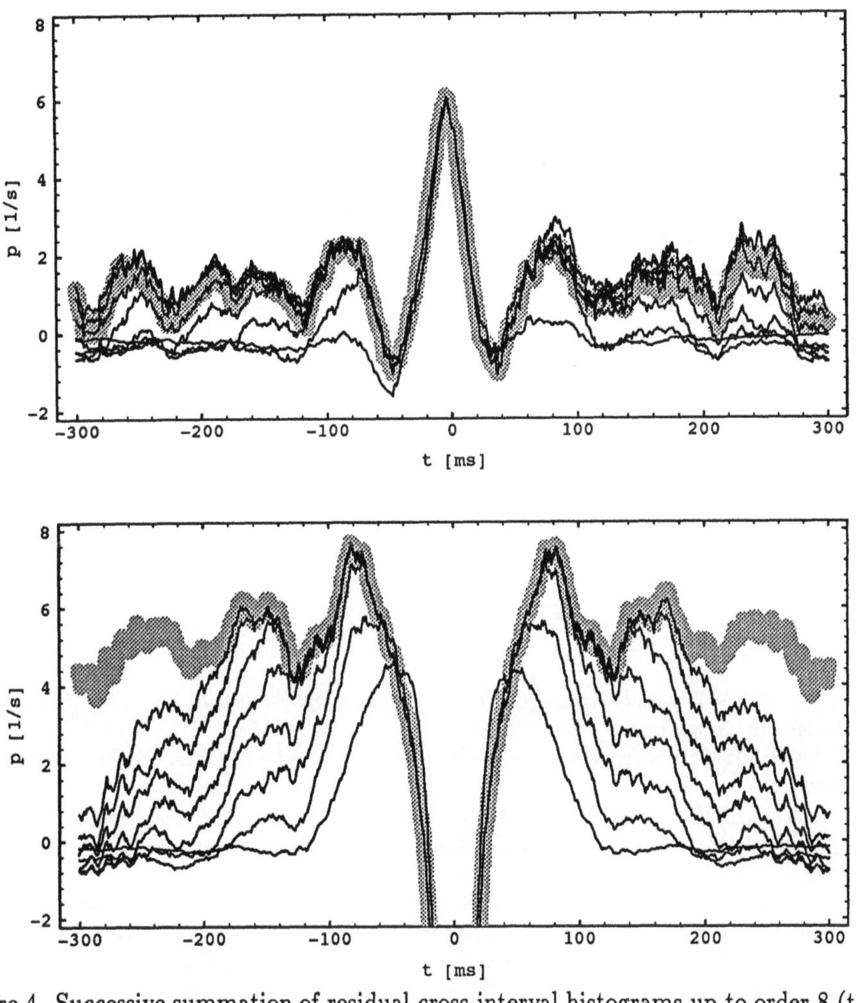

Figure 4. Successive summation of residual cross-interval histograms up to order 8 (top). The data are the same as before, vertical scaling is in counts/second. The result of this procedure adds evidence that the periodic bumps in the cross-correlation function are not just "noise" as might be inferred from the broad significance bands in Fig. 2. Rather, they indicate preferred times of occurrence for successive action potentials. A comparison with the net auto-interval histograms and the net auto-correlation function (bottom) reveals that the structure of higher order net cross-interval histograms cannot be induced by direct interaction alone. In particular, the difference in convergence of the summation of the two histograms suggests that more complex interactions may be involved.

add up the contributions of cross-intervals which have been kept apart so far. In this way, one can gradually reconstruct the average time course of the spike response of neuron A to an action potential of neuron B (see Fig. 4, top). One observes how the "bumps", which would hardly have been termed "significant" before, are slowly built up.

For an investigation to what extent this time structure is determined by neuron A's dynamics – for theoretical reasons, B's influence can be neglected – we employed the analogous procedure for auto-intervals, i.e., with reference points being the action potentials of neuron A itself (Fig. 4, bottom). One notices that bumps emerge with the same time structure as in the case of cross-intervals, but that it takes more higher order contributions until they are complete. This is true especially for longer time delays. The interpretation of this is twofold. First, it seems that the periodic structure in the cross-correlation function is already contained in A's spike train and could have been predicted from it. The difference observed in the convergence of the summation, however, tells us that there is a tendency for neuron A to selectively omit spikes in those parts of its spike train which are in the vicinity of a spike in neuron B.

To summarize, it appears quite promising to consider cross-intervals of different orders separately. A systematic evaluation of effects as described in the example may help to further unravel functional details of the dynamics of neuronal interaction in cortical networks.

REFERENCES

1. M Abeles (1982) *Local Cortical Circuits – An Electrophysiological Study.* Springer, Berlin, Heidelberg, New York.
2. M Abeles (1990) *Corticonics – Neural Circuits of the Cerebral Cortex.* Cambridge University Press.
3. E Ahissar, E Vaadia, M Ahissar, H Bergman, A Arieli and M Abeles (1992) *Dependence of Cortical Plasticity on Correlated Activity of Single Neurons and on Behavioral Context.* Science **257** 1412–1415.
4. A M H J Aertsen, G L Gerstein, M K Habib and G Palm (1989) *Dynamics of Neuronal Firing Correlation: Modulation of "Effective Connectivity".* J. Neurophysiol. **61** 900–917.
5. A M H J Aertsen, E Vaadia, M Abeles, E Ahissar, H Bergman, B Karmon, Y Lavner, E Margalit, I Nelken and S Rotter (1991) *Neural Interactions in the Frontal Cortex of a Behaving Monkey: Signs of Dependence on Stimulus Context and Behavioral State.* J. f. Hirnforschung **32** 735–743.
6. K-H Boven and A M H J Aertsen (1990) *Dynamics of activity in neuronal networks give rise to fast modulations of functional connectivity.* In: R Eckmiller, G Hartmann and G Hauske (eds.) *Parallel processing in neural systems and computers.* Elsevier, Amsterdam.
7. D R Brillinger (1976) *Identification of synaptic interactions.* Biol. Cybernetics **22** 213–228.
8. D R Cox (1962) *Renewal Theorie.* John Wiley & Sons, Inc., New York.
9. R Eckhorn (1991) *Stimulus-Specific Synchronizations in Visual Cortex: Linking Features into Global Figures?* In: J Krüger (ed.) *Neural Cooperativity.* Springer, Berlin, Heidelberg, 184–219.
10. J J Eggermont (1990) *The Correlative Brain. Theory and Experiment in Neural Interaction.* Springer: Berlin, Heidelberg, New York.
11. A K Engel, P König, A K Kreiter, T B Schillen and W Singer (1992) *Temporal Coding in the Visual Cortex: New Vistas on Integration in the Nervous System.* TINS **15** 218–226.

12. E Fetz, K Toyama and W Smith (1991) *Synaptic Interactions between Cortical Neurons.* In: A Peters und E G Jones (eds.) *Cerebral Cortex,* Vol. 9, 1–47, Plenum Publ. Corp.
13. G L Gerstein, P Bedenbaugh and A M H J Aertsen (1989) *Neuronal Assemblies.* IEEE Trans. Biomed. Engineering **36** 4–14.
14. D Hebb (1949) *The Organization of Behavior. A Neuropsychological Theory.* Wiley, New York.
15. D Heck (1991) *Gibt es eine neuronale Schwellenfunktion?* M.Sc. Thesis, University of Tübingen (Germany).
16. D H Hubel and T N Wiesel (1962) *Receptive Fields, Binocular Interaction, and Functional Architecture in the Cat's Visual Cortex.* J. Physiol. **160** 106–154.
17. P I M Johannesma and H F P van den Boogaard (1985) *Stochastic Formulation of Neural Interaction.* Acta Appl. Math. **4** 201–224.
18. G P Moore, D H Perkel and J P Segando (1966) *Statistical analysis and functional interpretation of neuronal spike data.* Ann. Rev. Physiol. **28** 493–522.
19. I Nelken (1988) *Analysis of the Activity of Single Neurons in Stochastic Settings.* Biol. Cybernetics **59** 201–215.
20. H Neven and A Aertsen (1992) *Rate Coherence and Event Coherence in the Visual Cortex: A Neuronal Model of Object Recognition.* Biol. Cybernetics **67** 309–322.
21. D H Perkel, G L Gerstein and G P Moore (1967) *Neuronal Spike Trains and Stochastic Point Processes. I. The Single Spike Train.* Biophys. J. **7** 391–418.
22. D H Perkel, G L Gerstein and G P Moore (1967) *Neuronal Spike Trains and Stochastic Point Processes. II. Simultaneous Spike Trains.* Biophys. J. **7** 419–440.
23. S Rotter and A M H J Aertsen (1991) *Towards a Description of Networks of Spiking Cortical Neurons: Membrane Potential and Firing Probability.* In: N Elsner and H Penzlin (eds.) *Synapse – Transmission – Modulation.* Thieme, Stuttgart, 302.
24. S Rotter and A M H J Aertsen (1992) *A Stochastical Model for a Network of Spiking Cortical Neurons: Time, Correlation and Synchrony.* In: N Elsner and D W Richter (eds.) *Rythmogenesis in Neurons and Networks.* Thieme, Stuttgart, 422.
25. S Rotter, D Heck and A M H J Aertsen (1992) *A Statistical Description of Spiking Cortical Neurons at Two Levels of Abstraction: The Correspondence between Intra- and Extra-Cellular Observations.* In: N Elsner and D W Richter (eds.) *Rythmogenesis in Neurons and Networks.* Thieme, Stuttgart, 423.
26. H Sompolinski, personal communication.
27. K Tanaka (1983) *Cross-Correlation Analysis of Geniculostriate Neuronal Relationships in Cats.* J. Neurophysiol. **49** 1303–1318.
28. K Toyama, M Kimura and K Tanaka (1981) *Cross-Correlation Analysis of Interneuronal Connectivity in Cat Visual Cortex.* J. Neurophysiol. **46** 191–201.
29. K Toyama, M Kimura and K Tanaka (1981) *Organization of Cat Visual Cortex as Investigated by Cross-Correlation Technique.* J. Neurophysiol. **46** 202–214.
30. E Vaadia, E Ahissar, H Bergman and Y Lavner (1991) *Correlated Activity of Neurons: A Neural Code for Higher Brain Functions?* In: J Krüger (ed.) *Neuronal Cooperativity.* Springer: Berlin, Heidelberg.

Brain Theory / A. Aertsen (Editor)

Spike Generation in Cortical Neurons: Probabilistic Threshold Function shows Intrinsic and Long-lasting Dynamics

Detlef Heck[a], Stefan Rotter[a] and Ad Aertsen[b] [1]

[a]Max-Planck-Institut für biologische Kybernetik, Spemannstraße 38, 7400 Tübingen, Germany

[b]Institut für Neuroinformatik, Ruhr-Universität, P.O.Box 10 21 48, 4630 Bochum, Germany

Abstract

It is commonly assumed that action potentials are the means by which neurons transfer information and, hence, participate in computational and information processing tasks in the brain. Therefore, the rule that governs the generation of such action potentials (spikes) is one of the elementary principles in the nervous system. Generally, the occurrence of a spike is probabilistically related to the potential difference across the cell membrane, such that the firing probability increases with the trans-membrane potential; this dependence is usually described by the *neuronal threshold function.* Surprisingly, and in spite of its apparent importance, the detailed form of this treshold function for cortical neurons is still unknown. Therefore we decided to investigate this threshold firing in more detail on the basis of long intracellular recordings from cortical neurons *in vitro.* In order to quantify the results, we developed a statistical analysis procedure, specifically designed to capture the relevant dependences. This approach results in a *stochastic transfer function,* describing the dependence of the firing probability on the trans-membrane potential and on the time elapsed since the occurrence of that particular potential value. Our results indicate that this transfer function of cortical neurons shows intrinsic dynamic properties, which cannot be reduced to a single, instantaneous threshold function, as used in most theoretical descriptions and neural network models.

1. Introduction

Faced with the enormous complexity of the brain, expressed in the large number of different elements and their intricate electrical properties and interactions, it seems reasonable to seek to reduce this complexity by searching for general principles and to test their validity in experimental and theoretical models. Indeed, the invention of the computer and the corresponding enormous increase in computational power made the modelling of such complex systems a feasible enterprise. An important first step in this direction was made in 1943 by McCulloch and Pitts[6], who reduced the known anatomical and physiological

[1]Additional support was received from the German Ministery for Science and Technology (BMFT)

variety of nerve cells to a simple binary threshold device, interacting with its neighbours via inhibitory and excitatory connections (the 'synapses'). This formal 'McCulloch-Pitts neuron' has only two activity states, and switches from the inactive to the active state when the sum over all inputs reaches a fixed threshold. Inhibition has a strong influence on a 'McCulloch and Pitts neuron', such that activity of only one inhibitory synapse fixes the 'neuron' in the inactive state (compare shunting inhibition). Although much has changed in our understanding of neural function since then, the notion of a neuron as a computational element with a nonlinear threshold function, transforming the input from many incoming fibers into a single output stream, remains common to most model formulations until the present day, irrespective of whether these models are continuous ('analog') neurons[9] or discrete ('spiking') neurons with 'action potential' events[7].

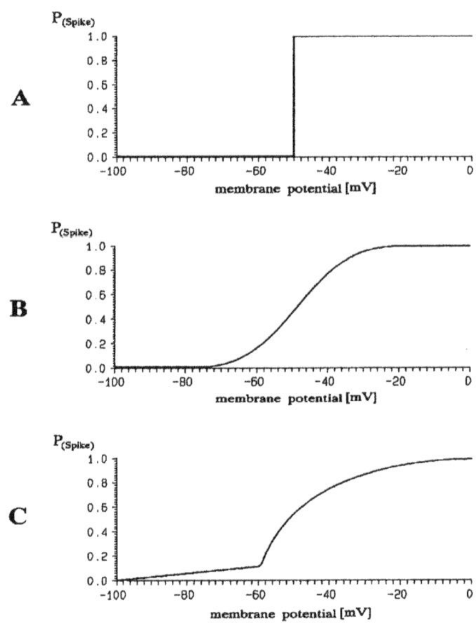

Figure 1: Three possible transfer functions for a single neuron model. A) A discrete threshold allows only two different output states. B) A sigmoid function defines a continuous mapping from the input to the output signal domain. C) A possible combination of A and B.

In many of these theoretical models, either one of two different types of threshold functions are assumed. One is the above mentioned 'hard', deterministic threshold proposed by McCulloch and Pitts[6] (Fig. 1A), the other is a 'soft', sigmoid shaped function introduced as a probabilistic descriptor by Little[5] (Fig. 1B)). Interestingly, and in spite of considerable early experimental work on peripheral nerves[11], the detailed form of the threshold function of cortical neurons is still unknown. This is all the more surprising, considering that a number of studies, both theoretical and experimental, have repeatedly emphasized the functional importance of the nonlinear threshold of central neurons[8,3] (see also the contribution of Freeman in this Volume). Recently, the form of the threshold function gave rise to a theoretical debate in the context of neural network models[10]. Here it was implicated in connection with the stability of neural activity, in particular at the low firing rates as typically observed in the neocortex[1]. In this context, an intermediate function was proposed[4], combining the properties of the 'hard' and 'soft' threshold (Fig. 1C).

Inspired by these considerations, we set out to measure the detailed form of the firing threshold function of neocortical neurons under conditions of steady, low-level firing. To this end, we measured simultaneously the time course of the membrane potential and the events of spike occurrences by intracellular recording from spontaneously active cells in an acute slice preparation. In a subsequent stage, we carried out a specifically designed statistical analysis of the relation between the membrane potential value and the probability of observing a spike event. In this context, it soon became clear that it would not be possible to describe the firing probability as a function of membrane potential alone. Clearly, other variables such as the time that passed since a particular potential value was attained, are expected to have an influence too. The result of this analysis, which will be described below, is a *stochastic transfer function*, which, unlike the usual instantaneous firing threshold, expresses the firing probability as a function of the trans-membrane potential and of the time elapsed since that potential occurred.

2. Methods

Acute coronal slices (thickness 400 μm) were prepared from the visual cortex of 4 to 6 week old rats (Lewis) according to standard procedures, and kept at room temperature in an artificial cerebrospinal fluid (ACSF[2]). For the experiment, the slices were transfered to a recording chamber (temperature 33 to 36 °C), where they lay fully submerged on a glass cover slip. Intracellular recordings were made with glass micropipettes (3 M potassium acetate) from cells in different layers of the cortical slice. The positions of the cells were estimated by visual inspection. If necessary, spontaneous spike activity of the cells was increased by adding 150 to 250 μM l-Glutamate to the perfusion medium and/or by injection of a constant, depolarizing DC-current.

Intracellularly recorded signals were digitized at a sampling rate of 10 kHz using a CED 1401 (CED, Cambridge, UK) laboratory interface and written to disk for later processing.

3. Firing Threshold as a Probability Distribution

The intracellularly recorded signals were analyzed off-line. First, from the digitized signal we extracted the times of spike events as the instants of the spike waveform maxima (Fig. 2A). From these we could easily compute the usual spike statistics, such as the interval distribution (Fig. 2B) and the autocorrelation function (Fig. 2C). These provide information about the firing mode of the cell (e.g. stochastic or quasi-periodic), as is illustrated for a typical recording from a cortical cell in Fig. 2. For the period this particular cell fired in a stochastic mode, more or less like a Poisson process with dead-time, as can be inferred from the exponential-like interval distribution and the more or less constant autocorrelation. In other cases, the firing pattern could be more regular, up to the point of being (quasi-)periodic.

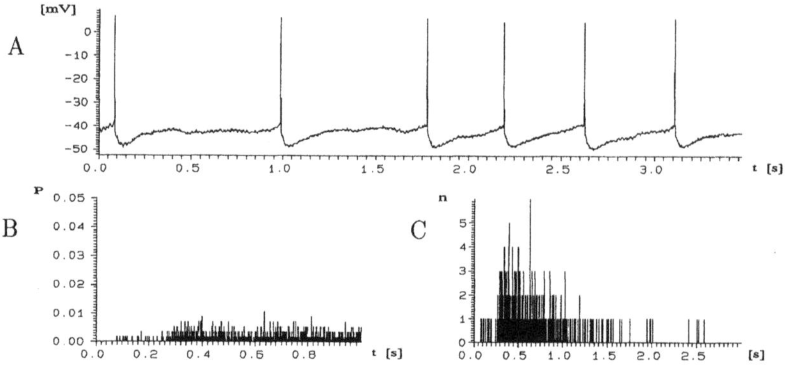

Figure 2: Typical segment of an intracellular recording from a cortical neuron in vitro (A), the autocorrelation function (B), and its interval distribution (C). This cell was recorded in layer II/III; a steady level of firing (mean inter-spike interval 638 ms, standard deviation 416 ms) was obtained by constant intracellular current injection (0.25 nA) and an addition of 200 μM glutamate to the medium. A total number of 548 spikes was collected during a total analyzed time of 350 sec. Observe that the firing of this cell is very reminiscent of a Poisson process with dead-time (refractoriness).

The probability of spike generation as a function of membrane potential and time was investigated by making the following analysis: First, every occurrence of a certain value of the membrane potential φ_i was registered and counted; this results in the count $n(\varphi_i)$ for that particular value. The collection of counts for all relevant values of φ is the potential distribution $n(\varphi)$. Second, whenever a particular membrane potential was found, we regarded this moment as time zero and determined the following two time intervals:

1) τ^- : the time elapsed since the last spike,
2) τ^+ : the time to go until the next spike,

for $\tau \leq$ some maximum value (in our application the duration of the longest inter-spike interval. See Fig. 3A.)

These two intervals were entered into two associated matrices of counts $n(\tau^-, \varphi)$ and $n(\tau^+, \varphi)$, with the rows representing the potential values φ and the columns representing the different time values τ. For each pair of doublets (τ^-, φ) and (τ^+, φ) found, the two corresponding matrix elements were incremented by one. This is illustrated in Figs. 3A,B, with the two members of each pair represented by the same symbol. The result of collecting all these pairs for all membrane potential values in the recording is two matrix histograms of counts, specifying for each potential value φ how often it was preceeded by a spike at τ^- ms, and how often it was followed by a spike at τ^+ ms.

From these matrix histograms we computed the transfer function of the neuron in terms of the firing probability as a function of membrane potential and time. It gives the probability of a spike Z to occur τ^+ milliseconds after the membrane potential attained

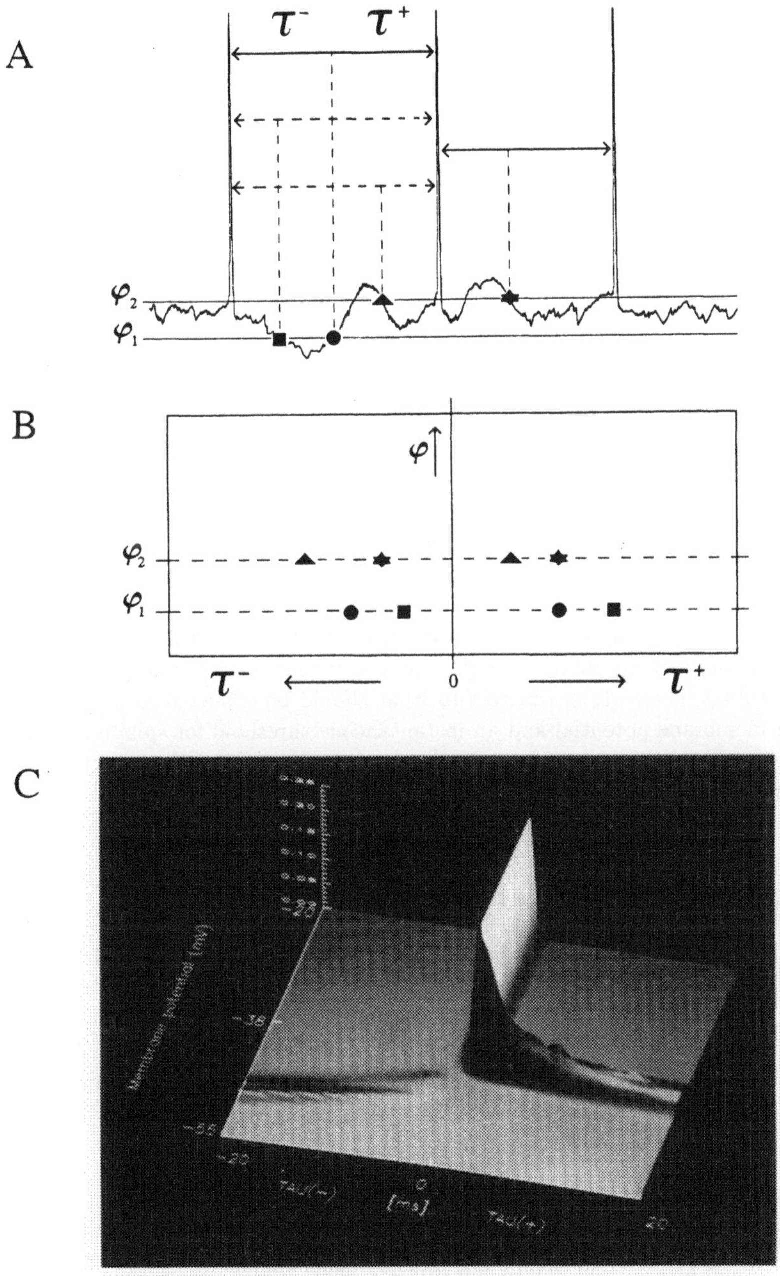

$P(Z^- \text{ at } \tau^-, Z^+ \text{ at } \tau^+ | \varphi)$

Figure 3: Illustration of the analysis procedure (A,B), and the resulting neural transfer function (C) for the cortical cell shown in Fig. 2. Further explanation see text.

a certain value φ (cf. the right-hand part of Fig. 3C).

$$P(Z^+ \; at \; \tau^+ | \varphi) = \frac{n(\tau^+, \varphi)}{n(\varphi)} \tag{1}$$

The analogous procedure for τ^- yields the probability of a spike Z to occur τ^- milliseconds before a certain value φ of the membrane potential (corresponding to the left-hand part of Fig. 3C).

$$P(Z^- \; at \; \tau^- | \varphi) = \frac{n(\tau^-, \varphi)}{n(\varphi)} \tag{2}$$

The graphical display of these probabilities is, consequently, a 3-dimensional graph with the firing probability plotted along the vertical axis, and membrane potential and time along the two horizontal axes. The two separate probabilities for τ^- and τ^+ can be combined into a single one: $P(Z^- \; at \; \tau^- \, , \; Z^+ \; at \; \tau^+ | \varphi)$. This can be plotted in a single picture, with the time running along the x-axis, and time zero, i.e. the time when the potential value occurred, in the middle (Fig. 3C).

The result of this procedure for the recording from the neuron in Fig. 2 is shown in Fig. 3C. Observe that, surprisingly, the spike probability distribution for this cortical cell is a non-separable function of φ and τ, covering a considerable region of the (τ, φ)-space, with an elongated tail extending to high values of τ. This result, which is quite typical for our recordings, is clearly in contrast to what should be expected for a neuron with a fluctuating membrane potential and an instantaneous threshold for spiking: in that case the probability distribution would essentially depend only on φ within a narrow range of τ-values immediately adjacent to the origin. The consequences of this will be described in more detail below.

4. The Influence of Membrane Potential and Time

In order to relate the above determined probability distribution to more conventional measures of the likelihood of spiking, we studied the behavior of the 'mountain range' in Fig. 3C at appropriately chosen cross-sections. This enables to further quantify the influence of each of the two variables φ and τ separately, keeping in mind of course the observed non-separability of these respective influences. (This is, in fact, the reason why we have to resort to cross-sections, rather than to the marginal distributions, which could be obtained simply by appropriate projection.)

4.1. Time-Dependent Threshold Functions

Let us first consider the influence of the membrane potential φ. This is, in fact, closely related to the original question on the form of the threshold function (cf. Fig. 1). Clearly, the 3-dimensional plot in Fig. 3C can be thought of to consist of a sequence of different threshold functions, each one refering to a different time τ^+ relative to the occurrence of a membrane potential. One can, for example, cut out a narrow slice, say 1 ms wide, from the 3-D probability distribution and, thus, arrive at the neural transfer function at time delay τ: $P(Z^+ \; at \; (\tau^+ = \tau \text{ to } \tau + 1\,\text{ms}) \,|\, \varphi)$.

Results of this procedure for three different choices of τ (at 0, 2, and 4 ms) are shown in Fig. 4. Not surprisingly, in view of the result in Fig. 3C, we observe that the firing probability is far from being instantaneous. Apart from the immediate response (Fig. 4A), considerable contributions arise for larger time delays (Figs. 4B,C). In fact, extending our analysis beyond the time window of 20 ms (Fig. 3C), we found that distinct nonuniformities in the probability curves in these time slices continue to exist (albeit gradually diminishing as compared to those in Figs. 4B,C) for delays of up to 200 ms or even more! This results obviously in a considerable amount of irregularity in the timing of spikes, to which we will return in the next section. Each of the threshold curves in Fig. 4 is characterized by a relatively smooth onset. The question whether this results from an intrinsically smooth threshold, or rather from a discrete, but fluctuating threshold that becomes blurred by the time averageing inherent to our analysis, cannot be decided at this point. Only the instantaneous response (Fig. 4A) shows saturation at probability one for high values of the membrane potential; curves at later times decrease again for increasing φ (Figs. 4B,C), reflecting refractoriness due to firing in earlier time intervals. Finally, there is a distinct tendency of the onset potential in the threshold curves to decrease with increasing time delay. This decrease saturates in the course of the first 5 ms, after which the threshold curves remain essentially the same, both in position and form, with only the amplitude very slowly decaying in the course of the ensuing few hundred milliseconds.

4.2. Voltage-Dependent Spike Latency Statistics

Another perspective on the spike probability distribution can be gained from choosing the orthogonal point of view. Making a cross-section along the τ-axis of Fig. 3C yields the distribution of spike latencies, relative to the time of occurrence of a particular membrane potential. A useful and compact way to characterize such a latency distribution is to specify its moments. Hence, we calculated for each delay distribution the first two moments - the mean latency and the standard deviation - and plotted these as a function of the membrane potential.

In addition, we calculated from them the coefficient of variation $\text{cv}=(\frac{\text{std.dev.}}{\text{mean}})$; this is a commonly used measure in spike statistics which, roughly speaking, measures the departure from an exponential distribution for which cv=1.

The resulting statistics for our cortical cell are shown in Fig. 4 (bottom row). Observe that the mean latency (Fig. 4D) and the standard deviation (Fig. 4E) show a distinct and similar dependence on membrane potential. They both start at low values around $\varphi = -35\,\text{mV}$, gradually increase when the membrane potential becomes more negative, to attain extremely high values (up to 500 ms) around $-46\,\text{mV}$. Both curves show a non-monotonicity for more negative potentials, for which we currently have no explanation. One possibility might be that this reflects certain properties of voltage-dependent ion channels.

Roughly speaking, these results imply that for more negative membrane potentials, spikes tend to occur both later and less precise in time. The relation between the two is expressed in their ratio, the coefficient of variation (Fig. 4C). Starting off closely above 1 at $-38\,\text{mV}$, it rapidly rises to a peak of 3.4 at about $-40\,\text{mV}$, after which it monotonically decays to a constant value around 0.7 for membrane potential values below about $-47\,\text{mV}$. Interestingly, for much of the range of membrane potential values the coefficient

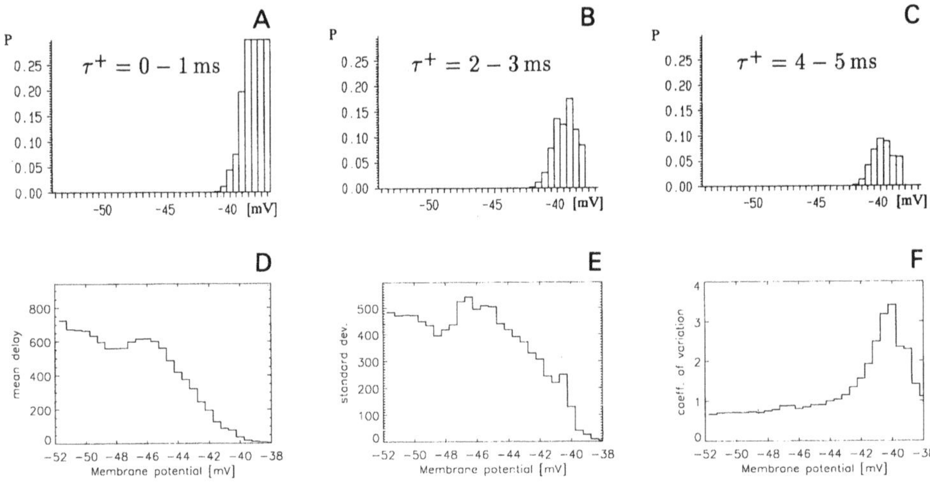

Figure 4: Firing probability functions (top row) and spike latency statistics (bottom row). Data are from the same cortical cell as in Figs. 2 and 3. Top row: Firing probability as a function of membrane potential for different time delays after that potential occurred: 0-1 ms (A), 2-3 ms (B), and 4-5 ms (C). Bottom row: Mean delay (D), standard deviation (E), and coefficient of variation (F) of the spike latency distribution.

of variation is 1 or higher. This implies that throughout this range, i.e. even for membrane potentials with a very high probability of firing, there is always an uncertainty as to when the spike will actually occur, which roughly equals the mean latency or is even higher. The result of this is evidently a considerable degree of stochasticity, as is confirmed by the spike train's Poisson-like behavior (cf. Fig. 2; see also the contribution of Rotter et al., this volume). The sharp peak in the relative uncertainty (Fig. 4F) seems to be primarily caused by a concurrent increase in the standard deviation (Fig. 4E), whereas the mean delay continues its smooth course in the corresponding potential regime (Fig. 4D). Apparently, there is a narrow voltage range (say, between -39 and $-42\,\mathrm{mV}$ in which firing is distinctly less precise in time than outside this range.

5. Discussion

We determined the stochastic neural transfer function of cortical neurons in vitro by statistical analysis of intracellularly recorded membrane potentials and spike sequences. This transfer function specifies the probability $P(Z^+ \; at \; \tau^+|\varphi)$ for the generation of a spike some time τ^+ after a certain membrane potential φ occurred. Appropriate cross-sections of this probability distribution can be related to conventional threshold curves and spike latency statistics.

We found that different cortical cells may have different transfer functions. This is expressed in the different shapes of the 3-dimensional probability distributions and, accordingly, in the different sets of threshold functions and spike latency statistics. However, one important feature shared by all cortical cells is that the neural transfer function is generally a non-separable function of membrane potential and time, extending over a varying, but in each case considerable region of the (φ, τ)-space. Consequently, single cortical neurons are expected to exhibit a considerable degree of randomness in the timing of their individual spike firings.

Occasionally we encountered neurons which, during the recording, switched spontaneously from a stochastic to a periodic firing mode, presumably related to small changes in the baseline membrane potential. Separate analysis of these different modes revealed drastically different transfer functions. Thus, cortical neurons may have different modes of firing, with each mode having its own specific transfer function.

In summary, our findings indicate that the transfer function of cortical neurons shows intrinsic dynamic properties. The non-separability of the different dependences implies that the transfer function cannot be reduced to a single, instantaneous threshold function, as used in most theoretical descriptions and neural network models.

References

1. M. Abeles et al., Firing patterns of single units in the prefrontal cortex and neural network models, Network, 1 (1990) 13.

2. L. J. Bindman et al., Comparison of the electrical properties of neocortical neurones in slices in vitro and in the anaesthetized rat, Exp. Brain Res., 69 (1988) 489.

3. F. H. Eeckman and W. J. Freeman, Asymetric sigmoid non-linearity in the rat olfactory system, Brain Res., 557 (1991) 13

4. D. Kleinfeld and H. Sompolinsky, in: C. Koch and I. Segev, eds., Methods in neuronal modeling. From synapses to networks, MIT Press (1989)

5. W. A. Little, The existence of persistent states in the brain, Math. Biosci., 19 (1974) 101.

6. W. S. McCulloch and W. Pitts, A logical calculus of the ideas immanent in nervous activity, Bull. Math. Biophys., 5 (1943) 115

7. N. Rochester et al., Tests on a cell assembly theory of the action of the brain, using a large digital computer, IRE Trans.Inform.Theory, IT-2 (1956) 80.

8. T. Sejnowski, in: G. E. Hinton and J. A. Anderson, eds., Parallel models of associative memory, (Lawrence Erlbaum Assoc., Hillsdale, New Jersey, 1981)

9. W. K. Taylor, Electrical simulation of some nervous system functional activities, Information Theory, Vol. 3 (ed. E. C. Cherry), (1956) 314, London: Butterworths.

10. G. Toulouse, Perspectives on neural network models and their relevance to neurobiology, J. Phys. A: Math. Gen., 22 (1989) 1959.

11. A. A. Verveen, On the fluctuation of threshold of the nerve fiber, Proc. 2nd Int. Meeting of Neurobiologists (1959) 282.

Brain Theory / A. Aertsen (Editor)

On spike synchronization

H. Glünder¶ and A. Nischwitz*

Institut für Medizinische Psychologie, Ludwig-Maximilians-Universität
Gœthestraße 31, D-8000 München 2, Germany

*Institut für Nachrichtentechnik, Technische Universität München
Arcisstraße 21, D-8000 München 2, Germany

Abstract

We start with historically founded reflections on the relevance of synchronized activity for the neural processing of information and we propose to differentiate between synchrony at the emitting and the receiving side. In the main part we introduce model networks which consist of chains of locally coupled and noisy spiking neurons. In the case of lateral excitation without delay as well as for delayed lateral inhibition these basic structures can turn homogeneous stimulations into synchronized activity. The synchrony is maintained under temporally varying stimulations thus evoking aperiodic spike fronts. Although we present some hypotheses, the question of how the nervous system deals with this network property remains to be answered.

1. INTRODUCTION

Half a century ago McCulloch and Pitts [1] stated that neurons are principally suited to perform Boolean operations. Undoubtedly, the authors were strongly influenced by the developing theory of automata and especially by the incredible perspectives of an effective mechanisation of the logical calculus. At that time, for instance, electronic AND-gates consisted of a resistive network for the summation of electric currents followed by an active thresholding device and therefore, they represented an attractive structural and functional analogue to nerve cells. However, much more importantly, this view marks the fundamental transition from regarding neurons as integrators to realizing them as coincidence detectors (cf. [2; 3]). Although coincidence detection is the principle of AND-gates, it did not become popular even in biological cybernetics and theoretical neuroscience. Instead, the generalized McCulloch/Pitts-(model)neuron which computes with a mathematical construct, namely instantaneous impulse rates, and not with impulses (action potentials or spikes), became standard – not only for most simulations of neural networks but also for interpretations of neurobiological experiments. Thus, until recently, only few scientists investigated the temporal fine structure of neural signals, i.e. of spike trains and bursts as well as the associated postsynaptic potentials, and hypothesized about its putative significance for the processing of neural information.

¶ The author is supported by the "Volkswagen-Stiftung" under the grant I/65 914.

2. THE RELEVANCE OF SPIKE SYNCHRONIZATION

In one of the early comments on this issue Wiener [4, chapter 10] points out that coincident or synchronized spikes at a neuron's input terminals will be much more efficient for the triggering of action potentials than asynchronous, for instance stochastic impulses. Owing to this basic functional property of nerve cells, Wiener concludes that synchronously oscillating nervous activity should be found in the brain – a notion which appears rather modern, although he concentrated on the α-rhythm while nowadays oscillations in the γ-band are favoured. Although Wiener's conclusion and its presently discussed versions are appealing, they lack stringency:

(i) In the same way as the statement of McCulloch and Pitts does not imply that neurons really act as AND-gates, the similarly reductionistic view of neurons as coincidence detectors does not imply the actual use of this faculty for neural processing (cf. the analysis of basic misconceptions in cybernetics by Taube [5] chapter 6).

(ii) Obviously, the type of coincidence detection considered here takes place at a neuron's axon hillock. Such somatic coincidences generally differ from synchronous input to a neuron. Consequently, action potentials that appear at the same time at various presynapses of a target neuron – thus representing synchronous input activity – need not cause coinciding excitatory postsynaptic potentials at its cell body (cf. [6] section 5). This discrepancy can be due to differences in conduction times as well as to delayed synaptic transmission which can be caused by molecular processes such as second messenger cascades. One may even conjecture that, within limits, neurons are able to produce somatic coincidences between non-coinciding action potentials by (learning) appropriate synaptic delays.

(iii) At least phenomenologically, synchrony need not be bound to periodic processes because aperiodic events may be synchronized as well. It should be realized that periodicity commonly refers to a *single* signal whereas synchronization exclusively concerns the (temporal) relation between *several* signals. Hence, it is somewhat surprising that the putative advantage of coincidence detection for the processing of neural information is quite often associated with oscillatory activity in the brain.

Of course, there is a rarely explicated reason for the association mentioned in comment (iii): The *generation* of synchronous spikes in neural populations by means of local cooperative processes, i.e. without central control (triggering or gating), is supported by short epochs of near to constant stimulation and consequently quasiperiodic firing. (This need not hold for already synchronized ensembles.) Whether this kind of short-term binding is recognized as oscillatory is a matter of taste. We prefer the aspect of fairly rapid variations in (synchronized) neural activity [7, section 4] rather than that of more or less stationary oscillations – or oscillations in the sense of slowly shifting and thus narrow spectral frequency bands. This emphasis appears justified by the notion that neurons are voltage controlled (stochastic) impulse generators (cf. [8]) which obviously serve the processing of time-varying signals.

According to remark (ii) there is generally little reason for the *emission* of synchronous action potentials from a neural population in order to optimally stimulate *coincidence detectors* (cell somata) if a *transmission channel* (axons, synapses and dendrites) with space-variant temporal properties must be assumed. However, one decade ago, a neural receiver mechanism was identified for which the emission of synchronized impulses could make sense, namely coincidence-detecting NMDA-type synapses [9; 10]. It can detect coincidences between action potentials that are transmitted at essentially the same dendritic site [11] but, in contrast to the somatic coincidence detection of tonic potentials, it is limited to small numbers of input signals. In short, one must be aware of what is or should be synchronized at which location.

Other reasons for the generation of synchronous events are their direct behavioural use, such as the synchronous emission of light flashes by populations of certain fireflies [4; 12], and the suspected relevance for general timing purposes [13; 14; 15].

We like to conclude this argumentation with yet another statement which anticipates the essence of our own investigations: We found that synchronization of neural spiking activity generally *must be expected* in populations of homogeneously stimulated neurons which are locally coupled – either in an excitatory space-invariant feedforward fashion or by delayed (recurrent) lateral inhibition. Because both are quite common interconnection schemes, especially in cortical structures, we conclude that synchronized action potentials should not be regarded as particular network states – at least unless these synchronizing mechanisms are commonly paired with desynchronizing ones, such as inhibitory and space-invariant forward couplings – or that our simulations turn out to be too simplistic.

3. NETWORKS OF LOCALLY COUPLED MODEL NEURONS

Our investigations started from the question regarding necessary conditions for the generation of synchronous impulses in populations of locally interconnected model neurons. In order to tackle this problem, we needed an appropriate, i.e. in larger populations still computationally manageable model neuron (unit) with spiking output. A chain of such units that are laterally coupled with their neighbours by either excitatory or inhibitory interconnections – with or without delay – was considered as a promising and simple enough network structure.

3.1 The model neuron

We use a model neuron with the subthreshold behaviour of a leaky integrator which can be characterized by its δ-impulse response

$$h(t) = e^{-t/\tau}/\tau \qquad \text{for } t \geq 0$$

with the time constant τ =10ms. We distinguish three kinds of input signals that are *linearly summed* by such units (all potentials are normalized to the threshold θ):

- the *feeding* input $e(t)$

  which represents stimulations from outside the network and changes the somatic resting potential by $u_e(t) = e(t) * h(t)$ ("$*$" denotes convolution). For most of the experiments reported here, we consider $E := e(t) = const.$ which simulates incoherent input via many weakly transmitting, e.g. "apical-dendritic" synapses (Figure 3) and changes the somatic potential according to the step response (Figure 2)

$$u_E(t) = E\left(1 - e^{-t/\tau}\right) \qquad \text{for } t \geq 0 \ .$$

- the *lateral* input $a(t) = \sum_\nu w_\nu \cdot p_\nu(t - \vartheta)$ ($w_\nu > 0$: excitatory; $w_\nu < 0$: inhibitory) is the weighted sum of impulse trains $p_\nu(t) = \sum_\kappa s(t - t_\kappa)$ that are transmitted with delay $\vartheta = const.$ from ν neighbouring units via, e.g. "basal-dendritic or somatic" synapses, i.e. from inside the network (Figure 3). (Times t_κ denote the spike positions in an impulse train.) This input alters the somatic potential according to $u_a(t) = a(t) * h(t)$. Figure 1 shows the assumed exponentially decaying action potential $s(t)$ (time constant $\tau_{AP} = 0.144\text{ms}$) together with its postsynaptic response at the soma $u_{PSP}(t)$ which is normalized to its *coupling strength* w.

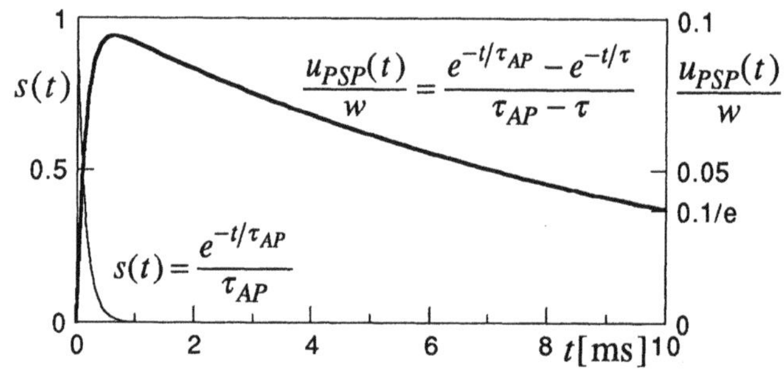

Figure 1. Action potential $s(t)$ and normalized postsynaptic potential $u_{PSP}(t)/w$

- the *noise* input $n(t)$ (individually computed for each unit) is a random process with uniformly distributed values from the range $\pm E/2$ which mimicks stochastic fluctuations $u_n(t) = n(t) * h(t)$ of the somatic potential.

In total, the change of the subthreshold somatic potential from the resting potential is

$$u(t) = u_e(t) + u_a(t) + u_n(t) = [e(t) + a(t) + n(t)] * h(t) \ .$$

(We neither consider synaptic habituation nor nonlinear synaptic transmission or interactions.) When the somatic potential exceeds the threshold θ, then an action potential $s(t)$ is triggered and 1ms later the somatic potential is set to the refractory potential for a period of 0.5ms before the integration can start again. The feeding input E is specified by the period T of the impulse train it evokes in a noisefree unit.

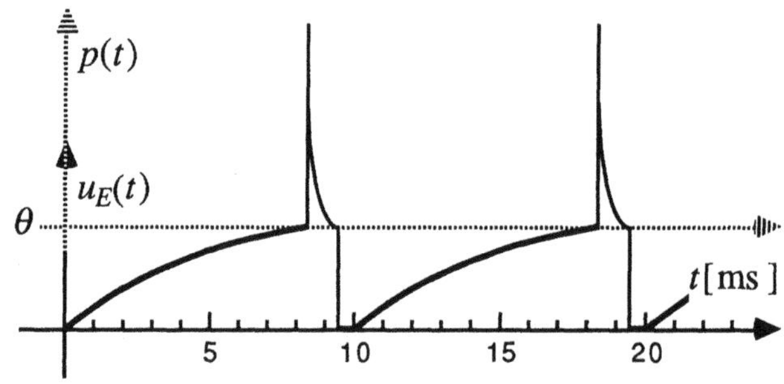

Figure 2. Typical time courses of the subthreshold somatic potential $u_E(t)$ as well as of the emitted spike train $p(t)$ with the period $T = 10\text{ms}$

3.2 The network

Figure 3 depicts the neighbourhood of a model neuron in our one-dimensional, single stage network. In order to avoid boundary problems in networks of manageable size the chain is cyclically closed and all its N units are coupled in the same way. We consider either purely inhibitory or excitatory interconnections without direct feedback from units onto themselves. Every unit receives input from its immediate $k \ll N$ neighbours on either side with a strength w_v that linearly decreases with the distance, i.e. with $|v|$. Unlike the coupling strength, the transmission delay is assumed constant which implies similar axonal conduction times as well as synaptic and postsynaptic processing (see point (ii) of section 2). In order to characterize the efficacy of interaction in the whole network, we introduce the *total coupling strength*

$$W = \sum_{\substack{v=-k \\ v\neq 0}}^{+k} w_v = const.$$

of every model neuron. Of course, there is a reasonable desire for global stability which in turn necessitates an upper limit of the excitatory coupling strength which we choose to $W_{crit} = +0.78$. For this *critical coupling strength* an unstimulated and noiseless unit begins to spike under synchronous unilateral input.

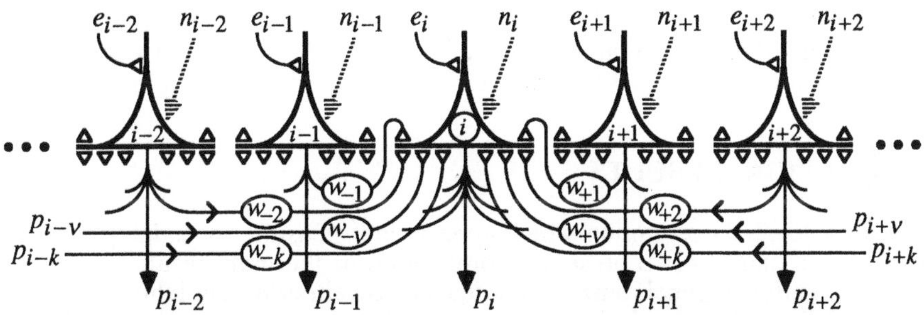

Figure 3. Local interconnection scheme of a unit in the one-dimensional network

Although we report on investigations of single stage or "single layer" networks, locally divergent forward coupling between "layers" – similar to the one proposed by Abeles [16, chapter 7] for the generation and transmission of synfire chains – appears more realistic for configurations without delay. Non-delayed lateral excitation can be directly achieved by forward coupling whereas non-delayed lateral inhibition requires the compensation of different delays between the central excitatory path and the divergent inhibitory ones which are due to the inevitable inhibitory interneurons.

3.3 Network simulation and measures of synchrony

We studied the *discrete* nonlinear dynamics of the networks, i.e. we performed simulations on a digital computer with the temporal resolution Δt. As a consequence, the zero delay in the lateral links can only be approximated, i.e. one must accept the mean intrinsic delay of $\vartheta_0 = \Delta t/2$. For the experiments reported here, all units of the network received the same feeding input for times $t \geq 0$. However, their initial somatic potentials ($t < 0$) were individually set to uniformly distributed random potentials from the "refractory potential to threshold"-range. Unless stated otherwise, the parameter values of the networks were choosen as follows:

$$\Delta t_s = 0.1\text{ms} \; ; \quad N_s = 64; \quad k_s = 8 \; ; \qquad E_s \text{ so that } T = 10\text{ms} \; ;$$

This standard setting turned out to be useful for most of the investigations and does not represent an extreme choice. The influence of deviations from these standard values on the quality of synchronization is explicated elsewhere [17].

In order to quantify the degree of synchronization we define the instantaneous *spike density* $S(t)$ which is given by the binarized spike aktivity of the network in a spatio-temporal window which is N units long and one spike duration $M \cdot \Delta t = 1\text{ms}$ wide, devided by the maximum possible activity in this window (Figure 5).

$$S(t) = \frac{1}{NM} \sum_{i=1}^{N} \sum_{j=0}^{M-1} P_i(t - j\Delta t) \quad \text{with } P_i(t) = \begin{cases} 0 & p_i(t) = 0 \\ 1 & \text{else} \end{cases}$$

Consequently, $S(t) = 1$ denotes an instant of perfect synchrony, i.e. all N units must have triggered action potentials at exactly the same time. Occasionally, we display the *envelope function* $\hat{S}(t)$ of the spike density. Finally, the *quality factor* η is defined as the mean over 50 runs of the maximum spike density that is determined during intervals of 50ms. We display $\eta_{50} = \max_{0<t\le 50}\{S(t)\}$ and $\eta_{200} = \max_{150<t\le 200}\{S(t)\}$.

For an assessment of the quality factor we provide the *reference quality* η_{ref} which results from "synchrony by chance" in the uncoupled ensemble but, since the reference quality depends on the impulse rate $r(W)$, which in turn can be converted to an equivalent reference quality, η_{ref} can also be specified for coupled networks.

4. SIMULATION RESULTS

After this detailed description which appears indispensible for judging the consequences of our findings, we present a compilation of the main results in Figure 4. Obviously, significant synchronization is feasible by either purely inhibitory or excitatory lateral links. (Figure 5 shows examples of corresponding spike densities.) In both cases, synchronization is based on the nonlinear characteristic of the somatic integration: The efficacy of a postsynaptic potential in delaying/accelerating the triggering of an action potential is higher for somatic potentials near the threshold than for small depolarizations. Therefore and with respect to the spike emission, advanced/retarded impulses are more strongly retained/impelled than late/early ones.

On the whole, a complementary behaviour is observable with respect to delays:

ϑ	$W<0$	$W>0$
$\vartheta_0 = 50\mu s$	desynchronization: $\eta_{200} < \eta_{ref}$	excel. sync.: $\eta_{200} > .9$ for $W > +.2$
$\vartheta_2 = 2ms$	good sync.: $\eta_{200} > .75$ for $W < -1$	no significant sync.: $\eta_{200} \approx \eta_{ref}$

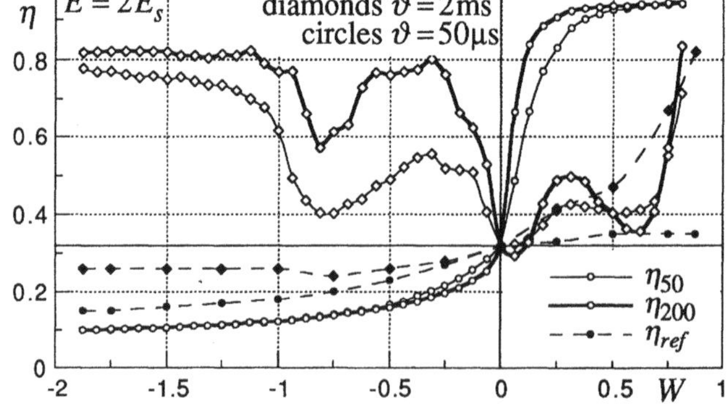

Figure 4. Synchronization quality $\eta(W)$ as a function of the total coupling strength W for transmission delays $\vartheta = 50\mu s$ and $\vartheta = 2ms$

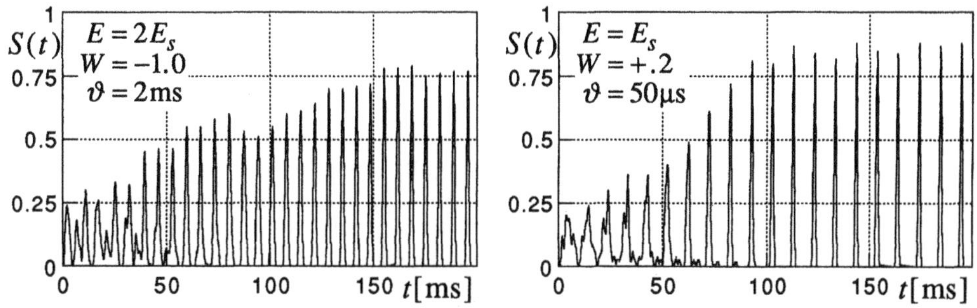

Figure 5. Examples of instantaneous spike densities $S(t)$ (note the different inputs)

Compared to the impulse rate $r_0(W_{crit}) \approx 195\text{s}^{-1}$ for the non-delayed transmission, delays considerably increase the impulse rate to $r_2(W_{crit}) \approx 400\text{s}^{-1}$ which means that with delay additional impulses are generated while without it retarded action potentials are accelerated up to synchrony. A similar effect holds for inhibitory links with $r_0(W=-1.87) \approx 60\text{s}^{-1}$ and $r_2(W=-1.87) \approx 142\text{s}^{-1}$. ($r_{ref} = r(W=0) \approx 180\text{s}^{-1}$)

Actually, synchronization depends on the amount of the delay. For $W = -1.0$ and $E = 2E_s$ we observed a continuous increase from desynchronization $\eta_-(\vartheta_0)$ to the maximum $\eta_-(\vartheta_2)$ which is sustained apart from 1ms dips at multiples of 6ms. However, for $W = +.2$ and $E = E_s$ we found a steep decrease from excellent synchronization $\eta_+(\vartheta_0)$ down to the reference level at $\vartheta = .5\text{ms}$ which is maintained except for peaks around multiples of 6ms that are less pronounced for η_{50} and decrease with ϑ.

Responses to a *linear downward sweep* of the feeding input are plotted in Figure 6. The envelope of the spike density $\hat{S}(t)$ indicates a consistently high degree of synchrony over the whole range of stimulation. Furthermore, this experiment nicely reveals the nonlinear transfer characteristic $r(e)$ of a coupled model neuron.

Figure 6. Spike density envelope $\hat{S}(t)$ and impulse rate $r(t)$ for a linear sweep of the feeding input $e(t)$

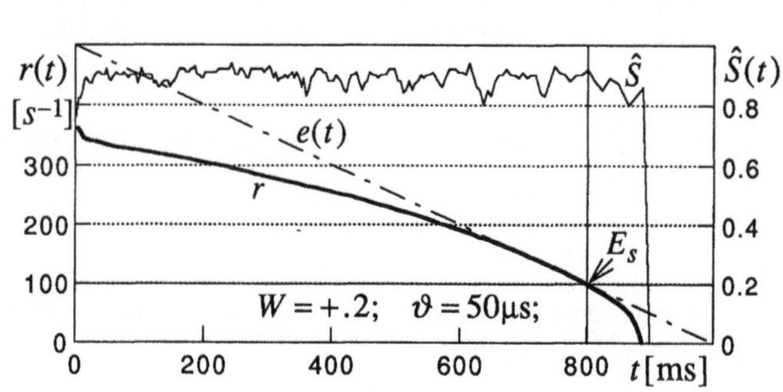

5. DISCUSSION

According to our findings, synchronized spiking activity results almost inevitably from homogeneously stimulated and especially excitatorily forward coupled or delayed (interneurons!) lateral inhibition networks. Contrary to the present euphoria over this *highly ordered* spatio-temporal binding, we are not certain as to its relevance for the processing of neural signals (see section 2): If a significant number of neurons really act as somatic coincidence detectors for *complex* spatio-temporal activity patterns, why then are synchronizing circuits required? However, aside from coincidence-detecting synapses, synchronous spike emission may turn out advantageous for reasons of selforganization.

After this general consideration we return to the issue of synchronized aperiodic spike fronts. In addition to the demonstration in Figure 6, we found that even abrupt variations of the global stimulation do not significantly disturb an already existing synchrony. – Moreover, spike synchronization happens stepwise and therefore, the time needed to reach a desired degree of synchrony depends on the spike rate.

Compared with related work on excitatorily coupled networks we have shown that neither nonlinear synapses [18] nor different time constants for feeding and synchronizing inputs [19] are a requisite for good synchronization. Although fully interconnected networks [20; 21] cannot really be compared they show states similar to those occuring in our experiments. In this context we would like to mention that our preferred total coupling strength $W = +.2$ means a rather *weak coupling*.

Finally, we must indicate a limitation of temporally discrete simulations arising from the fact that discrete nonlinear systems are not necessarily approximations of their continuous originals. Hence, all such simulations must be taken with extreme caution. Up to now we could only show a convergence for decreasing increments Δt.

6. REFERENCES

[1] McCulloch, W.S. and Pitts, W.H., *Bull. Math. Biophys.* **5** (1943) 115-133.
[2] Abeles, M., *Isr. J. Med. Sci.* **18** (1982) 83-92.
[3] Nischwitz, A., in: Yuan, B.Z., Zhao, K.H. and Dayhoff, J. (eds.) *Proc. Int. J. Conf. Neural Networks* (Publ. House Electronic Industry, Beijing, 1992) pp. 832-837.
[4] Wiener, N., *Cybernetics or Control and Communication in the Animal and the Machine* (MIT Press, Cambridge/MA, 1961).
[5] Taube, M., *Computers and Common Sense – the Myth of Thinking Machines* (Columbia University Press, New York/NY, 1961).
[6] Neven, H. and Aertsen, A., *Biol. Cybern.* **67** (1992) 309-322.
[7] Nischwitz, A., Glünder, H., von Oertzen, A. and Klausner, P., in: Aleksander, I. and Taylor, J. (eds.) *Artificial Neural Networks 2* (Elsevier, Amsterdam, 1992) pp. 851-854.
[8] Kling, U. and Székely, G., *Kybernetik (Biol. Cybern.)* **5** (1968) 89-103.
[9] Dingledine, R., *J. Physiol.* **343** (1983) 385-405.
[10] Jahr, C.E. and Stevens, C.F., *J. Neurosci.* **10** (1990) 1830-1837.
[11] Hounsgaard, J. and Midtgaard, J., *TINS* **12** (1990) 313-315.
[12] Mirollo, R. and Strogatz, S., *SIAM J. Appl. Math.* **50** (1990) 1645-1662.
[13] Winfree, A.T., *J. Theoret. Biol.* **16** (1967) 15-42.
[14] Pöppel, E., *Studium Generale* **24** (1971) 85-107.
[15] Pöppel, E., Ruhnau, E., Schill, K. and von Steinbüchel, N., in: Haken, H. and Stadler, M. (eds.) *Synergetics in Cognition* (Springer, Berlin, 1990) pp. 144-149.
[16] Abeles, M., *Local Cortical Circuits* (Springer, Berlin, 1982).
[17] Nischwitz, A., Glünder, H. and Klausner, P., in: Kohonen, et al. (eds.) *Artificial Neural Networks* (Elsevier, Amsterdam, 1991) pp. 1771-1774.
[18] Eckhorn, R., Reitboeck, H.J., Arndt, M. and Dicke, P., *Neural Computation* **2** (1990) 293-307.
[19] Hartmann, G. and Drüe, S., in: Eckmiller, R., Hartmann, G. and Hauske, G. (eds.) *Parallel Processing in Neural Systems and Computers* (Elsevier, Amsterdam, 1990) pp. 361-364.
[20] Erb, M. and Aertsen, A., in: Aertsen, A. and Braitenberg, V. (eds.) *Information Processing in the Cortex. Experiments and Theory* (Springer, Berlin, 1992) pp. 201-223.
[21] Deppisch, J., Bauer, H.-U., Schillen, T., König, P., Pawelzik, K. and Geisel, T., in: Aleksander, I. and Taylor, J. (eds.) *Artificial Neural Networks 2* (Elsevier, Amsterdam, 1992) pp. 921-924.

Part Four
The Brain: The Organization of Perception and Action

Brain Theory / A. Aertsen (Editor)

On the internal structure of cell assemblies

G. Palm

Department of Neural Information Processing, University of Ulm,
Oberer Eselsberg, W-7900 Ulm, Germany

Abstract

Based on estimates on excitation thresholds and contact probabilities of 'typical' cortical neurons, I try to develop a speculative picture of the appearance of a 'typical' cortical cell assembly, both with respect to space (sensory vs. non-sensory areas, blobbiness) and time (event coherence vs. rate coherence).

1. INTRODUCTION

In this short note I want to present some ideas concerning the fine structure of assemblies with respect to space and time. Cell assemblies as introduced by Hebb [1949] had been considered as groups of coactive neurons in the neocortex which are bound together by strengthened Hebb-synapses and which can be thought of as units for the representation of information. I have analyzed this idea from an information theoretical point of view [e.g. Palm 1980, 1991], where I was primarily interested in the total number of assemblies that can be stored and recalled and in their typical size. To this end I neglected possible temporal and spatial substructures and assumed perfect coincidence in timing and either fully or homogeneously randomly connected networks. The upshot of this work was

(i) that one can store many more assemblies than neurons,
(ii) that one can store and retrieve almost one bit of information per Hebb synapse and
(iii) that the typical size of a neocortical assembly should be around 10^4 (i.e. between 10^3 and 10^5).

The last point was the most speculative, since it really would have needed a more elaborate account of the internal structure of assemblies.

Internal structure in space: The cortical connectivity is not quite homogeneous, there is probably an almost homogeneous (in a statistical sense, only) local connectivity combined with a system of cortico-cortical projections which provides a typical cortical area with connections to and from about 4-5 other areas. This connectivity scheme suggests that one should think of an assembly as combined from a number of local areal subgroups. But what is the typical size and number of these subgroups?

Internal structure in time: The main question to decide here is the following: Within which time window do spikes have to coincide to be called coincident? It is clear that coincident firing of the afferents of a neuron enhances the chances to fire that neuron. One can easily estimate this enhancement from random firing at a certain frequency to coincident firing of the same number of spikes within a time window of a certain duration d [see Abeles 1991]. Coincident firing within a narrow time window can therefore be an effective way of making assemblies smaller.

To get a more quantitative handle on this problem we have to estimate the *excitation threshold* of a typical cortical neuron. This is the number of EPSPs that have to occur at the same time (say within 1 msec) at the neuron's input to drive its dendritic potential from resting value across the spike generation threshold. This is, of course, a simplified picture, but it may suffice for the following considerations. With this definition we can estimate [following Abeles 1991] this excitation threshold to be somewhere between 8 and 30.

2. ASSEMBLIES IN SENSORY AND NON-SENSORY AREAS; HIPPOCAMPAL GENERATION OF COHERENCE

We will now enter into some more elaborated speculations, concerning the size and appearance of cell assemblies, which are based on informed guesses about excitation thresholds and contact probabilities of cortical neurons based od anatomical observations [Braitenberg 1978, Schüz and Palm 1989, Braitenberg and Schüz 1991]. The basic idea is that a typical assembly can be thought of as a collection of local subassemblies, where each local subassembly is confined within the range which can be reached by a typical pyramidal cell, e.g. a cortical area of about 2 mm x 2 mm. Based on this idea we estimate the number of cortico-cortical and intracortical synapses which a typical neuron in the assembly receives from the assembly. We assume that these synapses are Hebb-synapses and we postulate that there should be at least about 5 to 8 of them converging on each neuron in the assembly. This number is based on estimates on the excitation threshold by Abeles [1991] and on the idea that a reasonable proportion of these excitatory synapses could be provided by nonspecific inputs. Our further considerations are related to, but different from the recent calculations of Wickelgren [1992] and Miller [1993].

With respect to timing we assume that within 2 mm of cortical distance a timing accuracy of about 4 msec can be kept [cf. Nelken 1992]. Thus we base our analysis on a time window of 4 msec. The following analysis may seem a little naive, since there are much more refined methods of calculation demonstrated for example by Abeles [1991], but such calculation would yield roughly the same results.

First let us consider a local assembly that is situated in a primary sensory area of the cortex like *V1*. Such an assembly can be formed by Hebbian strenghtening of sufficiently many synapses between the neurons of the assembly. This can easily happen if the neurons in the assembly are stimulated together often enough by the sensory input.

Once the assembly is formed, each neuron should receive coincident input through at least a certain number s of strengthened synapses from the others. In each time window (of 4 msec) only a certain percentage p of the neurons in the assembly are active. We have to assume a certain time window for firing coincidence that is effective in firing the next neuron, let us again say 4 msec in view of the typical duration of EPSPs.

If the assembly size is A, then on average pA neurons were active within the last 4 msec, thus the probability p' that a neuron gets at least s strengthened synaptic inputs can be estimated from a binomial distribution [as in Miller 1993] or simply from a Poisson distribution with mean pAp_l - where p_l is the probability of a local connection between two neurons within 2 millimeter distance. This probability p' should be larger than p.

Apart from the time windows we also have to specify the numbers p, p_l and s. Neurons that are active fire with a frequency which can vary greatly, but we assume here a value of 50 spikes per second for a typical cortical neuron. Thus for a 4 msec window we have p = 1/5. The excitation threshold of cortical neurons is assumed to be around 8 to 10 EPSPs within about 3-4 msec. Since we assume that our neurons already get some base-line of activation through the sensory stimulation and also through non-specific thalamic activation we will work with $s = 4$. The value for p_l can be estimated to be around $1/10$ or $1/20$; we take $p_l = 1/20$.

With these numbers the condition $p' > p$ is satisfied for A $\geq$ 230.

Thus we can assume that purely local assemblies in sensory areas should have about 250 neurons, which are all in an elevated state of activity when the assembly is activated, but not all active at exactly the same times, the "event coherence probability" p being about $p = 1/5$.

Next we want to consider global assemblies encompassing several areas. To bind neurons in several areas together, we need cortico-cortical connections. Between sensory areas that are cortico-cortically connected, this is easy once the subassemblies have been formed. In addition it would be good to have a special representation for the co-occurence of neural events in several sensory areas. This could be done by recruiting representative neurons in other non-sensory areas of the cortex. This idea is referred to as 'chunking' in the psychological literature. In particular Wickelgren [1992] has analysed this situation in the context of cell assemblies. The problem with this idea is that there is no sensory stimulation in such an association area, we only may assume an unspecific thalamic excitation.

Let us assume that $s = 6$ additional non-local synapses are sufficient to fire a typical neuron in an 'attentive' (pre-excited) association area. Even with this assumption it turns out to be hard to recruite sufficiently many neurons in such an area. We need an additional source of excitation. This source would be even more efficient if it would produce coincident excitation within our 4 msec time window.

We share with Robert Miller [1991, 1993] the idea that the hippocampus may be such a source of excitation (in particular to cortical association areas). Thus the hippocampus is thought of as sending out packets of coincident activation which are propagated from area to area across the cortex. We assume that about h neurons in the hippocamus being simultaneously activated by the theta-rhythm send axons to a typical cortical (association) area. They have a chance p_g to connect to one arbitrarily chosen neuron in that area. The cortico-cortical projection neurons are assumed to have roughly the same chance p_g of making a global connection, but only a fraction f of cortical cells takes part in the necessary cortico-cortical projection.

Thus the probability p' that a neuron in a non-sensory area gets at least s synaptic inputs via hippocampal and/or cortico-cortical projections from a other areas can again be calculated from a Poisson distribution, this time with mean $(pAa{\cdot}f+h)p_g$.

In this formula we take $p_g = p_l = 1/20$ and $a{\cdot}f = 1/12$. We assume that about 1/8 of all cortical neurons provide a cortico-cortical projection fiber and that there should be roughly as many projections converging on an area as diverging from an area. This number may be 4-5 and we should assume that at least 2-3 sensory subassemblies give input to our non-sensory population. Thus $a{\cdot}f$ may be between 1/8 and 1/16.

There should be about $p{\cdot}A$ neurons recruited within one time window in order to have a similar effect as a subassembly in a sensory area. Thus we require that $p' > (A/N)p$. Here the total number N of neurons below 4 mm^2 cortex is estimated as $N = 2{\cdot}10^5$. This requirement is satisfied for $h \geq 15$.

Thus the hippocampus can indeed provide the necessary input for recruiting chunking neurons in a cortical association area, if it sends out roughly $h = 15$ axons per 4 mm^2, i.e. 4 axons per mm^2 of non-sensory cortical areas. If all non-sensory areas would receive such a hippocampal input, then about $4{\cdot}10^5$ cells in the hippocampus would have to be activated coherently by the theta-rhythm. (If all cortical areas (sensory and non-sensory) were activated, this number would be about $6 \cdot 10^5$) I believe that these numbers are not unreasonable for the hippocampus [see also Miller 1991].

So we might consider this 'injection' of coherent activity packets into the cortex as one possible function of the hippocampus. As discussed in more detail by Miller [1991, 1993] and Kuroda [1991], this idea is well compatible with many experimental and clinical observations. For example anterograde amnesia, which is typically induced by hippocampal lesions in humans, is a loss of the ability to store new information which is typically complex and multi-modal and thus might require the recruitment of 'chunking' subassemblies in cortical association areas.

The emerging picture of cortical cell assemblies based on the above calculations is also very interesting. Cell assemblies may vary greatly in size depending on the number of areal subassemblies that they have. Up to now there is no easy way of estimating this number of subassemblies. Earlier calculations based on their information content [Palm 1986] hint in the direction that a total assembly should have somewhere around 10^4 neurons with a working range from a few thousand to several tenthousand. A sensory subassembly should normaly have at least around 250 neurons, which are not all firing synchroneously, whereas a typical non-sensory subassembly should have at least around 50 neurons, which fire synchroneously. I believe that each assembly should have at least one sensory subassembly and that large assemblies should have a reasonable fraction of sensory subassemblies, so the total number of subassemblies should be somewhere between 20 and 1000. This is still a small fraction of the about 40.000 local patches of 4 mm^2 size that would cover a human cortex. In large cortical areas (many mm^2) it is possible that several local subassemblies of about 2-4 mm^2 size belong to the same assembly. In sensory areas this will probably be the usual case.

In sensory areas the average activity should be higher than in non-sensory areas. Event coherence (v.s. rate coherence) should be more strongly expressed in non-sensory association areas. The higher activity level in sensory areas gives rise to an interesting possibility of information processing that has been discussed in the recent literature: In addition to rough coincidence of activation (time windows of 100 msec), termed *rate coherence*, the cortex may make use of fine coherence (time windows of a few msec as discussed in this paper), termed *event coherence*, to articulate some sub-structure in the local subassemblies [see Neven and Aertsen 1992]. This should be possible in all areas where the subassemblies are somewhat larger (at least 50 neurons per mm^2), mainly in the sensory areas. These coherently firing subgroups of sensory assemblies could be bound by coincident activation through cortico-cortical projections to different chunking assemblies in non-sensory areas. In this way the global structure of multi-areal assemblies provides a natural mechanism to 'read out' an interpretation of the sensory substructures. The global structure itself can be formed by Hebbian learning in cortico-cortical projections, provided that during learning the non-sensory neurons get enough coincident activation from some source - presumably the hippocampus.

3. HOW MANY ASSEMBLIES CAN BE LEARNED?

Based on the simple calculations of the last section we now want to estimate the total number of such assemblies that can be learned.

Obviously, if we first pick at random a certain number S of 4 mm^2 blobs from the sensory areas and a number C of 4 mm^2 blobs from the non-sensory areas, then about 250 neurons from each sensory blob and about 50 neurons from each non-sensory blob, it may well be that these neurons cannot be formed into an assembly by Hebbian learning. This will happen if there are not sufficiently many synapses connecting these neurons (compared to the above requirements).

In the last section we have focussed on *sensory-driven assemblies*, i.e. assemblies that originate in sensory areas, monitoring sensory coincidence, but possibly with some additional 'chunking' subassemblies which are formed in non-sensory areas. These non-sensory subassemblies can become the basis for more abstract conceptual assemblies [cf. Wickelgren 1992]. In the present calculation, however, we will only try to estimate the total number of sensory-driven assemblies. Therefore we will only pick the blobs and neurons in the sensory areas, which essentially determine the total assembly to be formed (including its non-sensory parts).

The probability P that a chosen set of neurons in sensory cortical areas can form an assembly, is just the probability that each cell in each local subgroup receives at least c excitatory synapses from its local neighbours in the assembly, where c is chosen so large that with these c synapses the chance for a neuron to be fired (i.e. to get at least s inputs) is above p (as before). This time this chance is calculated from a binomial distribution, or approximately from a Poisson distribution with mean pc. For our parameter-values $p = 1/5$, $s = 4$ we have to require that $c \geq 10$.

The probability for a sensory subassembly with 240 neurons, i.e. the probability that each neuron gets at least $c = 10$ synapses, can again be calculated from a binomial or from a Poisson distribution with mean $240 \cdot p_l = 12 \approx c$. Thus it is roughly 2^{-200}. Thus the expected number of learnable sensory assemblies containing S local clusters is

$$N_S = \binom{B}{S}\binom{N}{24+c}^S P \quad \text{with } P \approx 2^{-200S}.$$

Here N is the total number of neurons in a 4 mm^2 patch or blob of sensory cortex and B is the total number of such blobs. We take $N = 2 \cdot 10^5$ [cf. Schüz and Palm 1989] and $B = \mathit{Total\ area\ of\ sensory\ cortex} / 4\ mm^2 = 10^4$.

$$\text{Thus } \log_2 N_S = S(\log_2 B - \log_2 S) + 240S(\log_2 N - \log_2 240) - 200 \cdot S \approx 2{,}200\, S.$$

I will assume that a sensory assembly should have at least 3 local blobs. Adding up the numbers N_S from $S = 3$ to $S = Smax$ we obtain the total number L of learnable sensory assemblies. If the largest reasonable assembly has Smax local clusters, then clearly $L \approx N_{Smax} \approx 2^{1,800 Smax}$. These L assemblies are formed within 400 cm^2 or roughly 80 cm^3 of sensory cortex. This cortical volume should contain roughly $7 \cdot 10^{13}$ synapses.

We can try to estimate the total amount of information that can be embodied in the formation of sensory assemblies. Following my own earlier calculations, we can learn M of these L learnable assemblies as long as less than half the synapses are strengthened by Hebbian learning. As the total number of H Hebbian synapses we boldly take $H = 5 \cdot 10^{13}$. Thus $Smax \cdot 250 \cdot c \cdot M \leq H/2$ gives a safe estimate for the number of sensory assemblies that can be actually stored in a lifetime, yielding $M = 10^{10} / Smax$.

The information content of such an assembly is $log_2 L = 2{,}200\ Smax$ bits of learned information (plus $200 \cdot Smax$ bits of inborn information, if you like).

In total we can store $2.2 \cdot 10^{13}$ *bits* of information in $5 \cdot 10^{13}$ Hebb synapses giving a pattern storage capacity of about 44 % (bit per synapse). According to our own earlier calculations [Palm and Sommer 1992] we will actually be able to recall only 1/4 to 1/2 of this information by association, i.e. about $6 \cdot 10^{12}$ *bit* or 15 % (of a bit per synapse).

Since this value does not depend on *Smax,* we might have gone through the same argument with any value *S*, and it would perhaps be more reasonable to use an estimate of the average number *S* of sensory blobs in a typical assembly for the calculation of the average information content of an assembly and of the number *M* of storable assemblies.

On the other hand we should know *Smax* in order to determine the maximale size of a sensory assembly. But how can we estimate *Smax*? It seems reasonable to assume that even in a well stimulated sensory cortical area the activation is not uniform across the area, but somewhat 'blobby'. Furthermore, in a usual stimulation not all sensory areas are stimulated equally strongly. So let me assume that at most 1-5 % of all the 4 mm^2 local blobs are stimulated in a sensory-driven assembly. This means *Smax* is about 500. Thus the total number of neurons in an assembly will still be in the range between 10^3 and 10^5. The sensory-driven assemblies will probably be the larger ones (beween 10^4 and 10^5), whereas the more 'abstract' assemblies that can be formed mainly in association areas will typically contain at least about 40 small local clusters of neurons adding up to 10^3 to 10^4 neurons.

With respect to the total number of sensory assemblies our uncertainty is equally large, but there is no easy access to this question, anyway. If we take $Smax = 500$

and $S = 100$ for an average assembly, then there are $L = 2^{1.800 \cdot 500} = 10^{271,000}$ potentially learnable sensory assemblies and of these $M = 10^{10} / 100 = 10^8$ sensory assemblies can be actually formed by learning during a lifetime. In this case the information content of an average sensory assembly would be about $2 \cdot 10^5$ *bit*. The non-sensory assemblies would typically be smaller, but more.

In summary, we have derived a reasonable picture of a typical assembly and we have made some informed guesses about the size and total number of cortical assemblies, but we still have the greatest uncertainty about the simplest question: the size of a 'typical' cell assembly.

4. References

Abeles, M.: Corticonics. Cambridge University Press, Cambridge (1991)
Braitenberg, V.: in Theoretical approaches to complex systems (R. Heim and G. Palm, eds.) 171-188, Springer, Berlin (1978)
Braitenberg, V. and Schüz, A.: Anatomy of the Cortex. Statistics and Geometry. Springer, Berlin Heidelberg New York (1991)
Hebb, D.O.: The organization of behaviour. A neuropsychological theory. Wiley, New York (1949)
Kuroda, Y.: Concepts Neurosci. 4, 221-228 (1991)
Miller, R.: Cortico-Hippocampal Interplay. Springer, Berlin Heidelberg (1991)
Miller, R.: to appear in Memory Systems of the Real Brain (1993)
Nelken, I.: in Neural Network Dynamics (J.G. Taylor et al., eds.), 29-49, Springer, London (1992)
Neven, H. and Aertsen, A.: Biol. Cybern. 67, 309-322 (1992)
Palm, G.: Biol. Cybernetics 36, 19-31 (1980)
Palm, G.: in Brain Theory (G. Palm and A. Aertsen, eds.), 211-228, Springer, Berlin (1986)
Palm, G.: Concepts Neurosci. 2, 97-128, 1991
Palm, G. and Sommer, F.: Network 3, 177-186 (1992)
Schüz, A. and Palm, G.: J. Comp. Neurol. 286, 442-455, (1989)
Wickelgren, W.A., Concepts Neurosci. 5, 1-53 (1992)

Brain Theory / A. Aertsen (Editor)

Does the nervous system try to learn temporal invariants?

G.J. Mitchison

The Physiological Laboratory, Cambridge, CB2 3EG, United Kingdom

Abstract
A synaptic learning rule for extracting temporally invariant components from sensory data is described, and its behaviour is illustrated with some examples from stereoscopic vision.

1. INTRODUCTION

One theme of this meeting has been the temporal variation of neural signals. W. Freeman has discussed the significance this can have from a dynamical system point of view, and M. Abeles has shown how precise temporal patterns (synfire chains [1]) can be used for signalling within the nervous system. There is indeed good evidence, eg [2], that neural responses can show complex, internally generated, temporal patterning. However, my aim here is to describe briefly a different -- almost opposed -- approach, which encourages one to look for mechanisms which *reduce* the time variation in the responses of neurons to incoming signals. The idea is that much of what concerns us in sensory data changes slowly, and that it will be profitable for the nervous system to latch onto these time-invariant (or near-invariant) components. As an example, the visual changes caused by our movements will lead us to see a sequence of views of objects in our surroundings. A representation which extracts the invariant feature, the object itself, will obviously have advantages in economy over a set of representations, one for each view. Perhaps more important than this, however, is the idea that the task of tying together these views, a non-trivial problem for the nervous system, can be assisted by an algorithm which seeks time-invariant components.

2. ALGORITHMS FOR LEARNING INVARIANCES

Several authors [3-5] have devised algorithms of a neural character which could be used for extracting such components. One algorithm of this class [5] works as follows: The neuron is taken to be linear, so that its output y is the weighted sum of inputs x_i, $y=\Sigma w_i x_i$. The change in y over a small time interval -- perhaps a few hundred milliseconds -- is denoted by Δy. This is related to the changes, Δx_i, in the x_i by $\Delta y=\sum w_i \Delta x_i$. Suppose the goal is to minimize the average value, over many learning trials, of $(\Delta y)^2$. Altering the weights w_i by gradient descent to achieve this yields the rule $dw_i/dt=-\varepsilon \Delta y \Delta x_i$, where ε is a rate constant which should be chosen small so as to allow the changes to represent an average over many trials. This is like an anti-Hebbian synaptic learning rule, except that it is short-term variation in the input x_i and the output y which prompts changes in w_i rather than the values of these variables at one time. For this reason, rules of this type have been called *differential* synaptic learning rules [6-7].

Unfortunately, applying our differential synapse rule by itself will generally lead to a collapse of the weights to zero (reflecting the fact that a zero output does not vary). To avoid this trivial outcome, one can impose a conservation constraint on the weights, such as $\sum w_i^2=1$. This will generally force the system to find minimally varying solutions, but one often gets an embarrassment of riches: there are a great many possible sets of weights which satisfy these conditions. One way to select a useful, non-redundant set of receptive fields is to add a bias -- a small perturbing term -- to the differential synapse. For example, one can minimize instead of $(\Delta y)^2$ the sum $(\Delta y)^2+\lambda(\zeta-y)^2$, where ζ is the bias and λ is a constant which can be chosen to be small because its role is only to slide the state point within a space of solutions [5]. This is the form of the algorithm I shall be using here. It is equivalent to the synaptic modification rule

$$w_i^+=w_i - \varepsilon\{\Delta y \Delta x_i + \lambda x_i(y-\zeta)\} \quad (1)$$

The bias will usually be an input from another neuron. It does not have to be time-invariant itself; in fact, it can be one of the inputs x_i from the sensory array. What the learning rule does is to find an invariant response which is close to the bias.

3. STEREOSCOPIC VISION

Consider now an example of how this might work. There are reflexes which keep the gaze fixed when the head or body are moved, but even if the eyes individually remain exactly fixed on an object the angle made between the two lines of sight inevitably alters, and this produces a shift in stereoscopic disparity. The effect of this is, to a first approximation, to add a linear gradient of disparity across the visual field [8]. This can be seen from Figure 1, which shows that oblique viewing changes the angle which a surface makes to the loci of constant disparity (the Vieth-Müller circles) and thereby adds, locally at least, a disparity gradient. Oblique viewing also causes a multiplicative scaling of disparity, but we return to this later, assuming for the time being a purely additive effect of shifts in viewing angle.

Suppose first that disparity at a given retinal location is encoded by the firing rate of a neuron. Physiology suggests that it would be more appropriate to

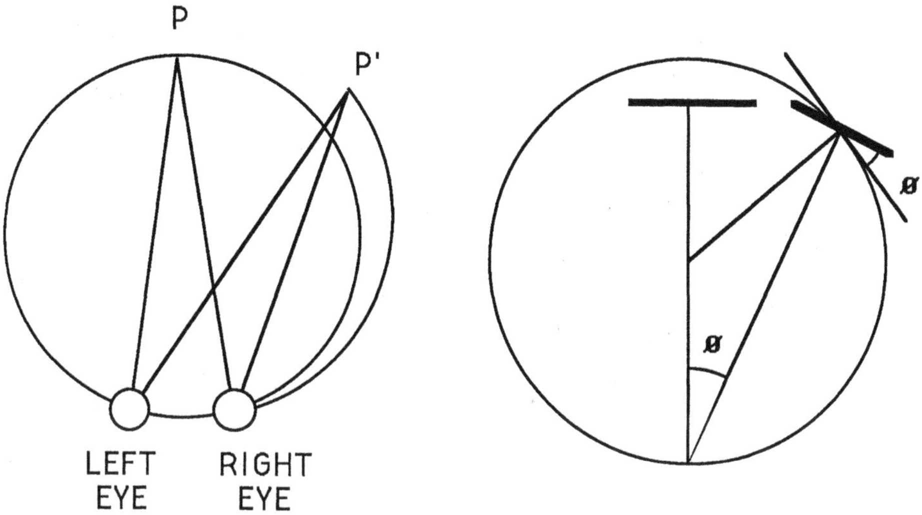

Figure 1
This shows the loci of constant disparity: the Vieth-Müller circles passing through the two eyes. If an observer rotates his head while maintaining fixation, so the point P takes up a position P' relative to the observer, there is a shift in disparity (because P' lies on a different circle from P) and also an added gradient (because a fronto-parallel surface is no longer tangent to its Vieth-Müller circle, but makes an angle ϕ to it, as shown on the right).

assume a population coding by neurons tuned to particular values of disparity [9-11], but rate coding is easier to understand. Applying the differential synapse algorithm, equation (1), in this situation, using a 5x5 array of units to represent the inputs, and with the unit at the centre of this array as bias, one obtains a set of weights w_i which define a balanced centre-surround receptive field (Figure 2). This is reminiscent of lightness perception, where gradients of illumination may be added to the (log) reflectance from surfaces. A suitable centre-surround filter will subtract these gradients and give a more faithful representation of the reflectances, as shown by Hurlbert and Poggio [12]. In much the same way, the centre-surround field learnt by our algorithm cancels disparity gradients due to changing viewing parameters. The psychophysical evidence shows a striking similarity between contrast phenomena in lightness and stereo [13], which in the former case are known to be underpinned by centre-surround mechanisms [14].

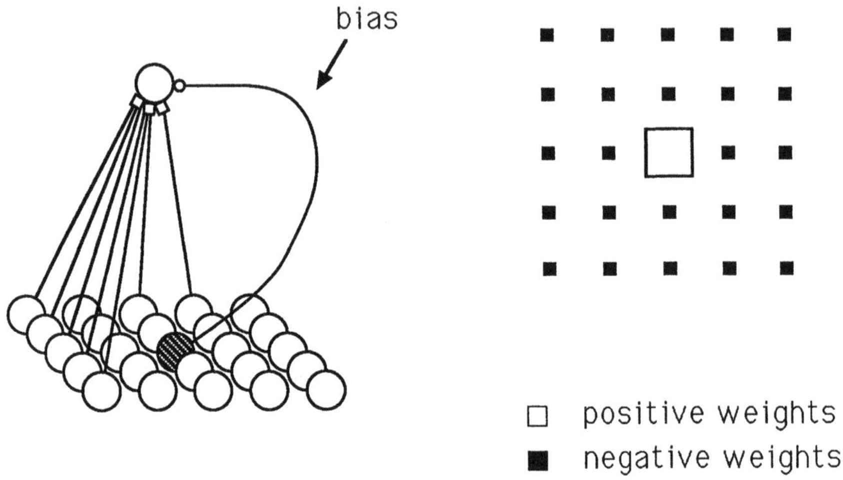

Figure 2
On the left is shown the output unit together with a 5x5 spatial array of rate-coded units which serve as inputs, the central unit (shaded) being selected as a bias. The disparities x_i in the training ensemble are assigned randomly, and the differences Δx_i are defined by a small added linear gradient across the 5x5 array in a randomly chosen direction. The weights which result from 10^6 learning steps of equation (1) with ε=.03, λ=.01 are shown as a Hinton diagram on the right.

Hurlbert and Poggio's algorithm [12] is formally equivalent to a special case of the differential synapse [5]. However, their algorithm needs to be supplied with the values of the underlying reflectances of the images in its training set, so that it can learn to cancel the added gradients. It should therefore be regarded as a means of mathematically constructing the ideal filter, rather than a learning procedure which the brain might use. By contrast, the differential synapse identifies the added components from their variation in time, and uses only information which is available to the visual system.

One important formal difference between stereo and lightness is that, in the case of stereo, the added gradients are generated by the observer (via the viewing geometry), rather than by an external agency (the lighting conditions). If the viewing parameters were exactly known, the gradients they cause could be cancelled. If they are not precisely known, a centre-surround may do a better job. This aspect of the use of data -- weighting it in accordance with its reliability -- is captured by the differential synapse. Figure 3 shows that a signal proportional to the value of the added disparity gradient at the centre of the 5x5 array, with added noise, entirely replaces the surround when the added noise is very small, but is only given small weighting when it is large.

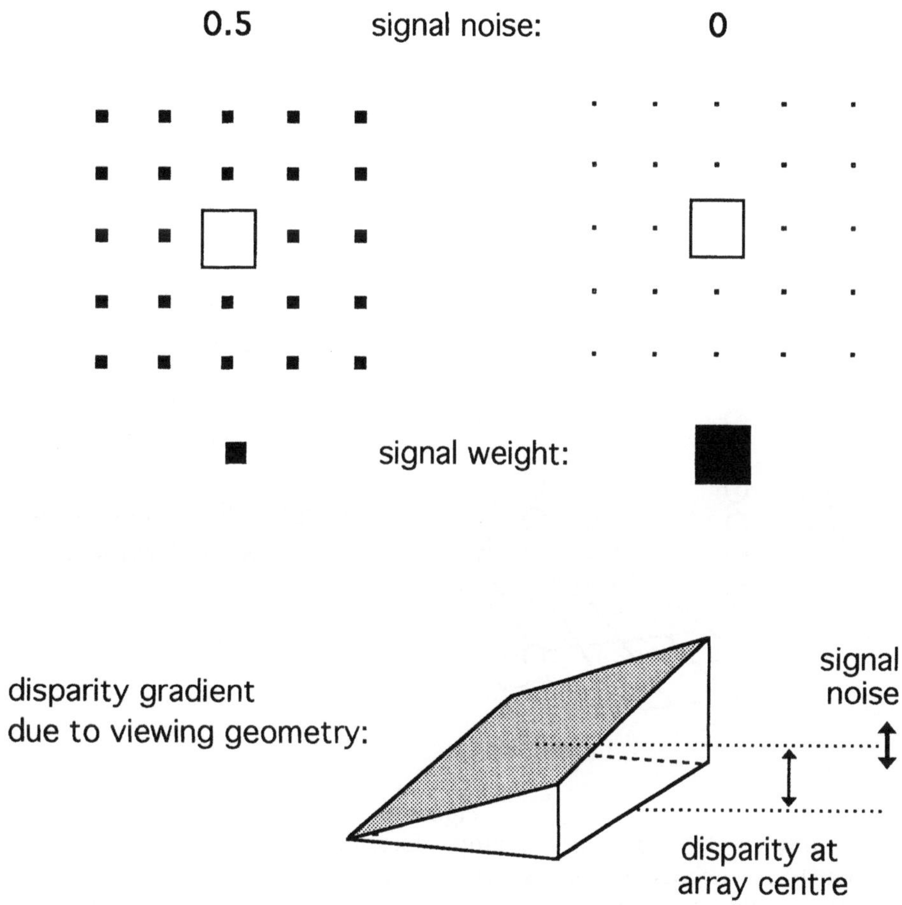

Figure 3
Simulations using equation (1) and a 5x5 spatial array of rate-coded inputs, as in Figure 2, but with one further input: a signal derived from the value of the added linear gradient at the centre of the array. If there is little added noise, this signal can be subtracted from the total disparity at the centre of the field to give a good estimate of the invariant component of disparity. But if the signal is noisy, the algorithm largely ignores it, relying instead upon the surround to get an estimate of the linear gradient.

We consider next a picture more compatible with the physiological evidence, where there is a population of disparity-tuned units, instead of rate-coded units, at each point of the spatial array (Figure 4).

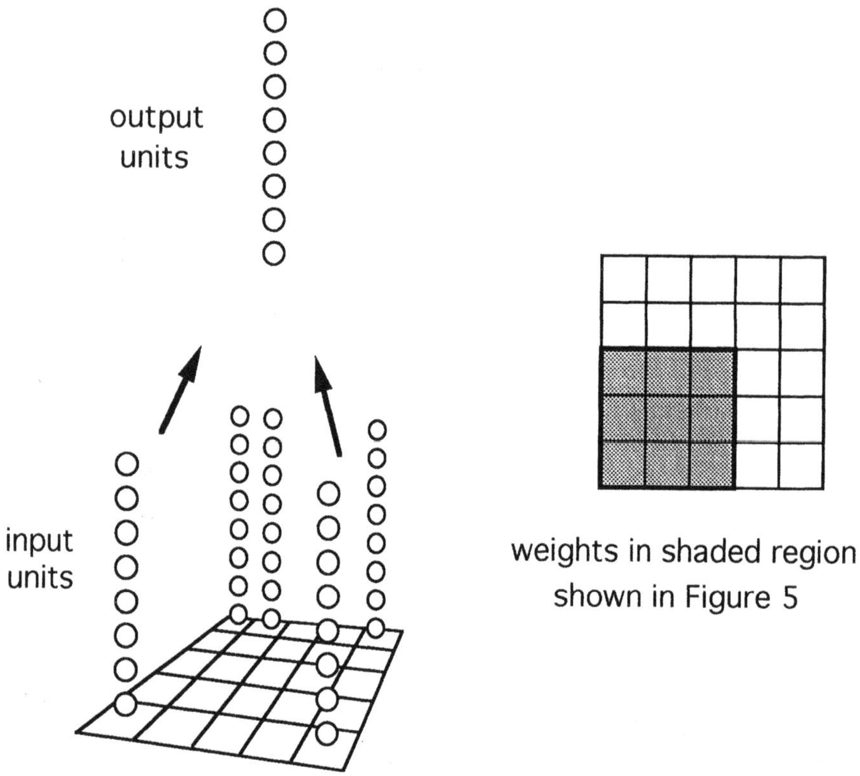

Figure 4
This shows a 5x5 array of retinotopic locations, at each point of which the disparity is encoded by N disparity-tuned units. There are N output units whose weights are trained by equation (1). Each output unit receives inputs from the full set of disparity-tuned units at all points of the 5x5 array. The bias of the i-th output unit comes from the i-th input unit at the centre of the array. As with the simulations shown in Figure 2, the disparities in the training sets are assigned randomly at each point of the 5x5 array, and the temporal changes in disparity are defined by a small randomly oriented linear gradient. The disparity $d_{\alpha\beta}$ at array location (α,β) is converted into an population-coded pattern by taking the activity of the i-th unit at (α,β) to be $\exp(-(d_{\alpha\beta}-\mu_i)/2\sigma^2)$, where μ_i is the mean of the i-th unit's Gaussian tuning curve and σ its standard deviation. The means are equally spaced, with separation σ.

Disparity at each point is represented by a set of units with Gaussian tuning curves centered at a set of regularly spaced disparity values (see legend to Figure 4). The output units are trained using the differential synapse, the i-th output unit being biased by the i-th input unit at the centre of the array. The algorithm learns the rather complex set of weights shown in Figure 5.

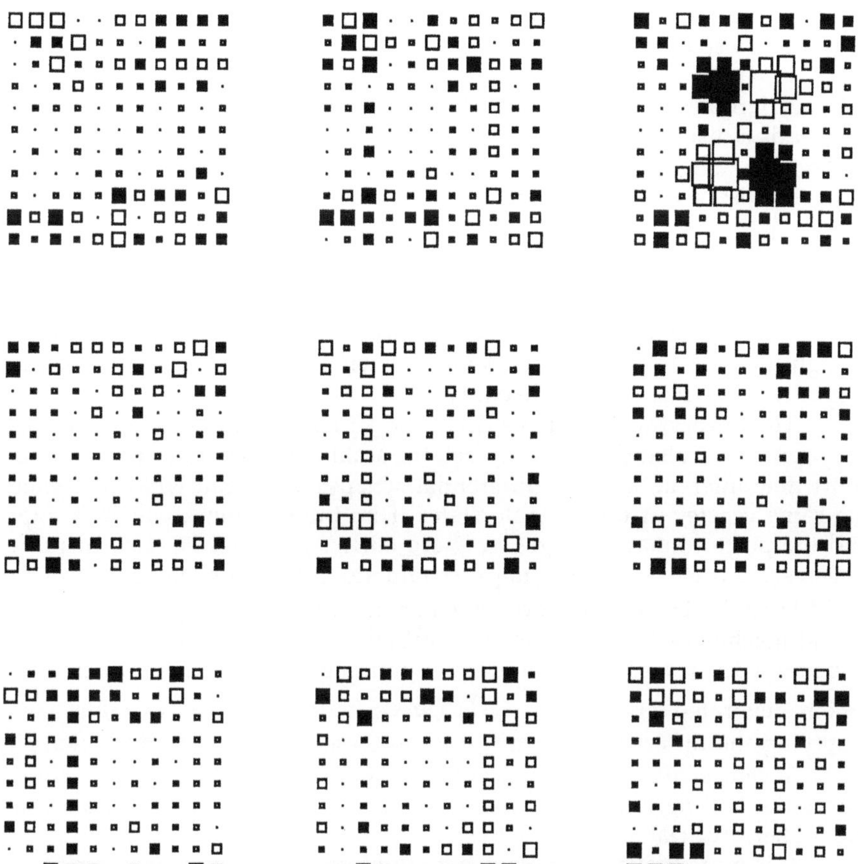

Figure 5
The synaptic weights learnt by equation (1) when disparity is coded by an ensemble of disparity-tuned units. Here there are N=11 units at each point of the 5x5 spatial array. For compactness, only the weights in the shaded region of the array in Figure 4 have been shown. The weights in the other quadrants can be inferred (approximately) from the spatial symmetry. At each of the 3x3 spatial locations the weights from the 11 inputs to the 11 output units are shown by an 11x11 Hinton diagram. (Constants as in Figure 2.)

If one converts the output of this network into a disparity value, by taking the average of the outputs weighted by the disparity peak of their units, then this turns out to be close to the disparity which would result from applying a centre-surround to the values of the input. We can therefore think of this set of weights as reproducing in a population coding the type of rate-coded centre-surround obtained previously (Figure 2). Note that this is a strikingly different pattern of interactions from the long-range inhibition and short-range excitation between neurons tuned to similar disparities which has been proposed by some authors [15-16].

Finally, we have to take account of a further significant difference between lightness and stereo, which is that a change of viewing parameters not only adds a disparity gradient but also introduces a multiplicative scaling by the divergence angle D (the angle subtended by the two eyes at a point). To see this, note the distance R to a point is approximately inversely proportional to D. Differentiating $R \sim 1/D$ we get $\delta R \sim \delta D/D^2$, where δD, the change in divergence for a shift in distance δR, is just the disparity. Thus changing the viewing distance alters the disparity, but (disparity)$.D^{-2}$ remains approximately constant as long as the distances between features in the visual scene are fixed. It is straightforward to show that oblique viewing adds a linear gradient not to disparity but to this rescaled disparity.

Thus everything we have discussed so far applies as long as we use the appropriately scaled disparity. In fact, the scaling is likely to change little over a short time interval, and the differential synapse algorithm will find essentially the same centre-surround stable states if no information about D is supplied and the proportionate changes in D are fairly small, 10-20%, say. In principle, however, the differential synapse could do something more interesting: it could learn the D^{-2} dependence, given an accurate signal for D and appropriate neural machinery to implement the scaling operation. For example, one could imagine a hybrid of a conventional summing unit and a product unit [17] which generated an ouput $y=D^{v}(\Sigma w_i x_i)$, with v and w_i variable weights. Applying the differential synapse rule to these weights, one finds that the w_i define a centre-surround structure and v converges to -2. Thus the algorithm learns the scaling and the cancellation of added gradients simultaneously in this rather artificial situation.

It is more plausible, of course, to think of the scaling being carried out on a population-coded representation of disparity. What happens if one supplies disparity in this form together with some representation, rate- or population-coded, of the divergence D, and applies the differential synapse algorithm using conventional linear units? One is then asking for the division by D^2 to be carried out by linear units, taking advantage of the properties of the population code. The result is, not surprisingly, that only an imperfect match to division is obtained, and only over a limited range of the variables. (This is true whether one uses the differential synapse or least squares' fitting of the desired output, and is therefore likely to reflect an inherent limitation of the architecture.) With suitable pre-processing, however, it may well be possible to learn more accurate scaling. Interestingly, there is evidence for an interaction between divergence angle D and disparity even in the earliest stages of stereo processing [18]. This may supply the repertoire of units needed for learning more exact invariance.

4. CONCLUSIONS

These examples show that the differential synapse algorithm can be used to learn a variety of types of receptive field. But the concept of temporal invariance itself is more important than the algorithm used to learn it. Furthermore, the way that temporal invariance has been imposed by this particular algorithm -- asking for the neuron's firing rate to remain as constant as possible -- is probably too simple-minded for biological reality. The significance of the firing of a neuron may often depend on the firing of other neurons, and it is therefore some function of the firing of a whole assembly which needs to be kept constant. This function could be made explicit as the output of a neuron which could then be trained by a rule like our equation (1). But it could also remain implicit, the term Δy in our learning rule being replaced by Δf, where f is some function of the firing rates of many neurons. Another possibility is that the difference Δy represents not the firing rate of a neuron but the post-synaptic depolarization in a region of its dendrites. In this case a neuron could learn an invariant property, but mix it with other, perhaps time-varying, components.

Thus temporal invariance mechanisms are not necessarily at odds with the kind of neural dynamics alluded to earlier, and a search for such mechanisms should not focus exclusively on constancy in firing rates of individual neurons. One might more profitably look out for differential synaptic learning rules, where it is the short-term differences in inputs, and of suitably defined output signals, which determines changes in synaptic strength. Note however that one expects these changes to be small, since the goal of the mechanism is to average over many learning trials rather than imprint any individual event. It would be important, therefore, to use a sustained train of strongly varying inputs as a stimulus. Despite the potential difficulties of such a search, it offers the considerable reward of revealing a powerful mechanism for detecting constancies of the sensory world.

5. REFERENCES

1 M. Abeles, Local Cortical Circuits, Springer-Verlag, Berlin, Heidelberg, NewYork 1982.
2 H.R. Dinse, K. Krüger and J. Best, Concepts in Neuroscience, 1 (1990) 199.
3 S. Becker and G.E. Hinton Nature 355 (1992) 161.
4 P. Földiák Neural Comp. 3 (1991) 194.
5 G.J. Mitchison Neural Computation 3 (1991) 312.

6 B. Kosko In J.S. Denker (ed.), AIP Conference Proceedings 151: Neural networks for computing (p. 277). New York: American Institute of Physics 1986.
7 A.H. Klopf In J.S. Denker (ed.), AIP Conference Proceedings 151: Neural networks for computing (p 265). New York: American Institute of Physics 1986.
8 G.J. Mitchison and G. Westheimer. In Vision: coding and efficiency, C. Blakemore (ed.) (p 302). Cambridge University Press, Cambridge, 1990.
9 H.B. Barlow, C. Blakemore, and J. Pettigrew J. Physiol. (Lond.) 193 (1967) 327.
10 D.H. Hubel and T.N. Wiesel, Nature 225 (1970) 41.
11 G.F. Poggio and B. Fischer, J. Neurophysiol. 40 (1977)1392.
12 A.C. Hurlbert and T.A. Poggio, Science 239 (1988) 482.
13 A. Brookes and K.A. Stevens, K.A., Perception 18 (1990) 601.
14 R.W. Rodieck, Vision Res. 5 (1965) 583.
15 R. Schumer, and L. Ganz, Vision Research 19 (1979) 1303.
16 S.R. Lehky and T.J. Sejnowski, J. Neuroscience, 10 (1990) 2281.
17 R. Durbin and D.E. Rumelhart, Neural Comp. 1, (1989) 133.
18 Y. Trotter, S. Celebrini, B. Stricanne, S. Thorpe and M. Imbert, Science, 257 (1992) 1279.

Brain Theory / A. Aertsen (Editor)

On the relation between cyto- and myeloarchitectonics in the human cerebral cortex

Bernhard Hellwig and Stefan Rotter[a]

[a]Max-Planck-Institut für biologische Kybernetik, Spemannstr. 38, 7400 Tübingen, Germany

Abstract

Anatomical maps of the cerebral cortex are based on areal differences in the layering and packing density of cell bodies or of myelinated fibers. Here we show that quantitative data on the cell body picture of a certain area permit to compute the corresponding myelin picture. Thereby only two simple assumptions are made: (1) A sigmoid curve represents the correlation between the cell body size of a neuron and the myelin content of its axonal arborization; (2) the majority of axon collaterals of pyramidal neurons - the most frequent cortical cell type - are horizontally directed and distributed slightly below the cell body. The results for three very different areas (the primary motor and visual cortices and a field in the parietal association cortex) are presented and compared with real myelin preparations. It is suggested that the cortical myelin picture is mainly produced by pyramidal neurons.

1. INTRODUCTION

In the cerebral cortex there are numerous areas with characteristic structural traits. Depending on the tissue elements considered two main types of cortical parcellation were proposed: cyto- and myeloarchitectonics. Cytoarchitectonics [1, 2] is based on differences in the distribution and the sizes of cell bodies (Figure 1). Myeloarchitectonics [3] relies on the packing density of intracortical myelinated fibers. Thereby attention is mainly focused on those horizontal fibers which are organized in bands known as the stripes of Baillarger (Figure 1).

Cyto- and myeloarchitectonics seem to be closely related. Structural changes from area to area as revealed in the cell body picture are paralleled by changes in the distribution of intracortical myelinated fibers [1]. However, the exact nature of this relation is not immediately obvious. Figure 1 shows that the outer stripe of Baillarger, situated in layer 4, corresponds to densely packed, small cell bodies, whereas the inner stripe of Baillarger, located in layer 5b, coincides with less densely packed, large cell bodies.

Degeneration studies indicate that the stripes of Baillarger consist of local axon collaterals of cortical neurons: Degenerating horizontal fibers are revealed only after intracortical lesions [4–7], but not after isolating the cortex from the white matter [7, 8]. It has been suggested [9, 10] that the most likely sources of these fibers are the pyramidal neurons. They are the predominant cell type in the cortex, and a large number of their local axonal

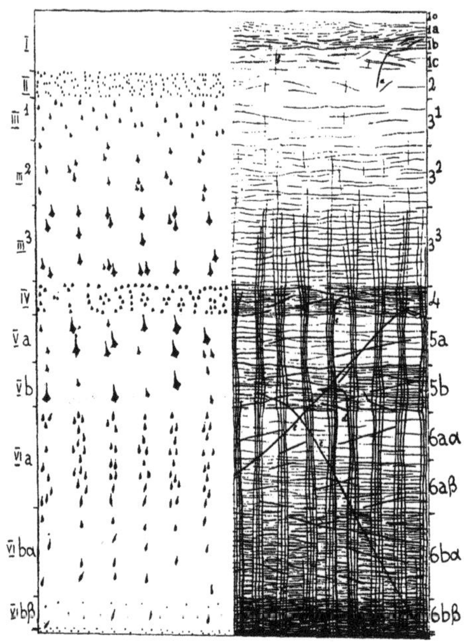

Figure 1.
Vogt's scheme [3] of cytoarchitectonics (left) and myeloarchitectonics (right) in a typical piece of association cortex. The stripes of Baillarger are the horizontal bands of myelinated fibers in layers 4 and 5b.

ramifications are horizontally directed (Figure 2).

In the present study it will be shown how the cortical myelin picture can be computed from cytoarchitectural data by making simple assumptions on the arrangement of horizontal axon collaterals of pyramidal cells and the rules of their myelination.

2. CYTOARCHITECTURAL DATA

Data were taken from a treatise by von Economo and Koskinas [2] on the human cerebral cortex which contains detailed descriptions of all areas. Three types of data were considered:

(1) Layer thicknesses: They are specified in a table on p. 797 of [2] as a percentage of the whole cortical thickness and were transformed into absolute values by multiplication with the absolute cortical thickness. In the descriptions of the various areas some of the layers indicated on p. 797 are further subdivided. In cases where the thickness of such sublayers is not precisely stated we either measured it on the photographs of the accompanying atlas or split the main layer into sublayers of equal thickness.

(2) Neuron sizes: They are specified in the descriptions of the various areas as the width of cell bodies, i.e. as their largest horizontal diameter. In cases where a range of

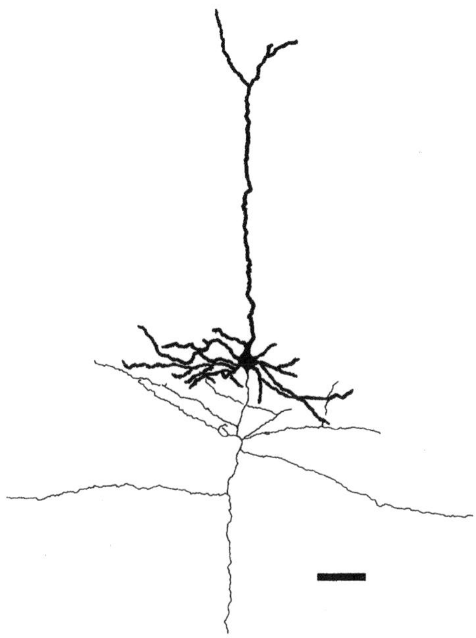

Figure 2.
Camera lucida drawing of a typical pyramidal cell (mouse cortex, Golgi stain). The axonal tree contains a number of horizontally directed collaterals, originating below the cell body. Bar, 50 μm.

neuron sizes is indicated we took a rounded average. In some layers sets of neurons with different (ranges of) cell body sizes are distinguished. We treated them separately. In area FA neuron sizes are sometimes described in qualitative terms (e.g. "small" cells). These terms are quantitatively defined on pp. 47-48 of [2].

(3) The number of neurons per 0.001 mm^3: The neuron density is specified in the descriptions of the various areas. In some layers sets of neurons with different cell body sizes are distinguished without indicating the neuron density for each set. In these cases we distributed the number of neurons per volume specified for the whole layer in equal parts to the different sets of neurons. The number of Betz cells (large neurons in layer V of area 4) was estimated from the photographs of the accompanying atlas.

In all cases measurements from the walls of sulci were given priority to those from the crests of gyri since in the sulcus wall the histological picture is less distorted by the rich folding of the human cerebral cortex.

3. TWO BASIC ASSUMPTIONS

Two basic assumptions are sufficient to transform cytoarchitectural data into realistic myelin patterns. The first assumption is that large neurons contribute more to the intracortical myelin content than small ones, and that this relation can be represented by the sigmoid curve in Figure 3.

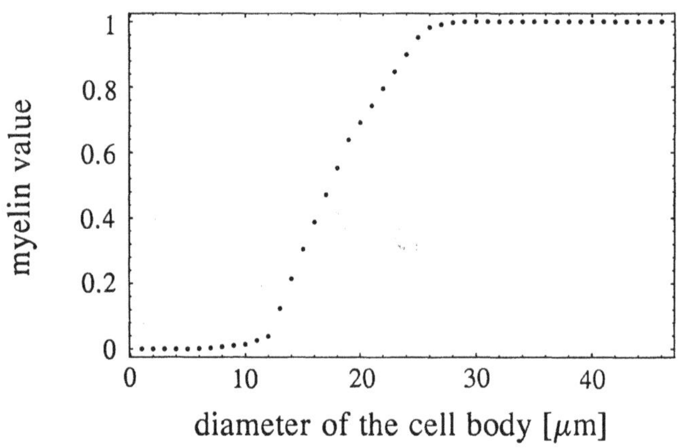

Figure 3.
A curve on the basis of which the neuron size, i.e. the largest diameter of a neuron's cell body, is transformed into a "myelin value". The larger the "myelin value", the more a neuron contributes to the myelin content of the cortex.

The second assumption is that the majority of axon collaterals of pyramidal neurons are horizontally directed and distributed slightly below the cell body. This is supported by studies on pyramidal neurons in various areas and in different species including man [11–14]. Therefore, we felt justified in using data from the rat visual cortex [11] (Figure 4) as a model of the distribution of horizontal collaterals with respect to the cell body and in applying them to the human cortex after appropriate scaling (see legend to Figure 4). Note that the second assumption concerns only pyramidal neurons. For the computation this means that the few neurons in layer I which are all of the nonpyramidal type [15] were discarded, and that below layer I, where the nonpyramidal cells account only for about 15 % of the whole neuron population [15, 16], all neurons were considered as pyramidal cells.

4. PROCEDURE AND RESULTS

We present here results from three areas covering the spectrum of architectonic variability: areas 4, 7 and 17 (according to Brodmann [1], FA, PE and OC according to von Economo and Koskinas [2]). The architecture of area 7, a field in the parietal association cortex, is paradigmatic for most neocortical areas [17]. Area 4 (primary motor cortex) and area 17 (primary visual cortex), on the other hand, represent two extremes in neocortical regional specialization.

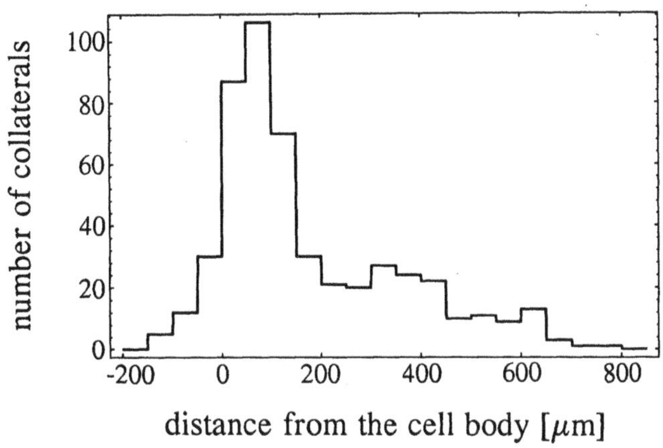

Figure 4.
Slightly modified diagram from a study by Paladino and Harth [11] on 39 Golgi-stained pyramidal neurons in the rat visual cortex. Distances (in a direction perpendicular to the cortical surface) were measured between the endpoints of axon collaterals and the cortical depth where the cell body is located. Negative distances indicate collaterals terminating above the cell body, positive distances mean that the endpoints of collaterals are situated between the cell body and the white matter. The diagram reveals that many collaterals terminate in a layer slightly below the cell body. Since the majority of axon collaterals displayed in the diagram is approximately horizontal [11], we used the diagram as a model for the distribution of horizontal collaterals with respect to the cell body. We scaled the diagram to each area in the human cortex, using as a scaling factor the thickness of the area considered divided by the thickness of the rat visual cortex (1200 μm).

Myelin patterns were produced in two steps. First, using data from von Economo and Koskinas [2], the size of the neurons was transformed into a "myelin value" by means of the curve in Figure 3, and the "myelin value" was multiplied by the number of neurons per volume. In cases where there are neurons of different sizes in one layer, this was done separately for each group of neurons and the results were added. The procedure yields, for each layer, the respective contribution to the population of horizontal myelinated fibers (Figure 5a).

In the second step of the computation, we took into account that the myelin is distributed over axonal arborizations: The diagrams in Figure 5a were convolved with the histogram of Figure 4 (Figure 5b). This simply provides for shifting the myelin into the appropriate position.

The densities of myelin thus obtained were represented as shades of grey (Figure 6a) in order to facilitate comparison with real myelin preparations (Figure 6b). The simulated myelin pictures are remarkably close to the real ones (cf. Figures 6a and 6b). Area 7 shows both stripes of Baillarger. In area 17, only one band of myelinated fibers is conspicuous, the so-called stria of Gennari. For area 4 the comparison is complicated by the fact that

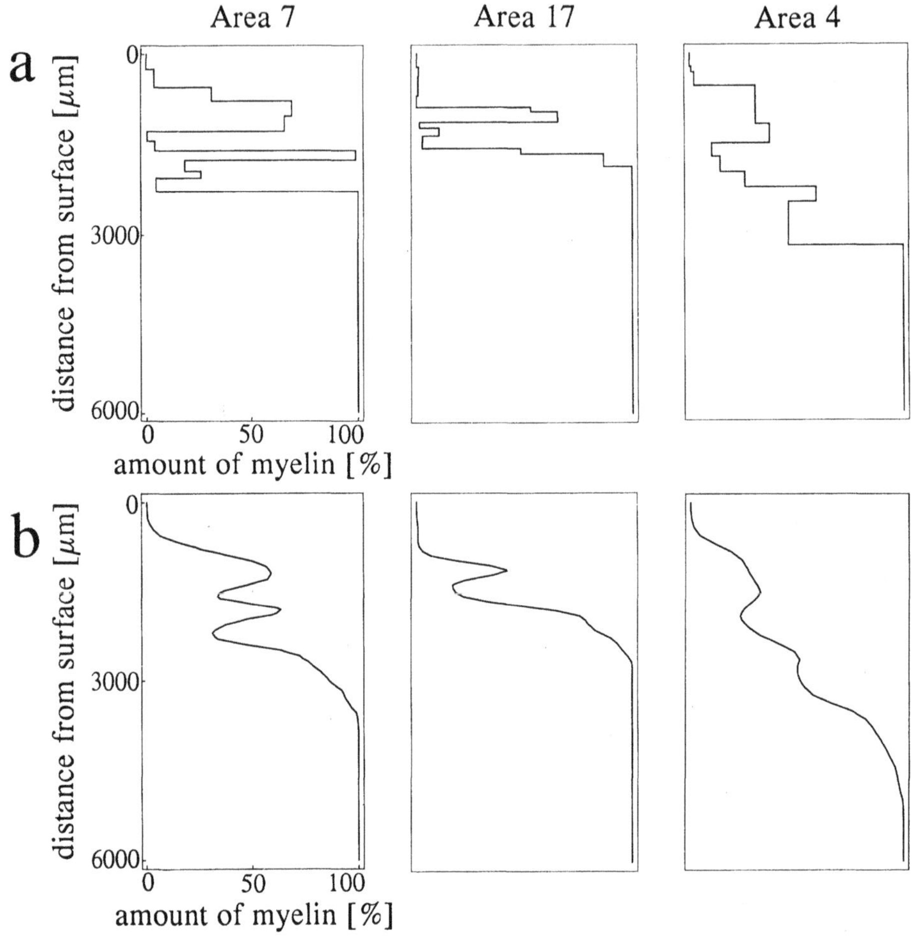

Figure 5.
Simulation of myelin pictures. **a**. The layer-specific amounts of myelin, computed in the first step of the simulation, are shown as a function of the cortical depth. The maximum amounts of myelin in the bottom parts of the diagrams (in area 7 below 2300 μm, in area 17 below 1900 μm and in area 4 below 3200 μm) correspond to the white matter which was added in order to facilitate comparison with real myelin preparations. **b**. The diagrams of Figure 5a were convolved with the histogram of Figure 4: In area 7 there are two myelin maxima above the white matter, area 17 shows only one maximum. In area 4 two myelin maxima are present, the lower one forming a plateau which is maintained up to the transition into the white matter.

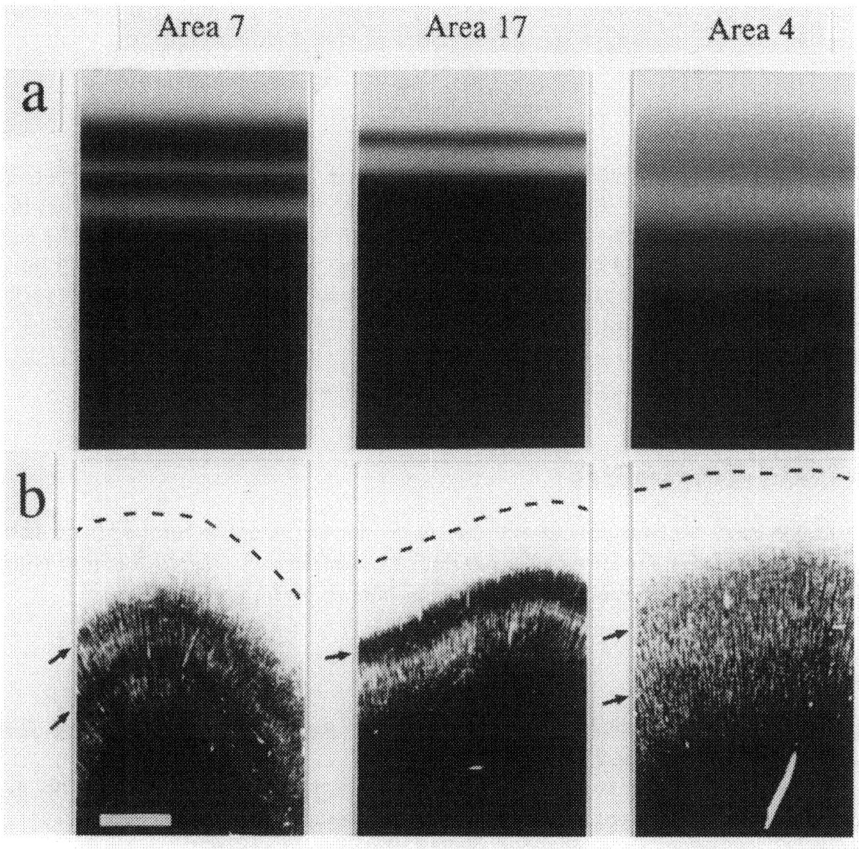

Figure 6.
Comparison of the simulation with real myelin pictures. **a**. Photographs from a computer screen representing the data of Figure 5b as grey values: white corresponds to 0 % myelin, black to 100 %. The amounts of myelin in between are linearly transformed into shades of grey. **b**. Myelin preparations (60 to 90 μm thick) from human cerebral cortex (Schroeder's technique [19], slightly modified). In area 7 two bands of myelinated fibers can be seen (arrows), the stripes of Baillarger. In area 17 there is only one stripe, the stria of Gennari. In area 4 (the preparation is taken from the anterior part of area 4) two myelin maxima are present (arrows), the lower one grading into the white matter. Note that real myelin preparations contain a conspicuous background of vertical fibers, which tend to blur the myelin patterns produced by horizontal fibers. Vertical fibers were neglected in the simulation. Dashed lines: cortical surface. Bar, 1 mm.

the real myelin picture is different in two subfields (cf. ref. 1, 2 and 18). The simulation (Figures 5b and 6a) is close to the anterior part of area 4 (Figure 6b).

5. CONCLUSIONS

The present findings support the assumption that the horizontal component of the myelin picture is mainly produced by axon collaterals of pyramidal cells, and that nonpyramidal neurons and afferent fibers play a minor role in this respect. Our results suggest also that the distribution of horizontal axon collaterals of pyramidal neurons and the principles of their myelination are remarkably similar in different areas. In a more extensive paper it will be shown that the rules linking cyto- and myeloarchitectonics which were proposed above apply also to other cortical areas, including the primary somatosensory cortex, the speech centers and a number of association areas.

ACKNOWLEDGEMENTS

We thank Prof. V. Braitenberg and Dr. A. Schüz for valuable comments on the manuscript, M. Dortenmann for excellent technical assistance, S. Würth for improving the English and Prof. R. Meyermann for providing human brain tissue.

REFERENCES

1. Brodmann, K., Vergleichende Lokalisationslehre der Großhirnrinde (Johann Ambrosius Barth, Leipzig, 1909).
2. von Economo, C. and Koskinas, G. N., Die Cytoarchitektonik der Hirnrinde des erwachsenen Menschen (Springer, Wien, Berlin, 1925).
3. Vogt, C. and Vogt, O., J. Psychol. Neurol. 25, 279-461 (1919).
4. Fisken, R. A., Garey, L. J. and Powell, T. P. S., Philos. Trans. R. Soc. London Ser. B 272, 487-536 (1975).
5. Colonnier, M. and Sas, E., J. Comp. Neurol. 179, 245-262 (1978).
6. Gatter, K. C. and Powell, T. P. S., Brain 101, 513-541 (1978).
7. Creutzfeldt, O. D., Garey, L. J., Kuroda, R. and Wolff, J.-R., Exp. Brain Res. 27, 419-440 (1977).
8. Le Gros Clark, W. E. and Sunderland, S., J. Anat. 73, 563-574 (1939).
9. Braitenberg, V., J. Comp. Neurol. 118, 141-156 (1962).
10. Braitenberg, V., J. theor. Biol. 46, 421-447 (1974).
11. Paladino, A. and Harth, E., in: Computer Analysis of Neuronal Structures (ed. Lindsay, R. D.) 189-207 (Plenum Press, New York, 1977).
12. Ramón y Cajal, S., Histologie du système nerveux de l'homme et des vertébrés, translated by L. Azoulay (Consejo superior de investigaciones cientificas, Instituto Ramón y Cajal, Madrid, 1911).
13. Gilbert, C. D. and Wiesel, T. N., J. Neurosci. 3, 1116-1133 (1983).
14. Landry, P., Labelle, A. and Deschênes, M., Brain Res. 191, 327-336 (1980).
15. Peters, A. and Kara, D. A., J. Comp. Neurol. 234, 242-263 (1985).

16. Braak, H. and Braak, E., in: Progress in Brain Research (eds. Swaab, D. F., Fliers, E., Mirmiran, M., Van Gool, W. A. and Van Haaren, F.) 185-212 (Elsevier Science Publ., Amsterdam, 1986).
17. Bailey, P. and von Bonin, G., The Isocortex of Man (University of Illinois Press, Urbana, 1951).
18. Vogt, O., J. Psychol. Neurol. 15, 221-232 (1910).
19. Braak, H., Architectonics of the Human Telencephalic Cortex (Springer-Verlag, Berlin, Heidelberg, New York, 1980).

Brain Theory / A. Aertsen (Editor)

Dynamics in neural systems. Some remarks

Werner von Seelen

Institut für Neuroinformatik, Ruhr–Universität Bochum, ND–04,
Universitätsstraße 150, D–4630 Bochum, Germany

1 Introduction

The analysis of neural systems is known as one of the most difficult problems of modern science. The reasons for this state of affairs are manifold. Two important aspects are the high dimensionality of neural systems, which can be reduced only to a limited extent through boundary conditions, and the fact that the observed data contain the complexity of the world as well as the complexity of the system itself. Separating these two sources of complexity is difficult. In the few examples, where this segregation has been achieved, the degree to which simple systems in a complex environment can behave in a seemingly complex manner is surprising.

An undisputed insight among neuroanatomists and neurophysiologists is that neural data are represented in space and time. The relative importance of these variables for system description shifts from time to time. At present we are witnessing a renaissance of dynamic aspects, leading sometimes to the effect of reinventing the wheel. Older theoretical insights, for example from control theory, are overlooked in favour of popular interpretations, which are not always tested with respect to how far analogies hold. Because theories do not systematically lead to questions, which enable us to analyse unknown systems, but at best help to avoid mistakes and delimit descriptions, nobody can propose the "right" experiments, which are only definable *a posteriori*. Below I will discuss a few aspects, which seem to me worth considering and which might help to formulate experimental questions a bit more precise. In my view it is essential that systems considered here *process information*. This may sound trivial, but the implications of this fact are often ignored. To animate the discussion I have sharpened the viewpoints somewhat.

2 Evolutionary aspects

The theory of evolution is certainly one of the best theories in biology. The idea to employ this theory in the analysis of neural systems is an old, but nevertheless, a fascinating one. Knowing the construction principles of a system facilitates its functional interpretation significantly. Viewed from that aspect, attempts to organize networks driven by data from the environment according to aspects of evolution seem very much worthwhile. But even for those who do not want to apply these methods it is important to realise that the relation to information and the inherent tendency to optimise are conditioned on structure in the manner sketched below.

1. The mechanism of evolution leads to the incorporation of an "average" knowledge about the world into neural "blue prints" of the structure. Information processing that secures survival is a primary requirement for the structure of nervous systems. Attempts to analyse neural systems without considering the information that organizes the system are often characterized by a certain arbitrariness.

2. If one accepts the thesis that nervous systems organize behavior which favours survival, this behavior defines which information must be processed. The definition of information therefore remains variable and can rarely be uniquely determined. The structure, however, must afford variability.

3. The mechanism of evolution has the inherent tendency to optimise the behaviour-related information processing at all levels. That means that both the "modules" and their cooperation are adapted to classes of tasks. From this it may be expected that as a result of functional adaptation, neither a "universal" temporal code nor a general usage of dynamic parameters exist. The theory of (simple) optimal systems shows instead how the useful signal and the perturbations determine a system's dynamics. It would therefore not be surprising if, for instance, orientation specificity and depth estimation were to possess different temporal properties.

The consequences of these roughly sketched arguments are manifold. In all cases they impose the requirement to take into account the structure of signal space, the dynamic properties of the effectors and the structure of the processing system (e.g. layers). The dynamic requirements of the behaviour will gradually change the dynamic structure of the input signals in the different parts of the effector system.

3 Signals in a natural environment

Living beings perceive their natural environment not in a sense of reconstructing that enviroment, but in a sense of organising behavior. The signals in a natural environment are characterised by

a. extremely large data sets

b. incompletely defined or fuzzy information

c. primarily non-random perturbations

d. high variability of the boundary conditions that contribute to the definition of information

e. as a rule the information content decreases with increasing distances of features in space and time.

In particular, item c. encounters a largely undeveloped theory, because expection based noise supression – as in the technical sector – can be only of limited effectiveness in neural systems.

With respect to the dynamics of the signal space, arguments can be put forward for the following constraints:

1. The temporal course of events, that are directly perceivable from the environment, lies in the range of 1 to 20 Hz. This does not mean that shorter time scales are not used (e.g., directional hearing), but that their processing time falls into that range.

2. The information content of events decreases as the spatial or temporal distance increases.

3. Coding on the temporal axis, being a time-consuming process, is limited through the perceivable sequence of events in the sense assumed above. This type of constraint is more stringent for the high-dimensional phase of information processing (e.g., primary visual areas) than in the domain of symbolic information processing, especially since the process of perception (e.g., eye movements) also needs to be organized in limited time.

4. Because of the structure of the signals the temporal resolution should be different in the different modalities. For instance, in the auditory systems the length of typical segments must be shorter than in the visual system, where the stationary part of the world can serve as a "memory".

5. Motion perception, particularly important in the visual system, is always related to the motion of masses, which can be described by second order differential equation. These dynamics are easily modeled, even though the design of sensors can pose problems (e.g., ill-posed problems in visual processing).

Much more serious consequences than these dynamic properties of signal space are brought about by the enormous amount of data to be processed. If we assume that, for instance, the information flow in the human visual system is in the order of magnitude of 10^9 bit/sec, then the rate at which behavior is organized can only be understood if models are continuously confronted with the data stream to reduce the dimensionality of the data (Mountcastle: "we do not see the world, we construct it"). If such a picture comes close to reality, then the dynamics of the information processing system must obey an additonal constraint: On the various levels of information processing a dynamic "match" with adequately coded model knowledge must be realised. As a result, the observable signal dynamics in neural systems are structured not only through the environment and the intended behavior, but also through interpreting model comparisons. That means that in dynamic optical recordings, even in "early" visual areas, we should expect already a behaviour-related interpretation of the data.

4 System dynamics

Adaptation through evolution should lead to task-dependent dynamics. Coordination among subsystems necessitates, however, adquate coding. General thoughts are difficult to verify. It may be useful, however, to consider some boundary conditions to point out the direction. The following aspects might be important:

1. Because neural systems organise behavior, the time course of relevant events in signal space and the dynamic properties of the effectors impose a temporal structure on the processing steps. Such structure is necessary, for instance, to control posture, locomotion and grasping movements. This temporal structure as a system constraint has effects all the way to sensory signal processing.

2. If the relevant part of courses of events in terms of psycho-physical measurements lies in the range of 1 to 20 Hz, the system dynamics should lie in the same range to afford the necessary interaction of system and signal. This means, that systems work in the transient regime. Because models are necessary for data compression and prediction (cf. Section 3.), the negative time axis is fully usable (where the present marks the origin). At present it is unclear, if this form of modelled knowledge has an *a priori* temporal structure or is, instead, event-correlated and affects parameters of action cycles. Because of the nonlinear system structure it might be very helpful to analyse experimentally the small signal behaviour at different states of the system.

3. Spatially two-dimensional laterally coupled nets (e.g., cortices) are a prominent structure at least for the evaluation of sensory information. These nets evaluate spatio-temporal neighbourhood of signals and generate a trade-off between accuracy and robustness of information processing. Because the spatial axis may scale functional parameters (e.g., orientation), the spatial structure of information processing is not restricted to space-dependent signals, for instance, includes acoustics. Instead, the spatial axis organises a variable space in which the number of neurons involved determines the accuracy of the operation or – depending on the view point – represents a parallel implementation. If this is true, one should expect that the spatial constraints of the structure may lead to similar dynamic properties of systems with the same kind of spatial representations, independent of their modality. In the same way the temporal axis can be used for tasks other than organising temporal order. Because of its high resolution, the temporal dimension could be used to "conserve continuity" and to code parameters in higher areas.

4. Since each transport of data between areas is assigned to a coordinate transformation (topographic mapping), the spatio–temporal representations have to be interpreted on different scales of both variables respectively.

5. As a rule, the combination of different representations in one layer (e.g. orientation, binocularity, color in area 17) increases functional diversity at the expense of spatial resolution. However this does not imply that the dynamic properties in

the representations are the same all over the area, because of the various feedback loops.

These very general aspects need to be made more specific. This must be judged on the basis of the existing data, which determine the appropriate level of abstraction, and of the incomplete controllability and observability of neural systems, which allows only for approximation. However, the fact that information must be processed, has to constrain the approximation procedure. Hence ultimately the dimensionality of information must be reduced and adapted to the dimensionality of the effector systems.

5 Temporal processes in visual systems

All our experiments are based on the conviction that the determination of a simple task configures a "simple" brain. If this were not true, the interpretation of neural signals would involve every single neuron. This segregation appears unproblematic if we restrict our attention to a single sensor. Difficulties are encountered, however, since the definitions of simplicity as used the by experimentor and by the system rarely coincides. This difficulty is illustrated by the response to "simple" oriented stimuli in visual areas regarded as "high" ones. The only way out of this problem seems to be to make use of the definition of information by the experimental animal through its behavior. Problems defined by physics (e.g. the invariance problem) are often more difficult to decide.

We have a lot of detailed knowledge about the visual system. Areas are assigned partial functions, the number of which can be increased by cooperation. The "where" and "what" paths have been distinguished and the cooperation among cells in the form of cell assemblies, which are coupled possibly through oscillations or – more realistically – through temporal structuring of signals (coherency) (Abeles et al., this Volume). A number of contradictory hypotheses leads to only few experimentally testable predictions, so that the difficulties reside at the conceptual level.

There are two time scales in the visual system which are remarkable. On the one hand, experiments using masking, which are most effective in the range between 60 to 80 msec, show that estimates made by the system take place on that scale. On the other hand, decisions can be made after about 250 to 300 msec. Moreover, the damped oscillations of neural response (PSTH) in the range between 3 and 15 Hz last longer than signals and suggest their interpretation by the entire system. This idea is supported by short time correlations between pairs of neurons within one or across several areas (Vaadia and Aertsen, 1992). Moreover, it seems essential that the visual system is an active system, able to control the flow of incoming data, depending on the ongoing processing of information and expectations. Here the problem arises to center the fovea at the right time on the right place. The necessary variations of gaze and focalisation require a temporal ordering of signals. Even though the operations take place in different subsystems, temporally distributed excitation and inhibition warrant in-step processing, i.e. an event-dependent time structure. The necessary duration of partial operations does not leave free processing capacity on the time axis available for the coding of other parameters. This restriction does not apply to visually guided behavior such as grasping and locomotion because the moved masses and the reduced dimensionality of the visual representation operate at a larger time scale. At present it remains undecided if, for example, form and colour parameters are coded temporally, or if event-dependent structuring of the time axis according to the above mentioned time scales allows to structure information hierarchically, which must then be demodulated at the target structure.

The presently much discussed binding of variables and 40 to 80 Hz oscillations need more experiments. Here one has to answer the question *what* it is that is being bound. The realisation of logical conjugation seems to be useful for energetic reasons locally, but only if signal parameters are coded.

The pure phase coding is relatively sensitive to noise, so that a coherence time containing several events might be a more realistic approach, despite of the fact the the coherence time reduces the numbers of states to be represented. The linking of spatial features to patterns in the time domain is clearly possible (Neven and Aertsen, 1992), but it needs an experimental verification.

Another aspect leading to temporal structuring can be found in the circuitry of cortical neural nets. Experimental observations show that the concept of receptive fields must be extended by dynamical aspects. Because of the temporal dependence of every spatial location of a receptive field (cf. Dinse, this Volume). In general, space and time are not separable in the cortex. Functionally, this implies that movement-induced form changes of objects and spatial frequency-dependent dispersion of pattern contributions in subsequent layers can be kept minimal. If these observations are combined with the damped oscillations in PSTHs, a general interpretation is suggested. If we assume that the degree of both intra- and inter-area feedback in visual cortex is large, then the temporal structuring of neural signaling is implied.

Taking the simplest case of a layer having linear feedback, its output as a function of r can be characterized in Fourierspace by

$$Y(\vec{r}) = \frac{X(\vec{r})H_1(\vec{r})}{1 - H_R(\vec{r})} \tag{1}$$

with X being the input, H_1 the forward coupling and H_R the feedback. Since the coupling contains propagation times as well as temporal filters $A(i\omega)$ aside of the spatial contribution $\tilde{H}(\vec{r'})$, a Taylor expansion yields

$$Y(\vec{r}) = X(\vec{r})H_1(\vec{r})[1 - \exp(-i\omega t_o)A(i\omega)\tilde{H}_R(\vec{r'}) + \exp(-i\omega 2t_o)A(i\omega)^2\tilde{H}(\vec{r'})^2 - \cdots] \tag{2}$$

with $\vec{r'}$ being the position vector. (2) can be viewed as an implicit description of an iteration. It allows for the following interpretation.

1. Feedback can give rise to iterations in which the (observed) selectivity and accuracy increases with time. Thus, an initial estimate will improve, oscillations depending on the cortical structure may be observed as a side-effect. In relation to the space-dependent variables, the layers act as a time-dependent resolution pyramid.

2. If higher areas are included into the feedback process, a structurally dependent temporal order in the cooperation will come about.

3. Using feedback for estimation tasks minimizes a mean quadratic estimation error (von Seelen et al., 1987).

These characteristics are not altered if a nonlinearity is added to the system. This forms yet another time–dependent formation process that realizes cooperation dynamics. However, the information dependent linking of knowledge taken e.g. from associative memories is not included on this level.

The mechanisms sketched here describe aspects suited for furnishing boundary conditions and temporal restrictions for the analysis of dynamical processes, since they exploit the characteristics of the actuator's signal space as well as its intrinsic structure.

The dynamic properties of neural systems are an essential part of their various functions in a complex task. These tasks define the demands in the dynamic range. Thus it is reasonable to assign the dynamic aspects to the concrete problems of control, estimation, classification, predictive interpretations, etc. Then, in a second step more general questions such as cooperativity schemes of 2D representations or coding of knowledge at an operational level can possibly be decided. This more experimentally oriented aspect does not exclude a more abstract description of the problem: The definition of collective variables in differential equations dscribing their equations of motion appears to be a very promising approach (Schöner, 1991). It links sensors and actuators by a suitable parametrization of the input signals in terms of the same collective variables. The observable stable patterns correspond to trajectories in the pattern dynamics and describe a given task together with its intrinsic boundary conditions. Besides of giving new insights into the detailed structure, the relatively high abstraction level of this approach to neural dynamics, combining sensor information and task specifity, might reveal more general principles.

In summary this contribution poses more questions than it provides answers. This mainly results from the following uncertainty: Suppose we did not have any problems in performing even the most imaginative experiments, would we have the theoretical means to make these data 'understandable' to us?

Acknowledgements

I wish to thank Jörg Kopecz, Hartmut Neven and Gregor Schöner for their help in preparing the final version of this manuscript.

References

Neven H, Aertsen A (1992) Rate coherence and event coherence in the visual cortex: a neuronal model of object recognition. Biol. Cyb. 56, 37-49

Schöner G (1991) Dynamic theory of action-perception patterns: the "moving room" paradigm. Biol. Cyb. 64, 455-462

Vaadia E, Aertsen A (1992) Coding and computation in the cortex: single-activity and cooperative phenomena. In: Information Processing in the Cortex: Experiments and Theory, pp 81-121. Aertsen A, Braitenberg V (eds). Berlin, Heidelberg, New York, Tokyo: Springer

von Seelen W, Mallot H.-P., Giannakopoulos F (1987) Characteristics of neuronal systems in the visual cortex. Biol. Cyb. 67, 37-49

Brain Theory / A. Aertsen (Editor)
1993 Elsevier Science Publishers B.V.

About neural areas, assemblies, attractors, and noise

G. Toulouse

Laboratoire de Physique de l'Ecole Normale Supérieure, 24 rue Lhomond, 75231 Paris, France

The following remarks have been stimulated by the talks of Moshe Abeles, Emilio Bizzi, Walter Freeman, Stan Gielen and Tomaso Poggio.

1. WHICH AUTOMATA DESERVE THE APPELLATION: "NEURON" ?

Perhaps this is due to the specific importance of an "appellation contrôlée" in my country, but the fact is that I still meet distinguished French neurobiologists who become outraged whenever a threshold automaton (viz. an automaton that computes a linear summation of its inputs, followed by a sigmoid function of this sum) is called a formal neuron, or for short a neuron. For those colleagues, such an automaton is just an inadequate caricature of a real neuron. My standard response goes quietly as follows:
1) this appellation has been in usage for at least fifty years (so I carry no personal responsibility);
2) threshold automata (or linear discriminators, in equivalent terms) have proved interesting on their own merit. The reason seems to be that they combine, in a very economical way, the essential ingredients of linearity (in the weighted summation of the inputs) and non-linearity (in the sigmoid function). Presumably, even if brains were using entirely different components, one would nevertheless have come to study nowadays formal neural networks (under a different name, of course). In brief, the inspiration and the name came historically from biology, and the connection with biology is still a powerful incentive, but it is not the only one. Hence a formal neuron is not just a misrepresentation of a real neuron;
3) as a corollary, anything intermediate between a formal and a real neuron deserves also the appellation "neuron".

All fine. I was becoming used to endorsing this laxist position, till recently I began to feel myself cornered in the attitude of the offended purist. "A theory of how the brain might work", by T. Poggio et al., makes use of units that compute 'hyperbasis functions' (HBF) of their inputs. As their name already tends to suggest, these HBF are freely chosen among a vastly extensive class of functions (including Gaussian functions, and multiplicative operations on inputs). Such HBF-units are certainly useful for curve-fitting (indeed they have been chosen ad hoc); and smooth approximation of functions is clearly one of the important problems solved by brains (see below). But are HBF-units close enough to real neurons, to be granted also the appellation "neuron" ? Conservatively -- yet with a nagging feeling that this choice may be as unwise as the Minsky-Papert rejection of multi-layer perceptrons (remember their 'monster of vacuous generality') -- I shall keep below with the previously defined restricted comprehension of the word.

It may be that, in the future, any network, with whatever units and adjustable weights, will come to be called a neural network. The two distinctive features, with respect to other computing architectures, would then reduce to:
- parallel distributed processing,
- learning, by adjustment of inter-units connection weights.

The word 'connectionism', often used as a synonym for neural net studies, has been indeed placing full emphasis on the second aspect.

At present, commonly used "units" of brain theories range, in order of increasing complexity, from threshold automata, or "integrate-and-fire" neurons, or more sophisticated neuron models (including cable-theoretic dendritic effects, veto inhibitory synapses, ion channel dynamics, etc.) to entities that are grey boxes in a schema, or "agents" in a society of mind, and endowed with almost homuncular faculties.

There is probably a cognitive lesson to be drawn -- beyond 'ethical frownings' -- from the misgivings that arise between scientists favouring different approaches. Let me attempt to sketch briefly three different strategies, each one with its own criteria of 'honest risk':

i) statistical physics looking for emergent collective properties of assemblies of simple automata; the results are not predetermined before the start of the study; the calculation must be feasible; it is an art of the soluble, where surprises are cherished; the hope is that the results obtained have some degree of universality, and will thus prove relevant in experimental situations, perhaps not yet investigated; the risk is that this hope has been misplaced, in a class of networks that is either too narrow or too vast;

ii) neuroanatomy trying to guess what is the function of some brain parts; the difficulty is to bring together all the available evidence, coming from data that are diverse and often dubious; it requires 'esprit de finesse' more than 'esprit de géométrie'; brain complexity and plasticity conspire to make rare the cases of right guesswork; even wrong-but-useful testable predictions are uncommon;

iii) artificial intelligence looking for devices that perform desirable computations; any structure, components, rules are admissible to achieve the goal purpose; the quality test, in this opportunistic game, comes from selection in overall performance by comparison with previously existing devices; the risk is to miss some essential hints from the fabric of human intelligence.

2. APPLICATIONS OF FORMAL NEURAL NETWORKS

The field of artificial neurocomputation is so fast moving, and so encompassing, that it is difficult to reach a balanced overview. Yet it seems one can list four domains of application where neural nets have been found useful in practice:

i) approximation of function,
ii) pattern recognition,
ii) adaptive control,
iv) optimisation.

Note that with my somewhat restrictive definition, neural networks are not omniconvenient. For example they are not efficient tools for general arithmetic calculations, as noticed long ago by von Neumann [1]. Boundaries between the classes of problems listed above are not always clear-cut, but some distinctions are useful in practice. For subsequent purposes, I wish to make a point about one difference between classes i) and ii).

In problem class i), the typical objective is to find a smooth function interpolating between given examples (values of the function are given for some values of the variables). In problem class ii), the typical objective is to map an input space into a discrete number of output values. In each case, one builds an input-output function from a set of examples, but the emphasis in the first case is on smoothness of the function (like in curve-fitting), whereas in the second case it is on discreteness (the goal function has plateaux separated by sharp boundaries). This smooth-vs-discrete distinction has consequences on the choice of appropriate network structures. For problem class i), a feedforward net, such as a multi-layer perceptron, is convenient. For problem class ii), a self-coupled net, such as an attractor neural network, has interesting advantages.

3. ATTRACTOR NEURAL NETWORKS

First, a reminder. Attractor neural nets [2], in contrast with feedforward nets, have many feedback connections. The working of an attractor neural net can be simply described in terms of the geometry of its state space. Any activity state of the network is represented by a point in state space. The neurodynamics divides this state space into attractor basins, so that the initial state (driven by the inputs into a point inside one of the basins) evolves toward the corresponding attractor (that defines the output state). Thus such a network functions naturally as a classifier, if it has been arranged that each attractor corresponds to one of the recognizable patterns.

An implementation of a pattern recognition task by such an attractor neural network presents three advantages:
a) robustness, due to the distributed representation of the computation over all the neurons of the network,
b) associativity, due to the convergence of the neurodynamics, that allows for completion of an incomplete input,
c) buffer property, due to the stability of the attractor state (in feedforward nets, no such possibility for persistent latency is available).

And now, for fairness, let me list the accompanying disadvantages (on the rear side of the advantages):
a') redundancy, due to the distributed representation (there can be no robustness without redundancy),

b') inflexibility, due to the discrete, quantized, character of the attractors,
c') trapping, so long as there is no additional mechanism to move the network out of a stable attractor state.

These basic notions will serve later in the discussion of cell assemblies.

4. PERCEPTION AND ACTION

1) In first analysis, perception is discrete and action is smooth.
Many ancient observations of the Gestalt school about perception fit rather well within the conceptual frame of attractor neural nets [3]. As one example: ambiguous stimuli, that lead to erratic flipping between two distinct percepts, may be considered as falling on the boundary between two attractor basins.
In contrast, many motor acts tend to be conspicuously smooth, continuous.
Referring to earlier discussion, it is tempting to draw this sketch: motor areas perform approximation of function (smooth interpolation from learnt examples), while perceptual areas are strongly involved in pattern recognition (discrete classification).
[This opposition should not be exaggerated. Perception of a continuous, time-varying, stimulus is possible. And there are motor patterns.]
2) Consistently, many neurons in perceptual areas have sharply-tuned receptive fields [4]; while most neurons in motor areas seem to have broadly-tuned motor fields [5].
3) The mammalian brain may be seen as an arched architecture, with (ascendant) perceptual pathways culminating in the entorhinal cortex (connected to the hippocampal loop), and (descendant) motor pathways. There are aspects of symmetry between the two pillar structures of the arch: on the perceptual side, the hierarchy of sensory receptors, primary and secondary areas, associative (parietal and temporal) areas, mirrors the hierarchy of muscle effectors, primary and secondary motor areas, associative (prefrontal) areas, on the motor side. And there is also an aspect of antisymmetry, because the dominant sense of information flow is ascendant on the perceptual side, descendant on the motor side.
In this light, it is interesting to observe and compare what systems of coordinates are used in perceptual and motor areas. Peripheral (retinotopic) coordinates seem to be kept till very high up in the hierarchy of visual areas; whereas all cortical motor areas, including the primary motor area M1, seem to ignore superbly their own peripheral, i.e. muscular, coordinates (the transformation of coordinates is left to spinal circuits). So, at least for that particular aspect of information processing, the (functional) antisymmetry appears to dominate over the (structural) symmetry.
Perhaps, the key to understand the nature and role of the retroprojections (from higher to lower areas, on the perceptual side) is to be found on the motor side; and vice versa, for the projections from lower to higher areas. That would argue in favour of a cortical theory with a deeper mix of sensori-motor considerations.

5. LOCALISED VERSUS DISTRIBUTED REPRESENTATIONS

The debate is still open between advocates of cardinal cells (localised representations à la Barlow) and advocates of cell assemblies (distributed representations à la Hebb). It is worth clarifying the notion of cell assembly, by distinguishing three acceptions.

1) Population coding.

Population coding has been much attested in motor areas. High motion accuracy is achieved by the simultaneous excitation of a large population of broadly-tuned neurons. Typically such a population might code for a displacement vector of the hand. Smooth modification of the excited population leads to smooth modification of the coded vector.

2) Multi-layer perceptrons, or also synfire chains.

Those are essentially feedforward structures, that are able to subserve information processing tasks, with associative properties (completion, convergence).

3) Attractor neural networks

In section **3**, the simple prototype case of persistent attractors was reviewed. However, if one wishes to cumulate all the advantages of cell assemblies, one is tempted to look for "attractors with internal structure". Either in the direction of "flexible templates" [6], i.e. continuously tunable attractors that would combine convergence and stability with flexibility (allowing for smooth landscaping inside an attractor basin, such as mental rotations [7]). Or in the direction of "complex reverberating attractors", i.e. attractors that are not states of simple persistent activity, but oscillating -- or even more complex -- with a frequency and a phase, that allow for temporal coding and synchrony effects. Such cell assemblies would begin to exhibit properties of space and time compositionality, lacking in previous neural network models [8]. Another motivation is the attempt to reconcile perception and action specific requisites (discussed in **4**).

One price to pay for such higher sophistication of cell assemblies should be clear: namely, increased interference between superposed assemblies, sharing the same neurons. There will be an unavoidable trade-off between attractor complexity and storage capacity.

6. CHAOS AND BIOLOGY

First, a generality on complex systems (such as hydrodynamics and biology). It has been a topic for long-standing debate [9,10,11] whether in such sciences a search for new universal laws was called for, or just a patient application of basic physical laws. Never far around, has been lurking the question: is this fundamental science, or applied science ? The notions of strange attractors and deterministic chaos are a good case in point. Both opposite camps may feel their predictions vindicated. In some sense, it is just an unfolding of classical mechanics (nothing revolutionary like relativity or quantum mechanics); but it is also a novel concept of wide interdisciplinary impact. Let us admit that it is original, perhaps more in the sense of a novel addition, than in the sense of a fundamental reconstruction of science foundations.

A recent article by David Ruelle [12] is well worth reading by potential users, such as brain theorists. Since the paper is exceptionally clearly written, I shall quote successively the abstract, and part of the conclusion.

"Under the name of chaos the mathematical ideas of differentiable dynamics have had a profound impact on physics. Some success stories are discussed. Attention is also directed to the excessive optimism of some currently attempted applications. It is shown how and why they should fail."

"The problem is to find out if various time evolutions observed in Nature correspond to low-dimensional deterministic dynamics. Various authors, using the Grassberger-Procaccia algorithm have claimed this was the case. A careful examination of the method indicates, however, that the 'dimensions' of the order 6 that are obtained are very close to the upper bound $2 \log_{10} N$ permitted by the Grassberger-Procaccia algorithm (N is the length of the time series used, of the order of 10^3). The 'proof' that one has low-dimensional dynamics is therefore inconclusive, and the suspicion is that the time evolutions under discussion do not correspond to low-dimensional dynamics."

Many measurements on brains, from macroscopic EEG to micro-electrode recordings, produce noisy time series. The practical question is: What is the difference between random noise and chaotic noise ? And what are the quantitative criteria to discriminate one from the other ?

At the moment [12], there is no criterion to distinguish between a high dimensional chaotic attractor and random noise. So the distinction is hardly worth considering. The interest therefore is for low-dimensional attractors. During the last ten years, many noisy time series (data obtained from meteorology, or economics, or biology, etc.) have been analysed with the standard Grassberger-Procaccia algorithm, and amazingly often, low-dimensional attractors were found. Ruelle's text gives a hint on the source of this 'mystery'. (Indeed, the reason why we know that there is chaos in meteorology comes from numerical study of theoretical models; direct data analysis alone would be inconclusive [13]). Now the period of enthusiastic exploration of raw time series should be followed by a more sober stage.

It remains that many periodic phenomena have been recorded in the brains (under the names of waves, oscillations, or rhythms); and that it is legitimate to investigate whether there are local or distributed pattern generators which are able to bifurcate from a cyclic attractor to a chaotic one. Much recent interest for chaos in brain theory stems from the search for a fast desynchronisation mechanism (i.e. a kind of noise control) [14]. The point is: once synchrony has been achieved, time-locked oscillators tend to be hard to desynchronise (desynchronisation of oscillators is discussed in two classical textbooks [15, 16]). In this context, deterministic chaos is appealing,-- provided it is strong enough to restore the asynchrony, and not just equivalent to a weak addition of extra noise.

Outside brain science, chaotic phenomena have been studied with some success in ecological populations studies. However it may be guessed that, more frequently than not, life is likely to hate chaos, and to avoid it safely, because unpredictability is dangerous for survival. Except in some occasions, as when a prey is followed by a predator, and may wish to switch on a chaotic generator to govern its escape. Indeed random action is often the best strategy in game theory.

7. REFERENCES

[1] J. von Neumann, The Computer and the Brain (Yale U. Press, 1958)

[2] D.J. Amit, Modeling Brain Function. The world of attractor neural networks (Cambrige U. Press, 1989)

[3] I. Rock, S. Palmer, *The legacy of Gestalt psychology,* Sci. Am. **12** (1990) 84

[4] W.T. Newsome, K.H. Britten, J.A. Movshon, *Neuronal correlates of a perceptual decision,* Nature **341** (1989) 52

[5] S.P. Wise, R. Desimone, *Behavioral Neurophysiology: Insights into Seeing and Grasping,* Science **242** (1988) 736

[6] D. Mumford, *On the computational architecture of the neocortex, II The role of cortico cortical loops,* Biol. Cyb. **66** (1992) 241

[7] A.P. Georgopoulos et al., *Mental rotation of the neuronal population vector,* Science **243** (1989) 234

[8] J. Fodor, Z.W. Pylyshyn, *Connectionism and cognitive architecture: A critical analysis,* Cognition **28** (1988) 3

[9] L. Brillouin, *Life, thermodynamics, and cybernetics,* Am. Sci. **37** (1949) 554

[10] P.W. Anderson, *More is different,* Science **177** (1972) 393

[11] J.J. Hopfield, *Physics, Biological Computation, and Complementarity,* in The lesson of Quantum Theory, Proc. of the Niels Bohr Centenary Symposium, eds. D. de Boer, E. Dahl, O. Ulfbeck (North-Holland, 1986) 295

[12] D. Ruelle, *Deterministic chaos: the science and the fiction,* Proc. Roy. Soc. Lond. A **427** (1990) 241

[13] D. Ruelle, private communication (1992)

[14] W. Freeman, *The Physiology of Perception,* Sci. Am. (1991) 34

[15] R.L. Stratonovich, Topics in the Theory of random Noise, 2 vol. (Gordon and Breach, 1963)

[16] P. Bergé, Y. Pomeau, C. Vidal, Order in Chaos (Wiley-Hermann, 1985)

AUTHOR INDEX

www.ingramcontent.com/pod-product-compliance
Ingram Content Group UK Ltd.
Pitfield, Milton Keynes, MK11 3LW, UK
UKHW050946290926
15470UKWH00005B/1678

9 780444 898395